ORGANISING COMMITTEE

G. J. Tatlock   (Chairman, University of Liverpool).

M. J. Bennett   (AEA Industrial Technology, Harwell,
                       Co-editor, Conference Proceedings).

H. E. Evans   (Nuclear Electric, Berkeley Nuclear Laboratories,
                       Berkeley).

H. M. Flower   (Imperial College).

P. J. Goodhew   (University of Liverpool).

J. A. Little   (University of Cambridge).

G.W. Lorimer   (Manchester Materials Science Centre,
                       Co-editor, Conference Proceedings).

S. B. Newcomb   (University of Cambridge).

J. Upton   (Conference Department, The Institute
                       of Metals).

MICROSCOPY OF OXIDATION

Proceedings of the First International Conference held at the University of Cambridge, 26-28 March, 1990

EDITORS: M J BENNETT AND G W LORIMER

Sponsored and organised by the Materials Science Division
of The Institute of Metals with the co-sponsorship of
The Institute of Physics and The Royal Microscopical Society

THE INSTITUTE OF METALS
1991

Book Number 500

Published in 1991 by The Institute of Metals
1 Carlton House Terrace, London SW 1Y 5DB

and
The Institute of Metals
North American Publications Center
Old Post Road, Brookfield VT 05036
U S A

British Library Cataloguing in Publication Data

Microscopy of oxidation.
1. Oxidation
I. Bennett, M. J. (Michael John) 1933- II. Lorimer, G.W.
III. Institute of Metals *1985-*
541.393

ISBN 0-901462-90-X

American Library of Congress Cataloging in Publication Data

Applied for

Compiled by *P i* c A Publishing Services from original typescripts
and illustrations supplied by the authors

Cover design: Jenni Liddle

Printed and bound in Great Britain

CONTENTS

PREFACE

The First International Conference on the Microscopy of Oxidation was held at Selwyn College, Cambridge, from the 26-28 March, 1990. The conference was sponsored and organised by the Materials Science Division of The Institute of Metals with the co-sponsorship of The Institute of Physics and The Royal Microscopical Society.

The direct observation and analysis of oxides, together with the identification of factors which influence oxidation reactions, are essential steps towards a fundamental understanding of the oxidation of alloys under service conditions. A fully quantitative description of the processes involved is still some way off, but major improvements in microscopy and analytical techniques have brought this goal one stage nearer. Hence it appeared timely to hold an international conference with the central theme of the application of microscopy in oxidation studies. The excellent response to the Call for Papers, the high quality of the papers and posters presented and the lively discussions which followed both the Oral and Poster sessions ensured that the meeting was a success.

It is hoped to hold the Second International Conference on the Microscopy of Oxidation in the Spring of 1993.

G.W. Lorimer
M.J. Bennett

SECTION ONE
Introduction

1 HIGH TEMPERATURE CORROSION OF METALS

P. Kofstad

Department of Chemistry, University of Oslo
P.B.1033 Blindern, 0315 Oslo 3, Norway.

ABSTRACT

The paper briefly summarizes fundamental aspects of transport processes and scale properties in growth of continuous scales. While early interpretations were based on the Wagner model and scale growth by lattice diffusion, it has been increasingly recognized that transport along grain boundaries and other easy diffusion paths often predominate in growth of highly protective scales. This, in turn, means that knowledge of the composition, structure and properties of grain boundary and interfacial regions are of fundamental importance in understanding high temperature corrosion. These aspects are illustrated by examples involving the effects of oxygen active elements and the growth of duplex oxide-sulphide scales. The influence of water vapour on the oxidation behaviour of metals and alloys - and particularly the loss of protective behaviour of technical steels - is also briefly discussed. There are no satisfactory interpretations of these phenomena, but as a possible correlation water vapour may significantly affect defect structures and defect-dependent properties of some oxide systems by serving as a source for proton defects in the oxides.

INTRODUCTION

High temperature corrosion constitutes a complex mixture of interrelated processes and reactions [1]. After the initial nucleation and a continuous film or scale of reaction products has been formed, the metal and the reacting gas are separated and the reaction proceeds through diffusional transport of reactant atoms or ions through the scale. In certain cases electron transport through the scale may alternatively be the rate-determining process. At the same time the gaseous reactant may dissolve in the metal substrate and will for alloys lead to internal corrosion phenomena such as, for instance, internal oxidation.

In addition to these transport processes other phenomena occur. Thus grain growth takes place in the reaction products, and if more than one reaction product is formed, these may react to form new phases. Furthermore, growth stresses are built up in the scales, and these are alleviated through various mechanisms which may include high temperature creep and deformation, cracking and/or spallation. Microchannels are also probably formed through these processes. It is also common that porosity and voids develop in the scales through the deformation processes and as a result of scale growth by outward transport of metal ions through the scale.

The field of high temperature corrosion has seen important developments over the last few decades, and have of course followed those in all related fields of materials and solid state sciences. After the theories of point defects in solids were developed and Carl Wagner had advanced his oxidation theory more than 50 years ago, it was for many years generally assumed in mechanistic interpretations that protective scales grow by lattice diffusion. On this basis numerous attempts were also made to change oxidation rates and behaviour through small dopant additions which were asssumed to be soluble in the lattice. The validity of the Wagner theory and the applicability of the principle of "the solid state diffusion theory of alloy oxidation" has been demonstrated for a few systems and today these approaches are still important for many interpretations of high temperature corrosion mechanisms.

Over the years it has been increasingly realized and demonstrated that the growth of highly protective scales involves lattice diffusion only to a small extent and that the predominating transport takes place along high diffusivity paths such as grain boundaries, interfaces and internal surfaces in the scales. When this is so, the important questions concerning the effects of impurities and dopants then become: to what extent do these become enriched or segregate at grain boundaries and interfaces and how do they affect grain boundary or interfacial properties and thereby important properties of the scales? [1].

The theme of this conference, "Microscopy of Oxidation", reflects recent developments. New and continually improved experimental techniques characterized by acronyms such as SEM, SAM, SIMS, TEM, XRD and XPS, etc. are increasingly being used to study the detailed microstructure and composition of scales, and particularly grain boundary and interfacial regions. In earlier years the themes of conferences were oxidation kinetics . It is interesting to note that at this conference there is not a single paper that has a title that contains the words "corrosion rates" or "kinetics". However, when one considers the overall corrosion phenomena, it is, of course, important to correlate all the microscopic aspects with transport properties and oxidation behaviour in general.

The title of this brief, introductory paper is "High Temperature Corrosion of Metals"- a vast topic which is impossible to review in a sensible manner in such a short space. As an alternative it is purposed to illustrate current important aspects and problems by the consideration of a few selected examples.

EFFECTS OF WATER VAPOUR ON THE OXIDATION BEHAVIOUR OF METALS AND ALLOYS

In numerous applications metals and alloys are exposed to steam or atmospheres that contain water vapour. It is well known that even small amounts of water vapour may for some alloys significantly affect their oxidation behaviour [1]. This may be illustrated in Fig.1 by some kinetic results on the oxidation of Fe-13%Cr at 980 °C in oxygen containing various amounts of water vapour ranging from 0.03 to 2 vol.% H_2O[2]. Within the time periods studied the alloys exhibit protective behaviour in the "dry" atmospheres, while increasing rates and loss of protectivity are observed as the water content in the ambient atmosphere is increased. In the dry atmospheres chromium is preferentially oxidized and accordingly the scales are enriched in Cr_2O_3. In the more humid oxygen thicker and less protective scales form which essentially consist of iron oxides (FeO, Fe_3O_4, Fe_2O_3) with inclusions of chromium oxide and the corresponding spinel in the inner part of the scales. This example is not a special case. It is well known that technical steels oxidize faster in humid than in dry atmospheres [1-3]. Even for the oxidation behaviour of unalloyed iron at high temperatures (> 900 °C) significant effects of water vapour have been reported; in this case the oxide composition has not been changed, but there are marked differences in the microstructure of the scales [4-7].

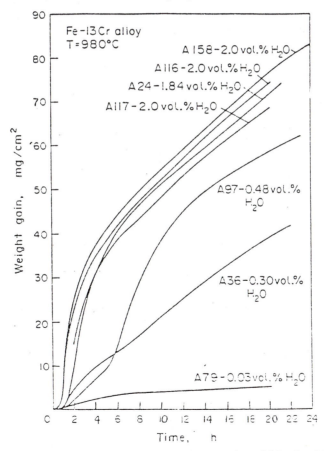

Figure 1: Oxidation of Fe-13Cr at 980 °C in air with H_2O contents ranging from 0.03 to 2 vol.% H_2O. (After Kvernes et al.[2])

The literature does not appear to provide any convincing interpretation of these effects. As a matter of fact, the problem often seems to be disregarded or "forgotten" in connection with high temperature corrosion. This also applies to studies of possible effects of water vapour on properties of ceramics; such studies have not received much attention. It has been and is common to study properties of oxide ceramics at low oxygen activities by equilibrating them in gaseous mixtures

and is common to study properties of oxide ceramics at low oxygen activities by equilibrating them in gaseous mixtures of hydrogen and water vapour. When so doing, it is generally assumed that hydrogen or water vapour have no significant effects on the defect structures and defect dependent properties of oxides. This may be valid for many oxide systems, particularly when the intrinsic defect concentrations are high and much larger than the expected effects of water or hydrogen. However, it has become increasingly evident that water vapour or hydrogen may greatly affect the defect structure and defect dependent properties of many oxides. A well known example of enhanced oxidation in sceam is that of oxidation of silicon [1,8]. In this case it is concluded that water molecules migrate along the channels in the quartz glass structure of the silica scales.

Gradually other examples are forthcoming which show important effects of water vapour on properties of crystalline oxides [9,10]. It is generally concluded that the water vapour serves as a source of hydrogen which dissolves as protons in the metal oxides, e.g.

$$H_2O = 2H_i^\cdot + 1/2O_2 \, (g)$$

The dissolved hydrogen ion is written here as an interstitally dissolved hydrogen ion, $H_i^\cdot$, (or proton), but in reality the hydrogen ions are associated with oxygen ions on normal lattice sites (HO_o). However, these alternative ways of writing the hydrogen defects are of no consequence in writing the defect reactions. For some oxide systems, particularly those with small concentrations of native, intrinsic defects, the concentration of dissolved protons may be sufficiently high that they may dominate the electroneutrality condition and thereby the defect-dependent properties under many conditions [9,10].

Yttrium oxide is one system where the effects of small amounts of water vapour/hydrogen defects on the defect structure have been demonstrated. Figure 2 shows the effect of water vapour at constant oxygen activity on the dc conductivity of

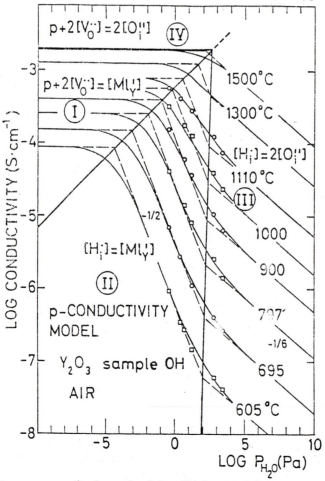

Figure 2: Effects of water vapour on the dc-conductivity of Y_2O_3 with Ca^{2+}-impurities at constant oxygen pressure (air). The figure also shows the interpretation of the results with delineation of various defect structure domains with different limiting/dominating electroneutrality conditions, e.g. $[H_i^\cdot] = 2[O_i^2]$, $[H_i^\cdot] = [Ml_Y]$, etc. (After Norby and Kofstad [11])

an yttria sample [11]. This contained small amounts of calcium and other lower valent cation impurities, and in this particular region the electroneutrality condition is concluded to be

$[H_i^\bullet] = [Ml_{Y'}]$

where $Ml_{Y'}$ is a lower valent cation impurity on an yttrium site with one negative effective charge. Figure 2 also shows the proposed defect structure model with different domains of limiting/dominating defect structure situations.

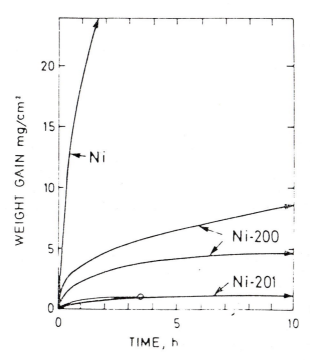

Figure 3: Corrosion of high purity nickel (Ni) and two commercial qualities of nickel (Ni-200 and Ni-201) at 700 °C in $O_2+4\%SO_2$ at a total pressure of 1 atm. (After Andersen et al.[22])

Time does not permit a detailed description and interpretations of the many results, and the interested reader is referred to the literature [9-10]. Protonic defects dissolved from water vapour may in certain regions completely dominate the electroneutrality condition and thus control the concentration of both majority and minority defects in the oxide. This is, in turn, means that even small amounts of water vapour can importantly affect properties dominated by majority defects (e.g. diffusion controlled growth of transport of reactants through growing scales) or by minority defects (e.g. high temperature creep) [9-12].

Although results of fundamental studies of effects of water vapour are limited, it is a rapidly expanding field and new results are gradually forthcoming. The creep rate of olivine (iron-magnesium silicate) at high temperatures (~1500 °C) is 2-3 times faster in H_2+H_2O than in $CO+CO_2$ mixtures with the same oxygen activity [13]. Furthermore, the sintering rates of magnesium oxide compacts are increased by the presence of water vapour in the ambient atmosphere and at high pressures the sintering rates have been reported to be proportional to $p(H_2O)^n$, with values of n ranging from 1 to 1.5 [14,15]. The reason suggested for this is that water interacts with the surface of MgO with formation of protonated vacancies [16].

In the interpretation of effects of water vapour on yttrium oxide (and a number of other oxide systems) it is proposed that protonic defects dissolve in the lattice. However, it would not be surprising if protonic defects also have particular effects on properties of surfaces and grain boundary regions, although the author is not aware of any cases where such effects on grain boundaries have as yet been estimated. At this stage one may only speculate, and in this respect it is interesting to return to the above examples on the oxidation of iron-chromium alloys. During an initial stage the oxidation is protective and consists to a large extent of chromium oxide. It is reasonable to conclude that the scale growth is controlled by grain boundary diffusion through the scale. However, in wet atmospheres the protection rapidly breaks down and iron oxides are formed on the scale surface. Does this mean that outward iron diffusion along grain boundaries in the initially protective scale is enhanced by the presence of water vapour and protonic defects?

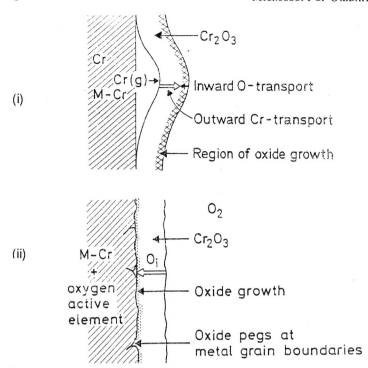

Figure 4: Schematic illustration of the growth of chromia scales in (i) the absence of oxygen active elements and with predominant outward transport of chromium through the scale, and (ii) the presence of oxygen active elements and with predominant inward transport of oxygen. In case (i) new oxide is formed at/near the outer scale surface and voids and porosity gradually formed at/near the metal/scale interface, while in case (ii) new oxide is formed at/near the metal/scale interface. (After Kofstad [1])

Regardless of the detailed interpretations, studies of these effects will be important for the understanding of properties of both ceramics and oxide scales.

SULPHUR AND HIGH TEMPERATURE CORROSION OF METALS

Reactions of metals in sulphur or sulphur-containing gases, and also other effects of sulphur in high temperature corrosion, have received wide attention in recent years, and examples may also illustrate interesting aspects of high temperature corrosion [1].

Sulphur gas is an agressive oxidant at high temperatures to common metals such as iron, cobalt, nickel, chromium and alloys based on these metals. The metal-sulphur reactions can generally be described as parabolic and the parabolic rate constants are orders of magnitude higher than for the corresponding metal-oxygen reactions at the same temperatures [1,17,18].

A major reason for these fast corrosion rates are that sulphides of nickel, cobalt and chromium can exhibit marked deviations from stoichiometry and this, in turn, means that they have correspondingly high concentration of lattice defects and thus exhibit high rates of diffusion. The major defects in these sulphides are cation defects and accordingly the sulphide scales on these metals and alloys grow by outward lattice diffusion of cations. As such, metal-sulphur reactions provide more examples of lattice diffusion controlled reactions than do metal-oxygen reactions.

An additional feature of these metal-sulphur systems is that they form relatively low-melting eutectics. For example, a Ni-S liquid solution with about 22 wt.% sulphur is formed at a temperature as low as 635 °C [1]. This thus means that the inner part of sulphide scales on nickel are liquid above this temperature.

When these metals are corroded in sulphur-containing gases such as O_2+SO_2/SO_3, the scales will under certain reaction conditions consist of duplex mixtures of oxides and sulphides, and the presence of sulphides may then greatly affect the corrosion behaviour. However, the specific effects are highly dependent on the distribution of the sulphides in the scales. This may, for instance, be illustrated by the reaction of nickel and dilute nickel-base alloys in gaseous mixtures of $O_2+SO_2/$

SO_3 at temperatures from 600 to 900 °C. Nickel corrodes rapidly under conditions when nickel sulphate is formed on the scale surface according to the reaction [19,20]

$$NiO + SO_3 = NiSO_4$$

Nickel migrates outwards through the scale and then reacts with the sulphate to form NiO and nickel sulphides, e.g.

$$9NiO + 2NiSO_4 = 8NiO + Ni3_{\pm x}S_2$$

As a result the scale consists of a duplex mixture of oxide and sulphide. In the inner part of the scale the sulphide must also consist of Ni-S liquid solution at temperatures above 635 °C. The reason for the rapid corrosion rates of high purity nickel under these conditions is in all probability that the sulphides wet the fine grains of NiO and form a three-dimensional sulphide network through the scale. Due to the high diffusion rates of nickel in the sulphides, the sulphide network serves as a high diffusivity path for outward transport of nickel and sustains the rapid rate of corrosion [19,20].

Similar studies were undertaken on two commercial grades of nickel, but could not reproduce the results on high purity nickel and as shown in Fig.3 the corrosion rates were slower. The commercial purity nickel contained small amounts of impurities of iron, manganese, silicon, and copper. Various dilute binary, ternary and quaternary nickel alloys with these metals were therefore prepared and tested under the same conditions. The corrosion resistance could be improved by small additions of iron, manganese and/or silicon, and for the ternary and quaternary alloys marked synergistic effects of the small alloy additions were also observed. The reaction products on the dilute alloys were similar to that for high purity nickel: sulphate was formed on the surface and the scale consisted of a mixture of oxide and sulphides. Thus, although the sulphides were present in the scales, in these cases they did not serve as a high diffusivity path for outward metal diffusion. It has been concluded that in these scales - which also contain small amounts of silicon, manganese and iron - the sulphide phase did not wet the oxide grains in the form of a three-dimensional sulphide, but rather the sulphide formed small particles at triple points of the oxide grains. A possible explanation of this is that the oxides of the small alloying additions, e.g. iron and or manganese silicates, are enriched or become segregated at the grain boundaries of the oxides and thereby completely change the properties of the grain boundary regions and the ability of the sulphides to wet the oxide grains [22].

Similar types of effects ought to be observed for many other systems and corrosion processes. In controlling these types of corrosion processes the important question then becomes: how can the properties of grain boundaries in protective scales be modified beneficially?

EFFECTS OF OXYGEN ACTIVE ELEMENTS

Although the effects of the so-called oxygen active elements have been known for more than 50 years, it is only in more recent years - and with the use of advanced instrumental techniques - that we have come closer to elucidating the mechanisms involved [1,23]. Many interpretations have been offered over the years and in many cases it has also been attempted to explain the effects by one single mechanism. It appears, however, that several effects are operative, but they generally appear to be related to interfaces and grain boundaries.

One important effect of the oxygen active elements is that they improve oxide scale adherence. Recently poor adherence of oxide scales have been proposed to be due to segregation of sulphur - which is present as an impurity in many high temperature alloys - to the metal/scale interface with a consequent embrittlement and reduced mechanical stability of the interface region. It has been suggested that yttrium - and other oxygen active elements - react with sulphur to form the corresponding sulphides and thereby eliminate the embrittling and weakening effects of the sulphur segregation [24,25]. If this is the sole mechanism, the reactive elements should really be termed "sulphur active elements". The importance of sulphur on the scale adherence for many alloys has been demonstrated in a number of investigations, but scale spallation also takes place in systems with extremely low sulphur impurity levels [26]. Consequently the sulphur effect does not appear to be the sole explanation.

It has also been demonstrated that the reactive elements have other effects. One important aspect as regards growth of chromia scales, for instance, is that the presence of yttrium changes the diffusional transport processes in the scales. While chromia scales - on unalloyed chromium and chromia-forming alloys without reactive elements - grow predominatly through outward migration of chromium, the effect of small additions of yttrium and cerium (and probably other reactive elements) is to make inward oxygen transport the predominant diffusional process [2,27,28]. Diffusional transport through growing chromia scales in all probability involves grain boundary diffusion, and it is then reasonable to look for effects of yttrium and other reactive elements on the properties of grain boundary regions of chromia scales. Several investigators have found that yttrium becomes enriched (segregated) at grain boundaries. It is not unreasonable to suggest that this changes the "defect structure" of the grain boundary regions so as to change these from metal to oxygen ion diffusing areas. Other detailed recent studies on the effects of small additions of cerium on oxidation of nickel have shown that cerium becomes enriched at grain boundaries of the NiO scales [29].

This change to predominant inward oxygen diffusion may also have important bearing on the adherence of the scales. When outward cation diffusion predominates in the growth of scales, new oxide is formed at or near the external surface and voids and porosity gradually develop at or near the metal/scale interface due to the reduced volume of the metal and the lack of sufficient plastic deformation of the scales. The voids and porosity reduce the adherence/coherence of the scales and increase the tendency for spallation and loss of protectivity. On the other hand, when inward oxygen diffusion predominates, new oxide is formed at the metal/scale interface provided the scale remains continuous and does not crack - and the tendency for formation of voids and porosity and loss of oxide adherence is greatly reduced [1].

CONCLUDING REMARKS

These are but a few examples of questions and problems that are being faced currently in the field of high temperature corrosion. They illustrate a key aspect of the field both as regards basic scientific questions and applications. Future developments will depend on the ability to characterize and model the scales and reaction products down to an atomic level and to correlate these results with transport and other defect properties of scales.

REFERENCES

1. P.Kofstad, "High Temperature Corrosion", Elsevier Applied Science, London/New York, 1988.

2. I.Kvernes, M.Oliveira and P.Kofstad, Corrosion Science, 17, 1977, p.237.

3. C.Fujii and R.A.Meussner, J.Electrochem.Soc., 111, 1964, p.1215.

4. A.Rahmel and J.Tobolski, Corrosion Science, 5, 1965, p. 333.

5. A.Rahmel, Werkstoffe u. Korrosion, 16, 1965, p. 662.

6. C.W.Tuck, M.Odgers and K.Sachs, Corrosion Science, 9, 1969, p.271.

7. N.Otsuka, Y.Shida and H.Fujikawa, Oxid. Met., 32, 1989, p.13.

8. B.E.Deal and A.S.Grove, J.Appl.Phys., 36, 1965, p.3770.

9. T.Norby, in "Advaces in Ceramics", Vol.23, Nonstoichiometric Compounds, The American Ceramic Society, Inc., 1989, p.107.

10. T.Norby, in Selected Topics in High Temperature Chemistry", Eds. Ø.Johannesen and A.Andersen, Elsevier, Amsterdam, 1989, p.101.

11. T.Norby and P.Kofstad, J.Am.Ceram.Soc., 69, 1986, p.780.

12. T.Norby and P.Kofstad, Solid State Ionics, 20, 1986, 169.

13. B.Poumelle and O.Jaoul, in "Deformation of Ceramics II", Eds. R.E.Tressler and R.C.Bradt, Plenum Publ. Corp., 1984, p.281.

14. B.Wong and J.A.Pask, J.Am.Ceram.Soc., 62, 1979, p.141.

15. J.A.Varela and O.J.Whittemore, Mater.Sci.Monogr., 14, 1982, p.439.

16. E.Longo, J.A.Varela, A.N.Senapeschi and O.J.Whittemore, Langmuir, 1, 1985, p.456.

17. S.Mrowec and T.Werber, "Gas Corrosion of Metals", published for The National Bureau of Standards and The National Science Foundation, Washington, DC, by the Foreign Scientific Publications Department of the National Center for Scientific, Technical Economic Information, Warsaw, Poland, 1978.

18. S.Mrowec and J.Janowski, in "Selected Topics in High Temperature Chemistry", Eds. Ø.Johannesen and A. Andersen, Elsevier, Amsterdam, 1989, p.55.

19. B.Haflan and P.Kofstad, Corros. Sci., 23, 1983, p.1333.

20. K.P.Lillerud, B.Haflan and P.Kofstad, Oxid. Met., Eng., 87, 1987, p.45.

23. "The Role of Active Elements in the Oxidation Behaviour of High Temperature Metals and Alloys", Ed. E.Lang, Elsevier Applied Science, London/New York, 1989.

24. J.G.Smeggil, A.W.Funkenbusch and N.S.Bornstein, Metall. Trans., 17A, 1986, p.923.

25. J.L. Smialek, in "Corrosion & Particle Erosion at High Temperatures", Eds. V.Srinivasan and K.Vedula, The Minerals, Metals & Materials Society, 1989, p.45.

26. K.N.Luthra and C.L.Briant, Oxid. Met., 26, 1986, p.397.

27. P.Skeldon, J.Calvert and D.G.Lees, Philos. Trans. R. Soc. London, 296, 1980, p.545,557,567.

28. G.T.Yurek, K.Przybylski and A.J.Garrat-Reed, J.Electrochem. Soc., 134, 1987, p.2643.

29. D.P.Moon, Oxid. Met., 32, 1989, p.47.

2 THE STUDY OF OXIDE SCALES USING SIMS

M. J. Graham

Division of Chemistry, National Research Council of Canada,
Ottawa, Canada K1A 0R9.

ABSTRACT

SIMS has been used to study the nature and composition of oxide films on metals and alloys ranging in thickness from ~1 nm-thick electrochemically-formed films to μm-thick oxides produced at high temperature. Emphasis is placed on thicker oxides where ^{18}O/SIMS has been used to investigate the extent of oxygen diffusion in Cr_2O_3, γ-Al_2O_3 and NiO, and the role of 'reactive elements' such as Ce in improving high temperature corrosion resistance.

Cr_2O_3 is found to grow primarily by cation diffusion, although a small amount of inward oxygen diffusion occurs down oxide grain boundaries. On pure Al, γ-Al_2O_3 grows by oxygen transport via local pathways through an outer amorphous oxide layer and subsequent oxygen incorporation at the periphery of the underlying growing γ-Al_2O_3 islands. Data for γ-Al_2O_3 scales formed on Fe-20Al at 800°C demonstrate oxide growth by outward cation diffusion. NiO is confirmed to grow predominantly by cation diffusion. ^{18}O/SIMS results for oxidized Fe-Cr alloys sputter coated with ~4 nm of ceria suggest that when Ce is located within the oxide scale there has been a change in oxide growth mechanism from predominantly cation to predominantly anion diffusion. TEM and EELS imaging microscopy have been used to complement the SIMS data, and have confirmed the presence within the scale of particles of $CeCrO_3$ which likely impede cation diffusion while allowing oxygen anion diffusion to proceed.

INTRODUCTION

SIMS is a powerful tool to study the nature and composition of oxides on metals and alloys. We have used the technique to examine the stability and growth mechanism of electrochemically-formed films and the influence of ions such as Cl⁻ incorporated within the oxide. We have also applied SIMS to examine the extent of oxygen diffusion in Cr_2O_3, γ-Al_2O_3 and NiO scales formed at high temperature and to study the role of 'reactive elements' such as Ce in improving corrosion resistance. In this last study SIMS has been used to determine the location of Ce within the oxide scale formed on ceria-coated Fe-Cr alloys and to study its effect on transport processes. TEM and EELS microscopy have been used to complement the SIMS data. The paper presents an overview of our SIMS work, with emphasis placed on the growth of thicker oxide films formed at high temperature.

EXPERIMENTAL

SIMS analysis has been performed with a Physical Electronics Inc. SIMS II system used in conjunction with a differentially-pumped 0.5 mm dia. xenon beam at 56° off normal, rastered over an area 2.5 x 3.0 mm, and electronically gated to analyze the central 10% of the crater area.

RESULTS AND DISCUSSION

Nature and Growth of Electrochemically-formed Oxide Films

In the past there has been considerable disagreement whether Cl⁻ is adsorbed on the oxide of a metal or alloy or is incorporated within an oxide film prior to pitting. We have shown using SIMS that anodic films formed on Ni contain up to ~4% Cl⁻ from solution [1] whereas films on Fe do not incorporate Cl⁻ [2]. Cl⁻ is also incorporated into films on Fe-26Cr alloys after electropolishing in perchloric-acetic acid. Figure 1 shows the SIMS profile of such a film. It is seen that the inner part of the film is Cr-rich and the outer part is Fe-rich; Cl⁻ is incorporated into the oxide to a level approaching ~20% in the middle of the film. This Cl⁻ incorporation into the film influences the open-circuit breakdown behaviour, but in the case of Ni was found to be not a precursor for pit initiation [1]. We have also used SIMS to determine the presence of any hydroxyl ions in passive oxide films on Ni, Fe and Fe-26Cr alloys [3,4]. No hydroxyl was present within passive films formed on Ni and Fe, while passive films on Fe-26Cr were found to contain a small amount: $0.6 \pm 0.1\%$ within the inner Cr-rich part of the film and $1.1 \pm 0.1\%$ within the outer Fe-rich part of the film. Non-passive films such as the

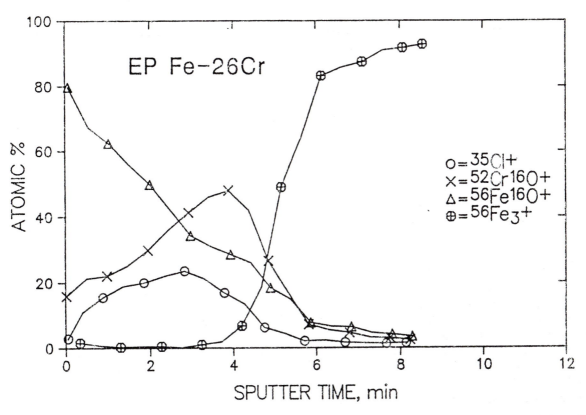

Figure 1: Positive polyatomic SIMS profiles for a ~1.8 nm-thick oxide film formed on Fe-26Cr after electropolishing (EP) in perchloric/acetic acid. (The $^{56}Fe_3^+$ profile indicates the position of the metal/oxide interface). Sputtering was by 1 keV xenon.

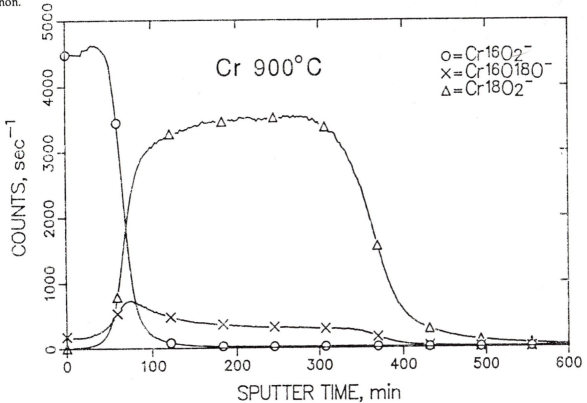

Figure 2: Negative polyatomic SIMS profiles for a 1.9 μm-thick Cr_2O_3 film formed on polycrystalline Cr at 900°C in 5 x 10^{-3} torr oxygen. 1.5 μm of oxide was produced in $^{18}O_2$ followed by 0.4 μm in $^{16}O_2$. Surface pretreatment was abrading to a 1 μm diamond finish and sputtering was by 4 keV xenon.

electropolish film on Fe-26Cr contained a considerable amount of OH^-, ranging from ~15% at a depth of 5 Å to ~2% at the metal/oxide interface [3,4]. SIMS has also been used to study the cathodic reduction, air stability and growth mechanism of oxide films formed in $H_2{}^{18}O$-enriched borate buffer solution [5]. Films formed at low potentials in the passive range were found to be unstable and thickened on exposure to the air. Films formed at higher potentials were air stable. ^{18}O/SIMS data for additional oxide grown at higher potentials showed that the oxide grew by inward oxygen diffusion.

Oxygen Transport in Cr_2O_3, γ-Al_2O_3 and NiO at High Temperature

SIMS analysis following sequential oxidation in $^{16}O_2$ and $^{18}O_2$ is an ideal way to study oxygen diffusion, although interpretation of the data is complicated by problems associated with interface broadening due to non-uniform sputtering and ion beam mixing. We have found that analyzing polyatomic species, e.g. of the type $M^{16}O_2{}^-$, $M^{18}O_2{}^-$ and $M^{16}O^{18}O^-$ rather than the usual $^{16}O^-$ and $^{18}O^-$ ions largely overcomes these difficulties [6]. It is possible using this polyatomic SIMS approach to distinguish grain boundary from lattice diffusion of oxygen, and to calculate diffusion coefficients.

(a) Cr_2O_3

Considering transport in Cr_2O_3, Fig. 2 shows a typical SIMS profile for polycrystalline Cr oxidized at 900°C first in $^{18}O_2$ and then in $^{16}O_2$. The $Cr^{16}O_2{}^-$ and $Cr^{18}O_2{}^-$ profiles show that the ^{16}O-oxide exists as a separate layer on top of the ^{18}O-oxide indicating that the major mass transport during oxidation is outward Cr diffusion. The mixed $Cr^{16}O^{18}O^-$ signal provides information on the extent of any inward oxygen diffusion; the profile in Fig. 2 is asymmetric about the interface between the $^{16}O^-$ and $^{18}O^-$-oxide layers, indicating penetration into the inner oxide. Two independent oxygen diffusion processes are in effect as illustrated in Fig. 3. The first is isotropic oxygen self-diffusion occurring at the interface between the oxide layers. This isotropic diffusion has no net transport and would not contribute to film growth. The second process,

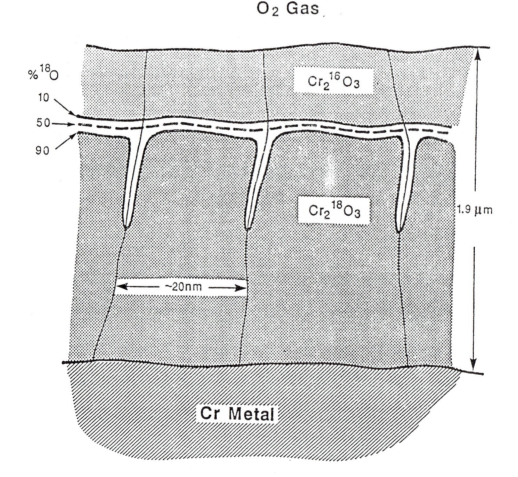

Figure 3: Schematic illustrating the extent of oxygen diffusion in Cr_2O_3 at 900°C as calculated from the SIMS data in Fig. 2. The 10% and 90% ^{18}O profiles indicate the extent of both isotropic diffusion and anisotropic diffusion down oxide grain boundaries.

associated with the tailing of the $Cr^{16}O^{18}O^-$ profile, is an anisotropic, localized inward oxygen diffusion assumed to be occurring down oxide grain boundaries. Oxygen diffusion coefficients can be calculated for the two processes from the area of the $Cr^{16}O^{18}O^-$ profile [7]. The diffusion coefficients for oxygen grain boundary diffusion are 2-3 orders of magnitude higher than for isotropic diffusion, but are still orders of magnitude too low to account for the observed oxidation rates. The major growth transport is by outward Cr diffusion; however ~1% of the oxide is created within the film as a result of inward oxygen diffusion.

(b) γ-Al_2O_3

Oxides formed on Al above 450°C consist of an outer amorphous Al_2O_3 film above γ-Al_2O_3 islands which form at the metal/oxide interface and protrude into the metal. Polyatomic SIMS data is therefore more difficult to interpret than on uniform Cr_2O_3 films, but results on single crystal Al surfaces can provide information regarding oxygen diffusion [8]. The results provide strong evidence that oxygen anion transport proceeds rapidly via local channels with some oxygen exchange in the amorphous Al_2O_3 film. The low level of the mixed polyatomic species $Al^{16}O^{18}O^-$ indicates that the anions are then incorporated at the peripheries of laterally growing γ-Al_2O_3 islands without much diffusion within the islands. The detailed morphology and kinetics of island growth depend on the substrate orientation [8]. More uniform films of γ-Al_2O_3 can be produced on Fe-Al alloys. Figure 4 shows the SIMS profile from an oxide formed first in $^{18}O_2$ and then in $^{16}O_2$ on Fe-20Al at 800°C. The latter-formed oxide is on the outside, clearly showing that the oxide has grown by outward cation diffusion. The total AlO$^-$ signal in Fig. 4 shows a pronounced increase near the oxide/metal interface due to a change in the level of sample charging. The $Al^{16}O^-$ peak at the oxide/metal interface corresponds to the initial air-formed film; a reverse oxidation ($^{16}O_2$ followed by $^{18}O_2$) does not show a $Al^{18}O^-$ peak at the same location indicating that inward oxygen diffusion is not occurring. Current studies on Fe-Al alloys are aimed at examining the mechanism of oxide growth at higher temperatures ($\geq$1000°C) where α-Al_2O_3 is formed.

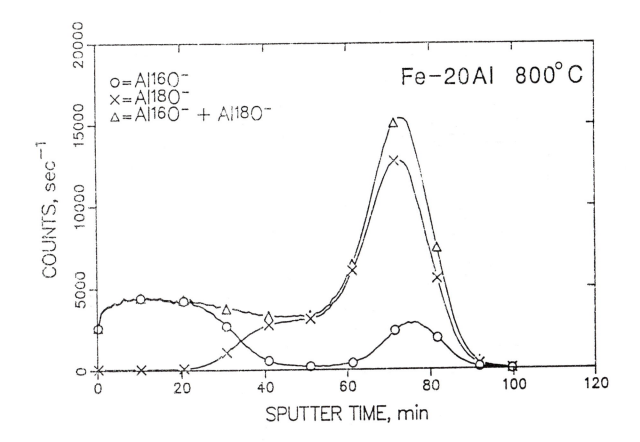

Figure 4: Negative polyatomic SIMS profiles for a 120 nm-thick γ-Al_2O_3 film formed on Fe-20Al at 800°C in 5 x 10^{-3} torr oxygen. 70 nm of oxide was produced in $^{18}O_2$ followed by 50 nm in $^{16}O_2$. Surface pretreatment was abrading to a 1 μm diamond finish and sputtering was by 4 keV xenon.

(c) NiO

For growth of NiO at 700°C, ^{18}O/SIMS experiments by Bishop et al.[9] and Atkinson and Smart [10] have confirmed that the rate controlling transport process is outward diffusion of Ni. Our SIMS data on single-layer NiO scales formed at higher temperatures are in agreement with these results. SIMS profiles for Ni oxidized sequentially in $^{18}O_2$ and $^{16}O_2$ at 900°C are shown in Fig. 5a. The profiles show, as in Fig. 2 for Cr oxidation, that the ^{16}O-oxide exists as a separate layer on top of the ^{18}O-oxide, confirming that the major mass transport during oxidation is outward Ni diffusion. The mixed peak, $^{58}Ni^{16}O^{18}O$ shows much less tailing into the inner layer than for Cr_2O_3 as the result of any inward oxygen diffusion. In Fig. 5b, the ~1 nm-thick ^{16}O-oxide film formed during surface pretreatment still exists in the region of the metal/oxide interface as seen by the small but distinct $^{58}Ni^{16}O_2^-$ peak, confirming growth by outward Ni diffusion. This clearly demonstrates the sensitivity of the polyatomic SIMS technique to detect 1 nm films under μm-thick external scales.

Oxidation of Ceria-coated Fe-Cr Alloy

The addition of small amounts of so-called 'reactive elements' such as Ce, Y, Hf, Th, or their oxide dispersions, greatly increases the high temperature oxidation resistance of Fe-Cr alloys under isothermal or cyclic conditions [11]. Beneficial effects also result from ion implantation of the active element [12], or from surface applied coatings [13]. We have applied very thin (4 nm) reactive element coatings by sputtering CeO_2 directly onto Fe-Cr substrates, and have used SIMS to determine the location of Ce with the oxide scale and to study its effect on transport processes. Additional experimental details may be found elsewhere [14].

SIMS profiles for short-term oxidation at 900°C of coated Fe-26Cr samples are shown in Fig. 6. The scale is principally Cr_2O_3 with a small amount of Fe in the outer part of the scale. The Ce in the oxide layer is detectable, and its maximum signal is observed to be away from the alloy/oxide interface. Longer-term oxidations show that the location of the Ce maximum moves further towards the oxide/gas interface. If the CeO_2 coating can be considered to be a stationary marker, its location towards the oxide/gas interface indicates that inward oxygen diffusion is becoming more predominant in governing oxide growth. $^{16}O_2$/$^{18}O_2$ experiments confirm that oxygen transport is occurring. Figure 7 shows data for a specimen oxidized at 900°C sequentially in $^{16}O_2$ and then $^{18}O_2$. In this longer-term experiment the location of the Ce maximum is now in the middle of the scale. There are ^{18}O maxima ($^{53}Cr^{18}O^+$) at the gas/oxide interface, within the scale at the position of the Ce maxima, and at the alloy/oxide interface showing that oxygen diffusion has occurred during oxidation. Similar results regarding oxygen diffusion on ceria-coated specimens have been obtained at 1100°C [14,15], on Fe-20Cr alloys [14,16] and on Fe-16Cr alloys [17]. Analogous results have also been obtained on Y_2O_3-coated alloys.

Examination of the SIMS data in Fig. 7 in the form of the fraction of ^{18}O associated with the Ce- and Cr-bearing species through the scale shows that a higher fraction is associated with the Ce-bearing species. If the Ce in the scale is located at oxide grain boundaries, the association of ^{18}O with Ce strongly suggests that transport of ^{18}O through the scale occurs along grain boundaries. The present SIMS lacks the lateral resolution to determine the precise location of the Ce or its form and distribution within the Cr_2O_3. However, comparison with the results of Cotell et al. [18] on the structure and location of Y within scales on Cr implanted with Y suggests that Ce would be expected to be segregated as a Ce-Cr oxide phase at grain boundary regions.

We have used TEM and EELS microscopy to attempt to identify Ce-Cr oxides in scales formed on ceria-coated Fe-Cr alloys. Figure 8 shows a bright field image and electron energy loss images for Ce from the same area of a 44 nm-thick stripped oxide film formed on Fe-26Cr at 900°C. The distribution of Ce-containing oxide is illustrated in Fig. 8b by the light-coloured regions. X-ray analysis in TEM shows the particles to have a Ce/Cr ratio ~1 suggesting that the particles are $CeCrO_3$ and this has been confirmed by electron diffraction measurements [19,20]. $CeCrO_3$ was also proposed to be the Ce-rich phase observed at Cr_2O_3 grain boundaries on oxidation of Ce-implanted Ni-30Cr alloys [21]. By analogy with the Y-implant work described above [18] where particles of $YCrO_3$ have been identified [22], it is proposed that particles of $CeCrO_3$, as well perhaps as Ce ions, retard the grain boundary diffusion of Cr cations to the extent that the mobility of oxygen anions along these grain boundaries exceeds that of the cations. Therefore, coating Fe-Cr alloys with a small amount of CeO_2 results in the formation of $CeCrO_3$ particles within the scale, which cause a change in oxide growth mechanism from predominantly cation diffusion for the uncoated alloy to predominantly anion diffusion.

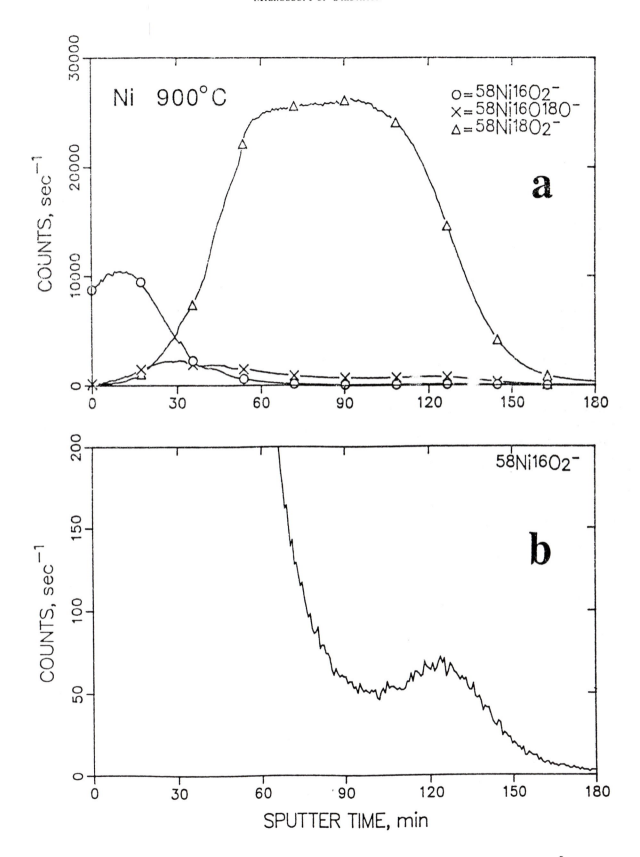

Figure 5: (a) Negative polyatomic SIMS profiles for a 3.4 μm-thick NiO film formed on Ni at 900°C in 5 x 10⁻³ torr oxygen. 2.6 μm of oxide was produced in $^{18}O_2$ followed by 0.8 μm in $^{16}O_2$. (b) Expanded $^{58}Ni^{16}O_2^-$ SIMS profile for the oxide in (a) showing the position of the prior ^{16}O-oxide at the metal/oxide interface. Surface pretreatment was electropolishing and sputtering was by 4 keV xenon.

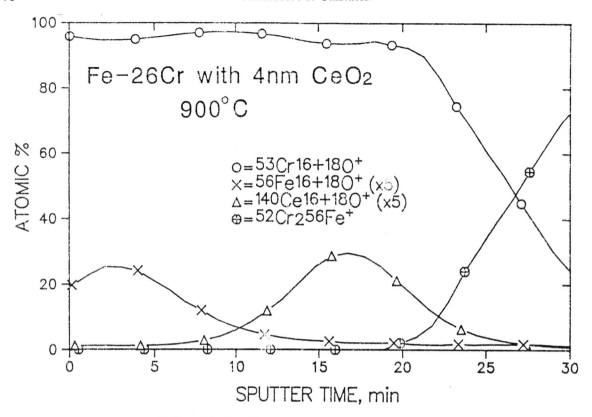

Figure 6: Positive polyatomic SIMS profiles for a 76 nm-thick oxide formed on Fe-26Cr coated with 4 nm CeO_2 after 1 min. oxidation at 900°C in 5 x 10^{-3} torr oxygen. (The $^{52}Cr_2$ $^{56}Fe^+$ profile indicates the position of the metal/oxide interface.) Sputtering was by 4 keV xenon.

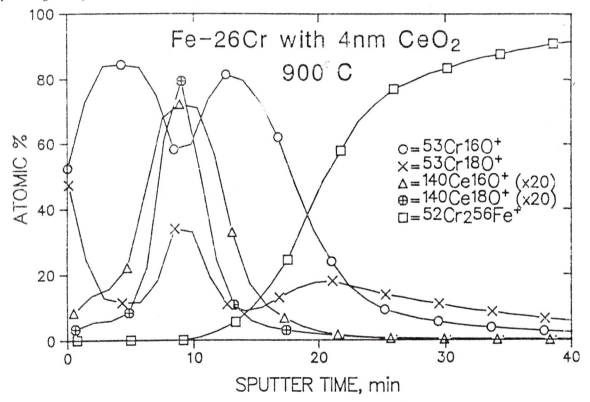

Figure 7: Positive polyatomic SIMS profiles for a 0.36 μm-thick oxide film formed at 900°C in 5 x 10^{-3} torr oxygen on Fe-26Cr coated with 4 nm CeO_2 after 0.7 h oxidation in $^{16}O_2$ followed 5.9 h oxidation in $^{18}O_2$; oxide thicknesses were 0.17 and 0.19 μm, respectively. (The $^{52}Cr_2$ $^{56}Fe^+$ profile indicates the position of the metal/oxide interface.) Sputtering was by 4 keV xenon.

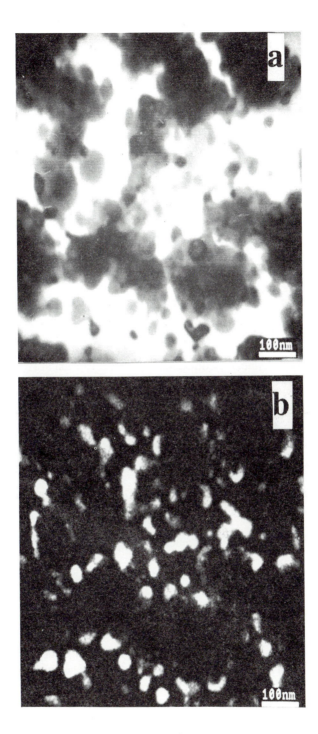

Figure 8: TEM bright-field image (a) and EELS images (884 eV) for Ce (b) from the same area of a 44 nm-thick oxide film formed on heating Fe-26Cr coated with 4 nm CeO_2 to 900°C in 10^{-3} torr oxygen, and stripped from the substrate.

ACKNOWLEDGEMENT

The author thanks R. J. Hussey, D. F. Mitchell, D. A. Downham, J. I. Eldridge, P. Papaiacovou, J. Shen, J. A. Bardwell, B. MacDougall and G. I. Sproule for their contributions to this work.

REFERENCES

1. B. MacDougall, D. F. Mitchell, G. I. Sproule and M. J. Graham, J. Electrochem. Soc., 130, 1983, p.543.

2. R. Goetz, B. MacDougall and M. J. Graham, Electrochimica Acta, 31, 1986, p.1299.

3. D. F. Mitchell, G. I. Sproule and M. J. Graham, Appl. Surf. Sci., 21, 1985, p.199.

4. D. F. Mitchell and M. J. Graham, J. Electrochem. Soc., 133, 1986, p.936.

5. R. Goetz, D. F. Mitchell, B. MacDougall and M. J. Graham, J. Electrochem. Soc., 134, 1987, p.535.

6. D. F. Mitchell, R. J. Hussey and M. J. Graham, J. Vac. Sci. Technol., A2, 1984, p.789.

7. R. J. Hussey, D. F. Mitchell and M. J. Graham, Werkst. und Korros., 38, 1987, p.575.

8. J. I. Eldridge, R. J. Hussey, D. F. Mitchell and M. J. Graham, Oxid. Met., 30, 1988, p.301.

9. H. E. Bishop, P. R. Chalker and D. W. Smart, Harwell Laboratory Report AERE R-12771,1987.

10. A. Atkinson and D. W. Smart, J. Electrochem. Soc., 135, 1988, p.2886.

11. J. Stringer and P.Y. Hou, Proc. Symp. on Corrosion and Particle Erosion at High Temp., Ed. V. Srinivasan and K.Vedula, TMS, 1989, p.383.

12. M. J. Bennett, J. A. Desport, M. R. Houlton, P. A. Labun and J.M. Titchmarch, Harwell Laboratory Report AERE R-13107, 1988.

13. P. Y. Hou and J. Stringer, J. Electrochem. Soc., 134,1987, p.1836.

14. R. J. Hussey, P. Papaiacovou, J. Shen, D. F. Mitchell and M. J. Graham, Proc. Symp. on Corrosion and Particle Erosion at High Temp., Ed. V. Srinivasan and K Vedula, TMS, 1989, p.567.

15. R. J. Hussey, P. Papaiacovou, J. Shen, D.F. Mitchell and M.J. Graham, Mater. Sci. Eng., A120, 1989, p.147.

16. P. Papaiacovou, R. J. Hussey, D. F. Mitchell and M. J. Graham, Corros. Sci., (in press).

17. R. J. Hussey, P. Papaiacovou, J. Shen, D. F. Mitchell and M. J. Graham, Proc. 11th Int. Corr. Congr., April 1990, Florence, Italy (in press).

18. C. M. Cotell, G. J. Yurek, R. J. Hussey, D. F. Mitchell and M. J. Graham, Proc. Symp. High Temp. Mat. Chem.-IV, Ed. Z. A. Munir, D. Cubicciotti and H. Takagawa, Vol. 88-5, The Electrochemical Soc. Inc., 1987, p.268.

19. M. J. Graham, D.A. Downham, R. J. Hussey, D. F. Mitchell and G. I. Sproule, Proc. 4th Berkeley Conf. on Corrosion-Erosion-Wear of Materials at Elevated Temp., Ed. A. V. Levy and J. Stringer, Jan. 1990, Berkeley, U.S.A. (in press).

20. D. A. Downham, R. J. Hussey, D. F. Mitchell and M. J. Graham, Proc. Symp. on High Temp. Oxid. and Sulphid. Processes, 29th Annual Conf. of Metallurgists, Aug. 1990, Hamilton, Canada (in press).

21. N. Patibandla, T. A. Ramanarayan, R. Ayer and F. Cosandey, Proc. Symp. on Corrosion and Particle Erosion at High Temp., Ed. V. Srinivasan and K. Vedula, TMS, 1989, p.585.

22. K. Przybylski, A. J. Garratt-Reed and G. J. Yurek, J. Amer. Ceram. Soc., 69, 1986, p.C-264.

3 APPLICATIONS OF LEED AND AUGER ELECTRON SPECTROSCOPY IN THE STUDY OF HIGH TEMPERATURE CORROSION MECHANISMS

H.J. Grabke and H. Viefhaus

Max-Planck-Institut für Eisenforschung, D-4000 Düsseldorf, FRG.

ABSTRACT

In the oxidation of alloys phenomena at the interface oxide/metal are important for the nucleation, adherence and morphology of oxide scales. Such phenomena can be observed by AES and LEED in the initial stages of oxidation or after laying bare the interface by fracture or spallation of the oxide. Three studies are reported:

1. In the system Fe-Al-Ti-C the formation of a thin epitaxial layer of Ti(C,O) beneath the alumina layer favours the nucleation of α-Al_2O_3 and improves its adherence.

2. Initial formation of epitaxial CrN layers in the oxidation of Fe-20Cr in N_2-H_2O-H_2 mixtures leads to an increased roughness and porosity of the Cr_2O_3 layer, the oxidation rate is enhanced compared to oxidation in the absence of N_2.

3. Processes causing void formation beneath oxide scales are favoured and accelerated by sulphur segregation from the alloy to the surface of the voids and cavities, no sulphur segregation was observed at coherent chromia/ or alumina/metal interface.

INTRODUCTION

The analysis of scales formed on high temperature materials in corrosive atmospheres is important to characterize their protective properties and to find out the mechanisms of their formation. Such layers may be composed of several phases, different oxides but also carbides, sulfides and nitrides, in stratified or columnar structures. Inclusions of other phases may be present in an oxide scale, also metallic inclusions, but also oxides or carbides in the metal alloy. Thus, information on the chemical composition and phase distribution are of interest, as well on the vertical as on the lateral distribution.

Generally, the thickness of the protective scales on high temperature alloys is in the range 1-100 μm. There are many methods of analysis for such 'thick' layers: metallography, X-ray structure analysis, electron microprobe, chemical analysis. However, also thin layers are of interest which are formed in the initial attack of atmospheres within a few seconds or minutes since their composition and structure may have great effects on the growth, morphology and adherence of the scale. Thin layers may also be formed in environments which are nearly inert. For studying 'thin layers' surface analytical methods have to be applied, and Auger electron spectroscopy (AES) is very convenient, as will be shown by three examples of investigations on mechanisms of alumina- and chromia-scale growth.

At present, mainly the high vertical resolution of AES is important which results from the small escape depth of Auger electrons: 0.4-2.5 nm. But in the near future also improved lateral resolution is available, and details of nucleation and scale inhomogeneities can be studied. By sputter profiling the Auger analysis can be extended to layers of about 100 nm thickness, in this range the sputter mixing which leads to levelling of concentration changes, is still tolerable. For the investigation of monolayers AES can be calibrated very exactly, however, the analysis of corrosion layers by taking sputter profiles can be only a qualitative or semi-qualitative method, but this will be sufficient for most purposes. AES gives information on the chemical composition, in some cases also on the chemical binding state.

In studies on the initial stages of oxidation LEED (low energy electron diffraction) can be a useful supplement for AES, since this method delivers structural information. The low energy electrons are reflected from the uppermost layer of surface atoms, diffraction leads to a diffraction pattern, characterizing the two-dimensional arrangement of the surface atoms and giving information on the periodicity of the surface structure and on the surface lattice constants. Especially, cases of epitaxy can be detected by LEED as will be shown by two examples.

OXIDATION OF Fe-Al-Ti ALLOYS

For the protection of high temperature alloys the formation of a continuous slow growing and well adherent oxide layer is desirable. A layer of α-Al_2O_3 is thermodynamically very stable under all conditions and slow growing, however, the nucleation of α-Al_2O_3 is difficult and the layers are often badly adherent and tend to spall off. Studies have been performed on the improvement of nucleation and adherence for alumina layers on ferritic and austenitic iron base alloys by various alloying additions [1-4].

Alloys have been prepared of the compositions Fe-6Al-M and Fe-27Ni-4Al-M where M = Ti, Nb, Y, Zr, V, W, B, Si... (concentrations in wt.%). One or more alloying elements have been added in the concentration range 0.1 to 1% and in some cases carbon.

The oxidation experiments have been performed at 1000°C in a H_2-H_2O atmosphere where the oxygen pressure corresponds to $p_{O_2} = 10^{-19}$ bar. After 1/2 h oxidation, the oxide layers were characterized by X-ray structure analysis, scanning electron microscopy and Auger electron spectroscopy (AES). Additional experiments have been performed in an ultra high vacuum-system, using LEED and AES to observe nucleation processes and epitaxial relations and taking AES sputter profiles for analysis of thin oxide layers [2,3].

The oxidation of ferritic Fe-6Al-M alloys leads to an initial formation of θ-Al_2O_3 before α-Al_2O_3 is nucleated.[1] The undoped Fe-6Al and most other studied alloys show badly adherent oxide layers. These layers spall off from the alloy according to growth stresses, in the isothermal experiment and even more during thermal cycling. In contrast, the oxide layers on alloys with additions of Ti, Zr, V, or Y are fine-grained and well adherent. Especially favourable oxidation properties were observed for the alloys Fe-6Al-0.5Ti and Fe-6Al-1Ti. The AES sputter profile of such alloy after 1/2 h oxidation at 1000°C indicates the presence of a layer of Ti-oxicarbide beneath the Al_2O_3 layer, some TiO_2 is present on the outer surface, Fig. 1.

The formation of the oxicarbide sublayer was simulated in an UHV experiment. The phenomena occurring on an Fe-6Al-0.4Ti-0.01C single crystal with (100) orientation were followed by AES and LEED. The surface of such specimen was cleaned by heating and sputtering under very good UHV conditions. After that the specimen was heated to 900°C and exposed to some oxygen at $p_{O_2} = 10^{-8}$ mbar. In this experiment, a simultaneous increase of the Auger peaks of Ti, O and C was observed while the Al at first stays unoxidized, see Fig. 2. Obviously induced by the chemisorption of oxygen, Ti segregates to the surface, also carbon diffuses to the surface and these elements together form a layer of Ti(C,O). Further, O_2 exposure leads to formation of α-Al_2O_3, the presence of Ti(C,O) favours the nucleation of this oxide. During these processes the LEED pattern stays unchanged showing always a clear 1x1 pattern. This observation indicates that the oxicarbide and the alumina form ordered structures with the same periodicities as the underlying Fe(100) surface. Such epitaxial growth is possible of Ti(C,O) according to

$$(100)_{Fe} \,/\!/\, (100)_{Ti\,(C,O)}$$
$$[100]_{Fe} \,/\!/\, [110]_{Ti\,(C,O)}$$

similar to the case of wustite growth on iron.[5] According to the LEED pattern the lattice constant of the first layer of Ti(C,O) corresponds to the lattice constant of the alloy. X-ray analysis indicates increasing values for the lattice constant in the range 0.4303...0.4315 nm for increasing thickness of the layer. Thus a gradual transition is possible from the lattice constant of the ferritic substrate to the lattice constant of the oxide. Compared to the lattice constant of pure TiC 0.4328 nm, the observed values are smaller, which most probably is caused by the presence of oxygen in the lattice. The average composition $Ti(C_{0.8}O_{0.2})$ may be deduced from the lattice constant.

The α-Al_2O_3 grows expitaxially on the oxicarbide according to

$$(111)_{Ti(C,O)} \,/\!/\, (001)_{Al_2O_3}$$
$$[110]_{Ti(C,O)} \,/\!/\, [110]_{Al_2O_3}$$

So the oxicarbide acts as a "graded seal" between the metal substrate and the oxide layer, the continuous transition of lattice constants between substrate and oxide layer allows an excellent adherence without growth stresses.

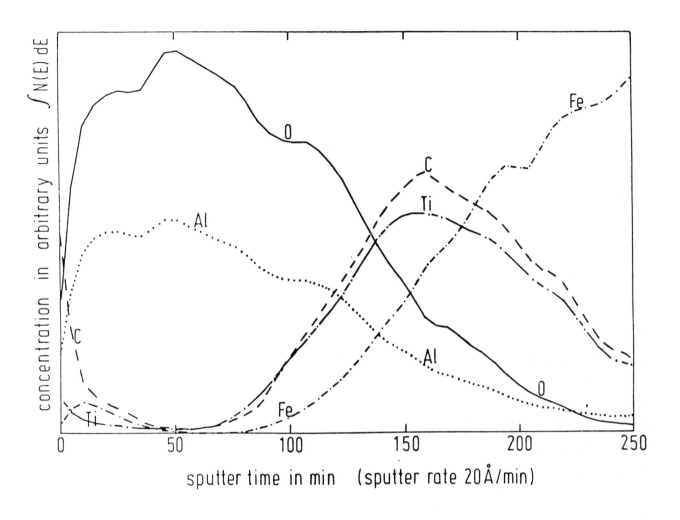

Figure 1: AES sputter profile of Fe-6Al-0.5Ti after oxidation at 1000°C in H_2-H_2O (sputtering rate 2 nm/min).

On the alloy Fe-6Al-0.5Nb an oxicarbide Nb(C,O) was observed, however, the presence of that oxide does not improve the adherence of the Al_2O_3 layer. On an alloy with 0.5% V and 0.01% C, no carbide or oxicarbide was detected. In contrast, the austenitic alloys with Ti and Zr formed continuous, well adherent α-Al_2O_3 layers, especially well behaved was the alloy Fe-4Al-0.5Ti-0.05C.

The oxidation of some of the alloys was also followed by gravimetric measurements in a microbalance. The alloys with Ti or Zr additions which form the well adherent oxide layers show a relatively high oxidation rate compared to Fe-6Al at least before oxide layer spallation occurs for the undoped alloy. The increased rate may be caused by increased inwards diffusion of oxygen at the grain boundaries of the fine-grained oxide layer on the doped alloys, but also by outward diffusion of Al^{3+} cations, since there may be a doping effect of Ti^{4+} and Zr^{4+} ions in the oxide layer on the doped alloys.

The good adherence and the protective properties of the scales on the Fe-Al-Ti alloys was also demonstrated by carbon permeation tests, in these tests after oxidation, samples were exposed at 900°C to an atmosphere containing radiocarbon C^{14} for long time and after that the specimens were analyzed for the uptake of radiocarbon. The carbon permeation through oxide layers on these Fe-Al-Ti alloys was unmeasurably small, obviously the pores and cracks of the alumina scale are sealed and closed by formation of Ti(C,O) with the intruding carbon. In contrast, there was a high carbon permeation in the case of undoped Fe-6Al, caused by cracking and spalling of the oxide layer.

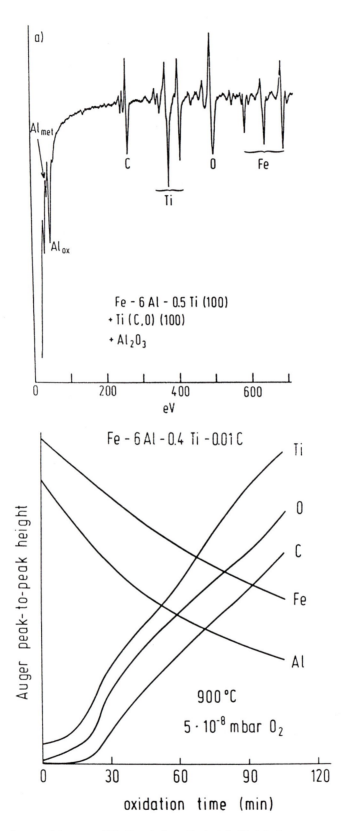

Figure 2: Simultaneous surface enrichment of Ti, C and O on Fe-6Al-0.4Ti-0.01C, induced by oxygen chemisorption at 900°C and $p_{O_2} = 5 \cdot 10^{-8}$mbar
   (a) Kinetics of the simultaneous chemisorption and segregation, recorded by AES.
   (b) Auger spectra of the layers of Ti(C,O) and Al$_2$O$_3$ formed by simultaneous chemisorption and segregation.

EFFECTS OF NITROGEN IN THE OXIDATION OF Fe-Cr ALLOYS

In the previous chapter it was demonstrated that the formation of intermediate layers can have positive effects in the formation and on the adhesion of oxide layers. On the other hand, formation of intermediate layers also can have negative effects in the formation of scales.

This was demonstrated in the study on the oxidation of Fe20Cr in H_2-H_2O mixtures at 900°C,[7,8] under these conditions usually a protective slow growing Cr_2O_3 layer is obtained. However, in the gravimetric measurement of the oxidation, differences were observed when using He or N_2 as a carrier gas. The samples were heated to the reaction temperature in hydrogen so that the clean oxide-free metal surface was exposed to the attack of He-H_2-H_2O or N_2-H_2-H_2O mixtures, respectively. The rate of mass increase is much faster in the presence of N_2 than in the nitrogen-free atmosphere, see Fig. 3. Especially during the initial period of oxidation in the presence of N_2 the oxidation rate is strongly enhanced, the layer must grow for some time to a larger thickness than without N_2 before a dense protective oxide layer is formed. After that initial period the parabolic law of oxide growth applies, but the growth rate is still faster than in the experiment in the absence of N_2. The parabolic constants of scale growth for the experiments in Fig. 3 are

$$He\text{-}H_2\text{-}H_2O\ k_p = 0.6 \cdot 10^{-13}\ g^2/cm^{-4}s^{-1}$$
$$N_2\text{-}H_2\text{-}H_2O\ k_p = 1.5 \cdot 10^{-13}\ g^2/cm^{-4}s^{-1}$$

Equivalent experiments were conducted adding 500 vppm HCl gas, this causes evaporation of $CrCl_2$ by the reaction

$$Cr_2O_3 + 2H_2 + 2HC = 2CrCl_2(g) + 3H_2O$$

The loss of mass by $CrCl_2$ evaporation leads to a retardation and at least to the stoppage of the Cr_2O_3 scale growth, as can be seen from Fig. 3. The evaporation is more pronounced in the case of oxidation in the presence N_2 (Fig. 3) indicating a higher reactivity of the Cr_2O_3 scale. The reason can be seen in scanning electron micrographs of the chromia scale (Fig. 4) formed in the absence and in the presence of N_2 - the latter scale shows enhanced roughness and porosity.

The differences in scale morphology are caused by processes in the initial stages of oxidation, as was demonstrated by AES investigation of layers formed in short time exposures (Fig. 5). After 1 min oxidation of Fe-20Cr in N_2-H_2-H_2O at 900°C the surface layer contains much more nitrogen than oxygen. Obviously, chromium nitride has been formed besides chromium oxide which is the stable phase. After 2 min, the nitrogen enrichment disappeared, a distinct oxide layer is perceptible and the nitrogen is distributed equally into the alloy. The observations demonstrate that in the initial state of attack both chromium oxide and nitride are formed and later on the nitride is overgrown by the oxide and dissolves into the alloy. However, the initial formation of chromium nitride obviously effects a marked roughness and porosity of the overgrowing Cr_2O_3 layer. The process is different in He-H_2-H_2O, no compound other than Cr_2O_3 is formed and a homogeneous, dense oxide layer can grow from the beginning.

It may be noted that these observations apply for the ferritic alloy Fe-20Cr, on the austenitic alloy Fe-20Cr-40Ni, no nitride was observed, only chromium oxide formation. The susceptibility of the ferritic alloy to nitride formation can be related to the epitaxy of ferrite and the CrN lattice[9]. This epitaxy favours the formation of CrN, as monolayer and as multilayer compounds on the ferritic alloy [9,10], whereas on the austenitic alloy there is no such epitaxy. Generally, nitride formation is possible on both alloys during attack of the virgin alloy surface, however, only on the ferritic alloy does this process play a portentous role.

EFFECTS OF SULPHUR ON THE ADHESION OF OXIDE SCALES

Some authors [11-14] have put forward the hypothesis that sulphur segregating at the interface oxide/metal can decrease the adherence of oxide scales. However, as yet it has not been proved convincingly that sulphur segregation can indeed occur at that interface for chromia or alumina forming alloys. To clarify this question, investigations have been performed on the alloys Fe-15Cr-160ppm S, Ni10Cr-9Al-90ppm S and on the intermetallic compound β-NiAl containing 110 ppm S. An oxidation chamber was constructed in which specimens could be oxidized at temperatures up to 1000°C and about 10^{-5}mbar O_2. The oxidation chamber was joined to an UHV-system by a gate lock so that the specimens could be transferred for sputtering and AES analysis before and after oxidation.

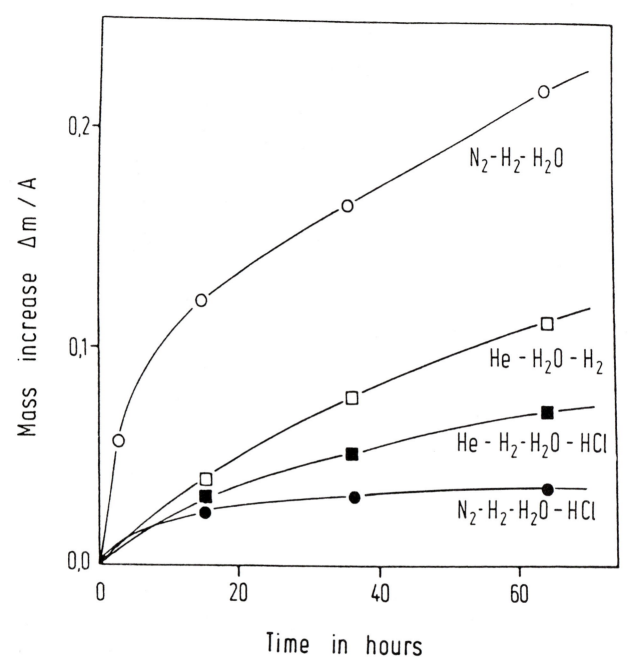

Figure 3: Effect of nitrogen in the oxidation of Fe-20Cr at 900°C, increase of oxidation rate in N_2-H_2-H_2O compared to He-H_2-H_2O and increase in formation of volatile $CrCl_2$ in N_2-H_2-H_2O-HCl compared to He-H_2-H_2O-HCl.

In the study of the alloy Fe-15Cr-160ppm at first the surface segregation on the clean metallic surface was observed at 800°C, and it was confirmed that after 5 min complete saturation of the surface with the segregated sulphur is attained. Afterwards the specimen was cleansed by sputtering at room temperature and oxidized for 10 min at 500°C in 10^{-5} mbar O_2. The next step was a 'segregation anneal' at 800°C for 5 min to establish segregation equilibrium and to attain sulphur segregation beneath the oxide - in case such segregation occurs. After that in an AES sputter profile no sulphur enrichment could be detected beneath the scale (Fig. 6a) a saturated monolayer of sulphur would have been seen clearly in the depth profile. The oxide layer had a thickness of about 30 nm and is composed of spinel in the outer part and of Cr_2O_3 in the inner region.

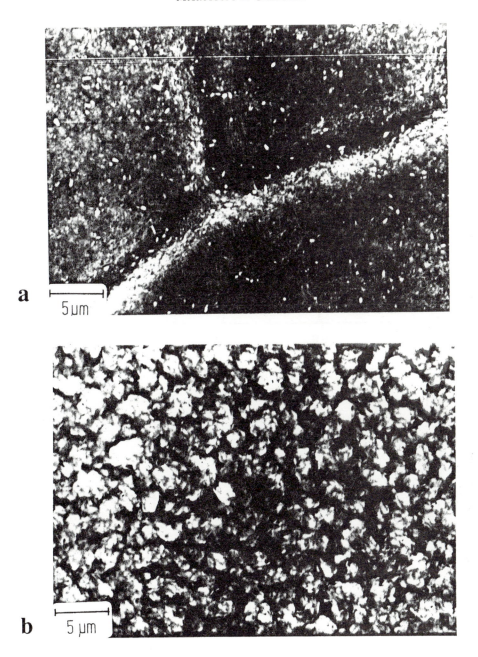

Figure 4: Scanning electron micrographs (top view) of the Cr_2O_3 scales formed on Fe-20Cr
    (a) after 60 h oxidation in He-H_2-H_2O and
    (b) after 60 h oxidation in N_2-H_2-H_2O.

The alloy Ni-10Cr-9Al-90ppm S had to be heated for 10 min to 900°C to attain saturation of the free metallic surface with segregated sulphur. Correspondingly, the following sequence was performed with that alloy: cleaning by sputtering, oxidation at 800°C in 10^{-5} mbar O_2, annealing for 10 min at 900°C and 10^{-9} mbar, after that the sputter profile was taken (Fig. 6b). A minor enrichment of sulphur was detected at the outer surface of the alumina layer, the sulphur concentration decreases towards the inner phase boundary, here at the interface Al_2O_3/metal is no marked sulphur enrichment.

The intermetallic phase β-NiAl with 90 ppm S was investigated in a similar way as the Ni-Cr-Al alloy. An Al_2O_3 layer of about 15 nm thickness was formed, and again in taking the depth profile no S enrichment was detected upon reaching the oxide/metal interface (Fig. 6c).

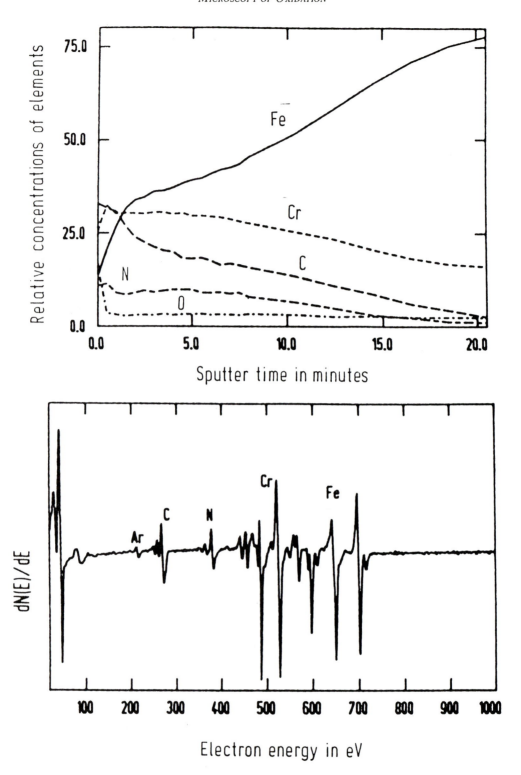

Figure 5a: AES study of Fe-20Cr after short time oxidation in N_2-H_2-H_2O at 900°C, Auger spectra of the surface and Auger sputter profiles. (After 1 min exposure, mainly chromium nitride on the surface.)

In these experiments with chromia and alumina forming alloys, the samples were oxidized to obtain a thin oxide layer and then annealed to attain segregation equilibrium, but no sulphur segregation was detected beneath the oxide layer. Now, these materials were also oxidized for long time to obtain thick oxide layers, afterwards the samples were introduced into the vacuum chamber and somehow the oxide scale was partially removed. This was possible by different methods,

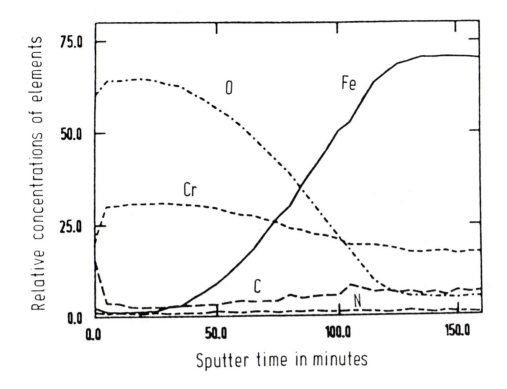

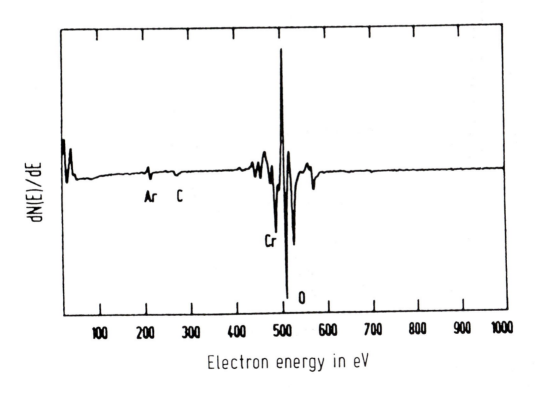

Figure 5b: AES study of Fe-20Cr after short time oxidation in N_2-H_2-H_2O at 900°C, Auger spectra of the surface and Auger sputter profiles. After 2 min exposure, the chromium nitride is overgrown by chromium oxide Cr_2O_3 and the nitrogen is dissolved into the alloy.

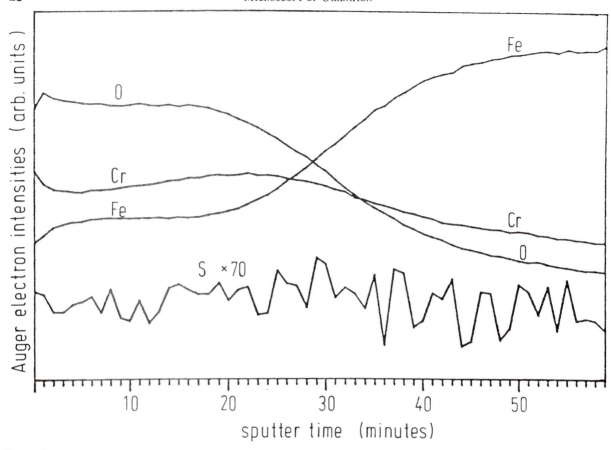

Figure 6a: Auger sputter profiles after short time oxidation of Fe-15Cr-160ppm, no sulphur enrichment.

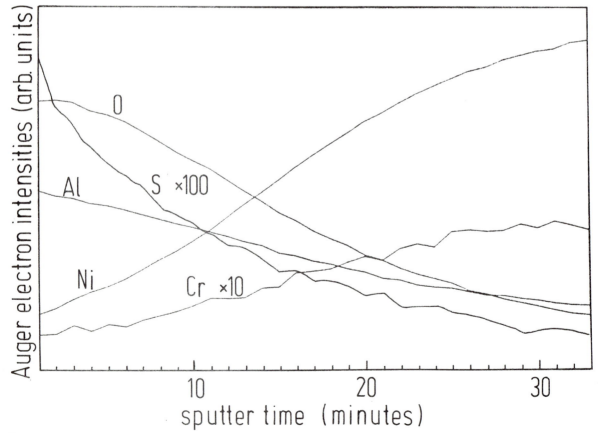

Figure 6b: Auger sputter profiles after short time oxidation of Ni-10Cr-9Al-90ppm S, a minor sulphur enrichment on the surface and in the scale.

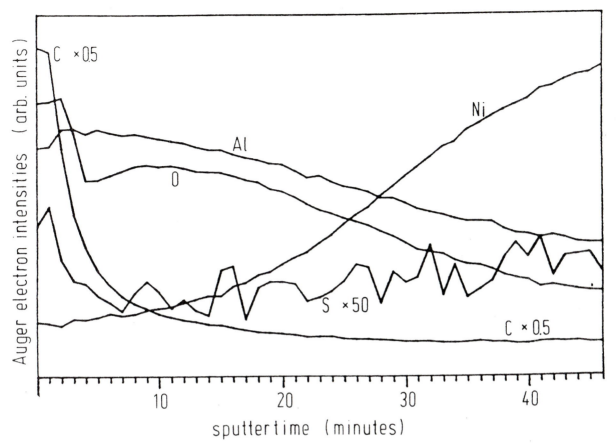

Figure 6c: Auger sputter profiles after short time oxidation of β-NiAl with 90 ppm S, no sulphur enrichment.

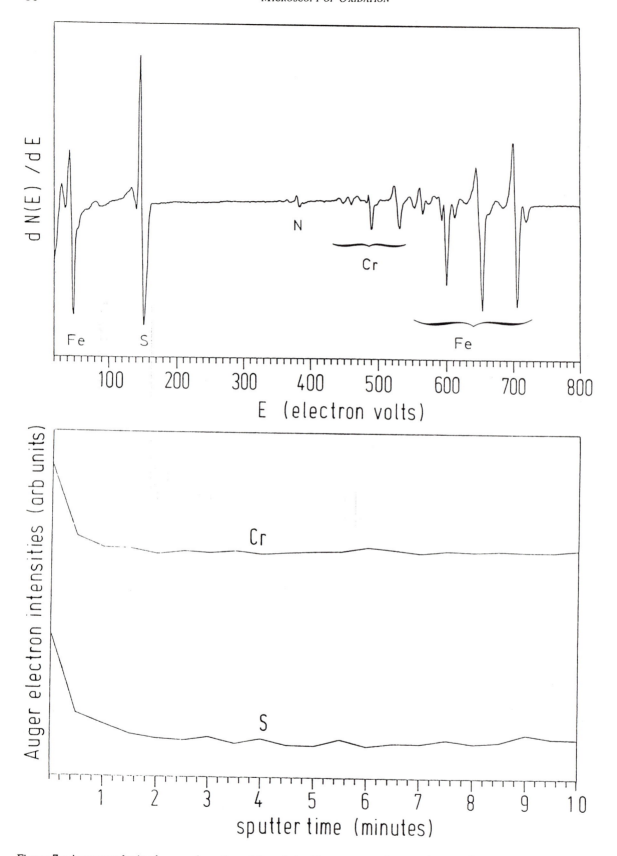

Figure 7a: Auger analysis of areas where the oxide scale had been removed on specimens after long time oxidation. Auger spectrum and sputter profile of sulphur, recorded on oxide free area of Fe-15Cr-160ppm S after 3 h oxidation.

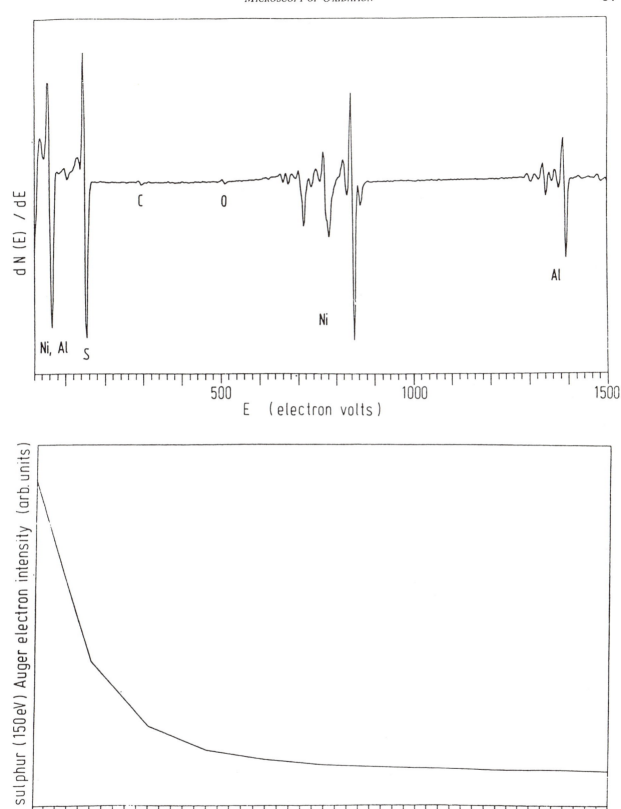

Figure 7b: Auger analysis of areas where the oxide scale had been removed on specimens after long time oxidation. Auger spectrum and sputter profile of sulphur, recorded on oxide free area of β-NiAl after 5 h oxidation.

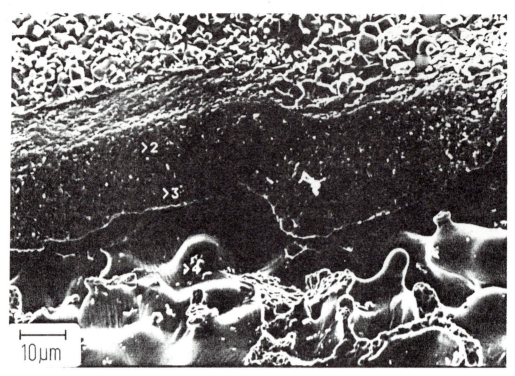

Figure 8a: Scanning electron micrograph of sample after long time oxidation, oxide scale partially removed. Fe-15Cr-160ppm S, after 3 h at 900°C and $6 \cdot 10^{-3}$ mbar O_2.

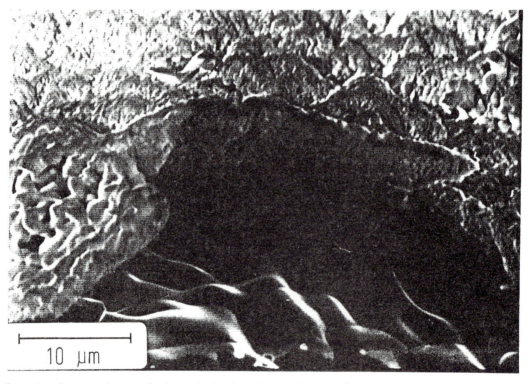

Figure 8b: Scanning electron micrograph of sample after long time oxidation, oxide scale partially removed. β-NiAl-90ppm S, after 5 h at 1000°C and 10^{-2} mbar O_2.

by bending or scratching the sample or by thermal cycling. In all cases sulphur was observed now by AES at spots where the oxide scale had spalled. This is demonstrated for Fe-15Cr-160ppm S which had been oxidized for 3 h in $6 \cdot 10^{-3}$ mbar O_2 at 900°C (Fig. 7a), the sputter profile indicates that the sulfur enrichment corresponds to a monolayer. However, when looking at such oxide-free spots of the alloy in the scanning electron microscope (SEM), one can see a lot of cavities beneath the oxide scale, the scale is linked to the metal only by small bridges (Fig. 8a). The suspicion arises that the sulphur is segregated not at the interface oxide/metal but at the surface of these cavities. By SEM investigation the presence of cavities at the interface oxide/metal was also demonstrated for the other materials, the Ni-Cr-Al alloy and β-NiAl, after long time oxidation and after removing the oxide scale. SEM photos are shown (Fig. 8b) of β-NiAl which had been oxidized for 5 h at 1000°C and at 10^{-2} mbar. Again cavities can be seen at oxide free areas, and sulphur segregation was detected by AES corresponding to one monolayer, see Fig. 7b. Obviously, sulphur does not segregate to an interface chromia/metal or alumina/metal, but the presence of sulphur accelerates the nucleation and growth of voids and cavities beneath these scales. There are several mechanisms in the growth of these oxides which may lead to formation of voids beneath the scale and these processes are strongly favoured by the highly surface-active sulphur, which decreases its surface energy upon surface segregation in the void.

REFERENCES

1. J. Peters and H.J. Grabke, Werkstoffe u. Korrosion 35, 1984, p.385.

2. J. Peters, H.J. Grabke and H. Viefhaus, Proceed. 10th Int. Congr. Reactivity of Solids, Dijon 1984, Eds. P. Barret and L.C. Dufour, Elsevier 1985, p.151.

3. H. Viefhaus, J. Peters and H.J. Grabke, Surface Interface Anal. 10, 1987, p.280.

4. K. Wambach, J. Peters and H.J. Grabke, Mat. Sci. Engng. 88, 1987, p.205.

5. J. Bardolle and J. Benard, Rev. Met. 49, 1952, p.613.

6. H.J. Grabke, K. Ohla, J. Peters and I. Wolf, Werkstoffe u. Korrosion 34, 1983, p.495.

7. D. Bramhoff, H.J. Grabke and P. Schmidt, Werkstoffe u. Korrosion 40, 1989, p.642.

8. D. Bramhoff, H.J. Grabke and P. Schmidt, The role of active elements, Ed. E. Lang, Elsevier Appl. Sci. 1989, p.325.

9. C. Uebing, H. Viefhaus and H.J. Grabke, Appl. Surf. Sci. 32, 1988, p.363.

10. C. Uebing, Surface Sci. 225, 1990, p.97.

11. A.W. Funkenbusch, J.G. Smeggil and N.S. Bornstein, Metall. Trans. 16A, 1985, p.1164.

12. J.G. Smeggil, A.W. Funkenbusch and N.S. Bornstein, Metall. Trans. 17A, 1986, p.923.

13. D.G. Lees, Oxid. Met. 27, 1987, p.75.

14. J.L. Smialek, Metall. Trans. 18A, 1987, p.163.

15. D. Wiemer, Untersuchungen der Grenzfläche Oxid/Metall bei der Oxidation von Modellegierungen und intermetallischen Phasen, Dr.-Thesis, Universität Dortmund, 1990.

SECTION TWO
Growth and failure of iron oxide scales

4 A SIMS INVESTIGATION OF THE OXIDATION OF IRON IN OXYGEN

M. Montgomery, P.L. Harrison* and D.G. Lees

Manchester Materials Science Centre,
Manchester University and UMIST,
Grosvenor Street, Manchester M1 7HS, UK.
*National Power TEC (formerly C.E.R.L.),
Kelvin Avenue, Leatherhead, Surrey KT22 7SE, UK.

ABSTRACT

Iron specimens have been oxidised sequentially at 550°C in natural oxygen and oxygen enriched with the oxygen-18 isotope at 0.1 atm. The oxides formed were magnetite (Fe_3O_4) and haematite (Fe_2O_3).

Examination of the cross-sectioned scales by scanning electron microscopy (SEM) showed that the magnetite consisted of a small inner columnar layer, a middle columnar layer and an outer porous columnar layer at the magnetite-haematite interface.

The location of the oxygen-18 in the cross-sectioned scales was investigated by means of imaging SIMS (Secondary Ion Mass Spectrometry). The results show that substantial cation transport occurred in the scale but that appreciable short-circuit transport of oxygen also took place either via grain-boundaries or micropores.

INTRODUCTION

A considerable amount of research has been carried out over the years on the oxidation behaviour of iron [1]. This is of technological importance because of the problems with corrosion of ferritic steels in gas-cooled nuclear reactors. The growth-mechanisms of magnetite in carbon dioxide have previously been studied using charged particle nuclear techniques by Taylor et al. [2] and Gleave et al. [3]. In the present work, imaging SIMS has been used to investigate the growth-mechanism of magnetite in oxygen as part of a wider study of the growth-mechanism of this oxide. The use of imaging SIMS in this way has been shown to be a powerful method for investigating oxide growth-mechanisms [4,5].

EXPERIMENTAL PROCEDURE

Coupons of pure iron were oxidised in 0.1 atm. oxygen. The samples were sequentially oxidised, that is, they were oxidised firstly in oxygen-16 and then in oxygen-16 enriched with the oxygen-18 isotope. The gas changeover was done at temperature and the samples were cooled under vacuum to prevent the migration of oxygen-18 down cracks which may form. They were oxidised in an apparatus that monitors the kinetics and the kinetics showed no rapid increases during the oxidation and were similar to those of other workers [6,7]. X-ray analysis was used to check for the presence of magnetite and haematite. The samples were then gold-coated, nickel-plated and cross-sectioned. The cross-sectioned samples were mounted in Woods Metal and studied with the optical microscope. They were then etched with 50% hydrochloric acid to reveal the grain structure and examined by means of the SEM. The cross-sectioned oxide was studied using imaging SIMS to locate the oxygen-18 within the oxide. A beam of spot size 0.3 μm was raster scanned over a defined area so that it just removed the surface atoms of the cross-sectioned scale. These atoms were detected by a Time of Flight Mass Spectrometer which was linked to a computer. In this way an image was built up which showed the positions of the oxygen-16 and oxygen-18 in the oxide.

RESULTS

In all the samples studied, both magnetite and haematite were present; the haematite layer could not always be seen optically but was detected by X-ray analysis. Magnetite formed a thick inner layer and haematite formed a much thinner outer layer. The magnetite consisted of a small columnar layer at the metal-oxide interface, a middle columnar layer and an outer more porous columnar layer.

Samples were oxidised for 6 h at 550°C. They were oxidised for the first 1.5 h in oxygen-16 and for a further 4.5 h in the oxygen-18 enriched gas. Figure 1 which is an SEM micrograph of an area analysed with SIMS shows the three layers in the magnetite. The total thickness of the oxide formed was about 12 μm. The imaging SIMS results for oxygen-16 and oxygen-18 are shown in Figs. 2a and 2b respectively. If oxidation had occurred by cation diffusion only, there would be a band of oxygen-18 enriched gas in the outer half of the oxide. There is a band of oxygen-18 at the gas-oxide interface as expected but one can also see some oxygen-18 throughout the scale. Point analyses at positions across the oxide gave a concentration profile of the enriched gas through the oxide (Fig. 3). This shows the presence of oxygen-18 as a band at the gas-oxide interface and it also confirms that there was oxygen-18 throughout the scale up to the metal-oxide interface. It also shows that the oxide at the gas-oxide interface does not contain 100% of the second gas, perhaps indicating that oxide was formed in pre-existing oxide.

Iron coupons were oxidised at 550°C in oxygen-16 for 46 hrs. and oxygen-18 for the final 19 h. From the kinetics which were monitored, 1/5 of the oxide should have grown in oxygen-18 enriched gas. The oxide formed here was 30 μm thick, of which 1.5 μm was haematite (Fig. 4). Imaging SIMS (Figs. 5a and 5b) showed that there was a band of oxygen-18 at the gas-oxide interface and oxygen-18 was present throughout the scale. The concentration profile (Fig. 6) confirms this result.

If diffusion was occurring, as this indicated, then if the time given for oxidation in oxygen-18 was increased relative to that in oxygen-16, there should be a build-up of oxygen-18 at the metal-oxide interface. For this reason, the following experiment was undertaken. A coupon of iron was oxidised for the first 0.5 h in oxygen-16, and then for a further 5.5 h in the oxygen-18 enriched gas at 530°C. The oxide formed was 9 μm thick (Fig. 7); the haematite was not optically visible although it was detected with X-ray analysis. SIMS analysis (Fig. 8a and 8b) showed a band of oxygen-18 at the gas-oxide interface as seen in the scales on the previous specimens. The oxygen-18 concentration profile (Fig. 9) through the scale confirmed this but, in addition, there was a peak at the metal-oxide interface.

DISCUSSION AND CONCLUSIONS

It has been concluded by previous workers [7,8] that magnetite grows predominantly by cation diffusion outwards, and this work is in agreement with that conclusion. In addition, the results from this work indicate that oxygen transport has taken place in the magnetite. The oxygen-18 profile is similar to the schematic graph given by Basu and Halloran [9] to describe a scale growing by both outward metal transport and inward oxygen grain-boundary transport with exchange.

Taylor et al. [2] also used oxygen-18 as a tracer to investigate the oxidation of Fe-9%Cr alloys at 580°C. A nuclear technique was used to locate the oxygen-18. They concluded that the growth of magnetite was by cation diffusion outwards and that there was no significant amount of anion diffusion inwards. However their experiments were conducted in 1 atm. carbon dioxide and not oxygen. Gleave et al. [3] conducted similar oxygen-18 tracer experiments in high pressure carbon dioxide at 500°C on low-silicon mild steel. They found that in the protective region of oxidation, new oxide was formed predominantly at the oxide-gas interface; however in one case, the tail of the oxygen-18 extended all the way to the oxide-metal interface. This gave a similar profile to Figs. 3 and 6. Further work carried out by Lees et al. [4] on this sample using the SIMS technique also showed the presence of oxygen-18 throughout the scale. The explanation given for this by Gleave et al. [3] was that oxygen transport took place by gaseous diffusion of carbon dioxide through micropores or cracks. In our work, we believe that short-circuit oxygen transport takes place via either grain-boundaries or micropores, and further investigation of these two possibilities is required. However, a direct comparison with our results cannot be made as the gaseous environments and pressures are different.

ACKNOWLEDGEMENTS

We would like to acknowledge the financial support given by the Science and Engineering Research Council in the form of a research grant and by the C.E.G.B in the form of a CASE studentship, and the support given in many forms by the Central Electricity Research Laboratories, Leatherhead.

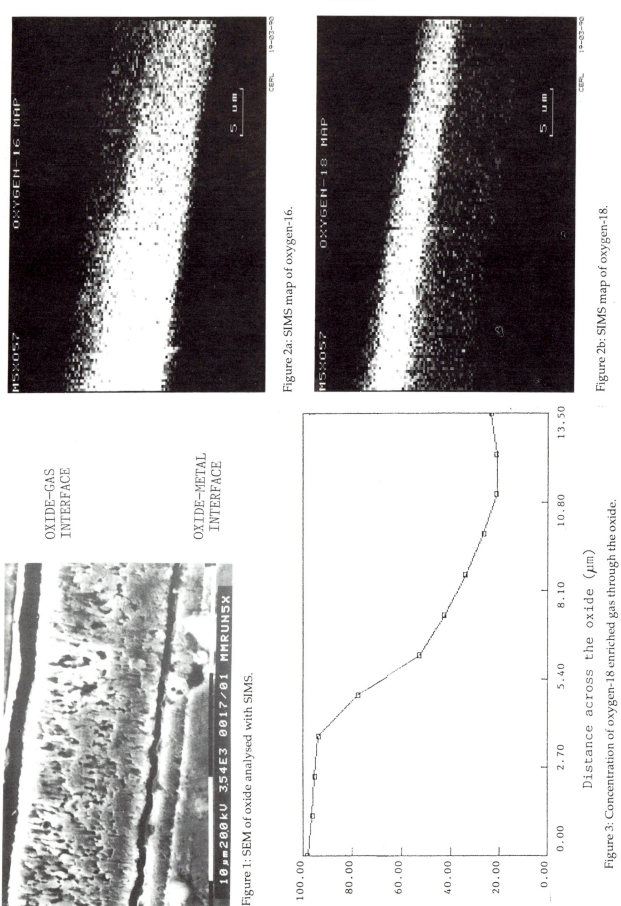

Figure 2a: SIMS map of oxygen-16.

Figure 2b: SIMS map of oxygen-18.

OXIDE–GAS
INTERFACE

OXIDE–METAL
INTERFACE

Figure 1: SEM of oxide analysed with SIMS.

Figure 3: Concentration of oxygen-18 enriched gas through the oxide.

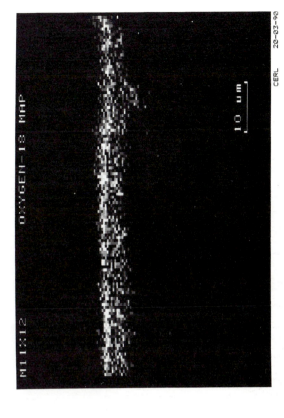

Figure 5a: SIMS map of oxygen-16.

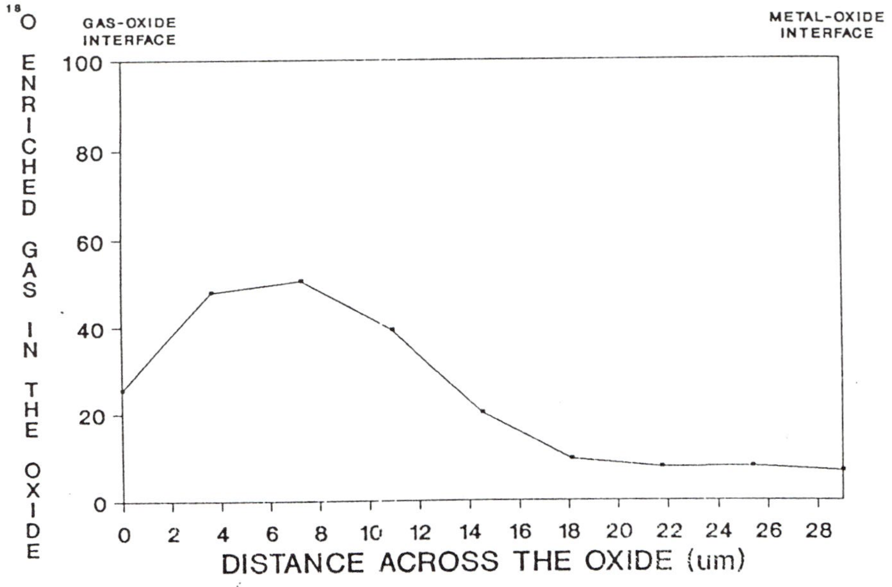

Figure 5b: SIMS map of oxygen-18.

Figure 4: SEM of oxide analysed with SIMS.

Figure 6: Concentration of oxygen-18 enriched gas through the oxide.

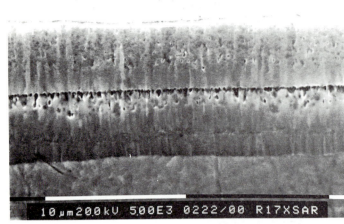

OXIDE–GAS
INTERFACE

OXIDE–METAL
INTERFACE

Figure 7: SEM of oxide analysed with SIMS.

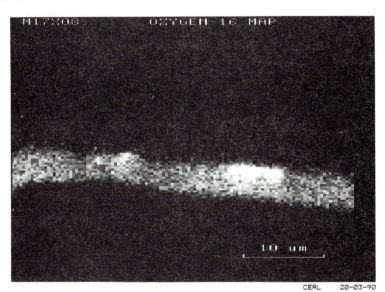

Figure 8a: SIMS map of oxygen-16.

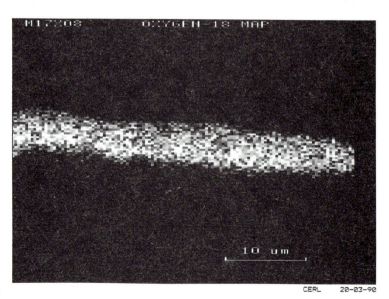

Figure 8b: SIMS map of oxygen-18.

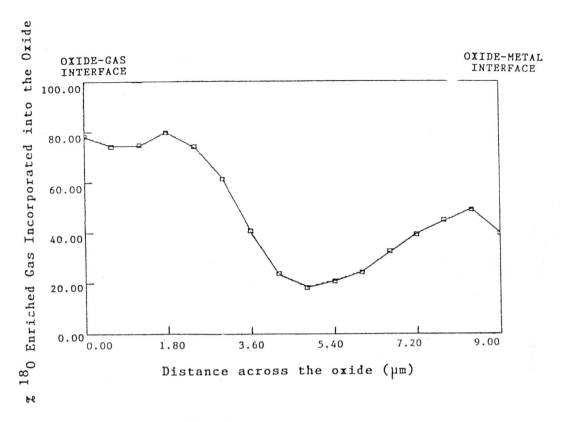

Figure 9: Concentration of oxygen-18 enriched gas through the oxide.

REFERENCES

1. P. Kofstad, High Temperature Corrosion, Elsevier Applied Science, 1988.

2. M.R. Taylor, J.M. Calvert, D.G. Lees and D.B. Meadowcroft, Oxid. Met. 14, 1980, p.499.

3. C. Gleave, J.M. Calvert, D.G. Lees and P.C. Rowlands, Proc. Royal Soc. Lond. 379, 1982, p.409.

4. D.G. Lees, P.C. Rowlands and A. Brown, Inst. Phys. Conf. Ser. No.90, 1980, p.57.

5. D.G. Lees, P.C. Rowlands and P. Humphrey, J. Mat. Sci. Tech. 4, 1988, p.1117.

6. R.J. Hussey, G.I. Sproule, D. Caplan and M.J. Graham, Oxid. Met. 15, 1981. p.421.

7. M.H. Davies, M.T. Simnad and C.E. Birchenall, Trans. AIME 3, 1951, p.889.

8. A. Brueckmann and G. Simkovich, Corr. Sci. 12, 1972, p.595.

9. S.N. Basu and J.W. Halloran, Oxid. Met. 27, 1987, p.143.

5 COATINGS FOR RUSTY STEEL: USE OF ANALYTICAL TECHNIQUES TO ELUCIDATE CORROSION BEHAVIOUR

N. L. Thomas

ICI Chemicals and Polymers, The Heath, Runcorn, Cheshire, UK.

ABSTRACT

A range of complimentary analytical techniques, including EPMA, SEM, EDX, TEM and LAMMS, have been used to investigate problems associated with protection of poorly prepared steel substrates by paint coatings. In particular, work has focussed on characterisation of rusts produced by natural weathering, coating failure mechanisms and the excellent performance of red lead oil-based paints.

INTRODUCTION

There is considerable commercial interest in paint coatings which will give good performance when applied to wire-brushed rusty steel substrates [1, 2]. This is because minimum surface preparation requirements prior to painting are difficult to achieve in many practical situations, due to the inaccessibility of structures or to cost and time constraints. In order to develop paint systems which are tolerant to poorly prepared surfaces it is necessary to gain a fundamental understanding of why many coatings fail on such substrates and why some systems, such as red lead oil-based primers, can give remarkably good performance. This paper describes how a range of complementary analytical techniques has been used to investigate this problem.

CHARACTERISATION OF RUST

It is important to have a well characterised rusty steel surface if results of coatings performance are to be meaningful.

Morphology of Natural and Synthetic Rusts

Figure 1 is a back-scattered scanning electron micrograph of a polished cross-section of wire-brushed rusty steel. The sample was taken from a panel which had been allowed to rust for 3 months at an industrial/marine site at Widnes in Cheshire. Although the original surface of the panel was quite smooth, after rusting for 3 months the steel surface has a rough topography. There are clearly defined regions where the steel has undergone more extensive corrosion: these are the so-called anodic sites and are the areas of active corrosion. Note that the rust is readily distinguished from the steel in Fig. 1 because of the atomic number contrast in the back-scattered electron image.

Semi-quantitative X-ray diffraction was used to identify the phases present and to give an estimate of the amorphous content of the rust. The major phase (>50%) in Widnes rust was found to be poorly crystalline lepidocrocite (γ-FeOOH). A small amount of magnetite (Fe_3O_4) was detected together with trace amounts of goethite (α-FeOOH) and haematite (Fe_2O_3). It was estimated that there was less than 15% of amorphous material present.

The rust produced on natural weathering develops as a compact, adherent layer, which is difficult to remove by wire-brushing. Its morphology is quite different from that of synthetic rusts produced in accelerated weathering tests, as illustrated in Fig. 2. The natural rust (Fig. 2a) is much more solid and compact than the porous, spongy rust grown in a salt-spray cabinet (Fig. 2b).

All coating tests were carried out on panels rusted by exterior exposure at the industrial/marine site at Widnes.

The Presence of Contaminants

Atmospherically grown rusts contain contaminants in the form of sulphate and chloride salts, which become active sites for further corrosion processes [3, 4]. To investigate the concentration and distribution of such salts, rusty steel samples were sectioned and polished (using isopropanol lubricant) and examined using Wave-length dispersive X-ray Analysis (WDXA) in a Cameca Camebax Electron Micro Probe (EPMA) at 20kV.

Figure 1: Polished section of wire-brushed rusty steel.
SEM backscattered image.

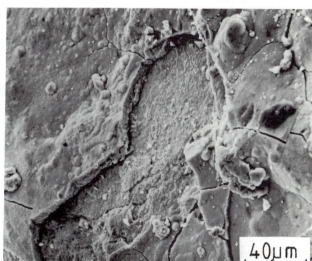

Figure 2a: Exterior exposure rust-plan view.

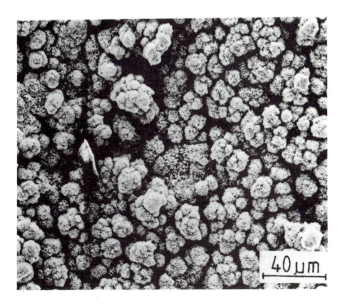

Figure 2b: Salt spray rust-plan view.

The major contaminants in Widnes rust were S and Cl, present at between 0.5 and 1.5 wt%. Elemental maps revealed that they are not uniformly distributed. Figures 3(a) and (b) are Fe and S maps respectively from a typical area, and show how the S is particularly concentrated at the steel/rust interface. Because the contaminants are deeply embedded in compact rust they are not easily removed by wire brushing.

Detection of Millscale

The sensitivity of the back-scattered signal to small differences in average atomic number allows differentiation between rust phases. This is illustrated in Fig. 4, where there is a finger of rust which shows up brighter than the surrounding material. An Fe scan across the sample revealed that this was an iron oxide phase with a higher Fe content than the surrounding rust, which is mainly lepidocrocite (γ-FeOOH). It was deduced that this was a layer of millscale, which is largely magnetite (Fe_3O_4), and has a higher average atomic number than lepidocrocite and therefore shows up brighter in the back-scattered image.

The ability to detect millscale and differentiate between Fe_3O_4 and γ-FeOOH was very useful when analysing coating failures.

ELUCIDATION OF FAILURE MECHANISMS

There are several factors governing the performance of coatings on rusty steel.

(i) Contamination of rust by aggressive ions is viewed as the most serious problem [5]. Mayne [6] has shown that the life of a paint system applied to rusty steel depends on the ferrous sulphate content of the rust. Breakdown occurs where nests of ferrous sulphate exist in the rust below the paint coating. Because of the presence of aggressive ions at the rust/metal interface, surface tolerant coatings are often designed to have very good barrier properties to prevent the ingress of water and oxygen [2, 7].

(ii) Failure finally occurs when the coating is stretched and cracked by underlying rust blisters that have grown at anodic sites. Hence strong, flexible coatings perform better than weak, brittle ones. A rust blister which has caused local bulging of the coating is shown in Fig. 5. This was a sample of clear resin on rusty steel exposed in a salt spray cabinet. Black blisters of magnetite were seen to grow underneath the resin, implying that the cathodic reduction of lepidocrocite to magnetite is taking place, as proposed in the mechanism of atmospheric rusting due to Evans [8]. Note how the fresh rust in Fig. 5 appears to be forming at the rust/paint interface.

(iii) Good wetting and penetration of rust is believed to be important [9] and will depend upon the choice of binder and solvent and the viscosity of the paint system. The effect of viscosity on the penetration of rust by chlorrubber resins was investigated. Coated rusty steel samples were cross-sectioned and polished and chlorine maps were obtained using Energy dispersive X-ray Analysis (EDX) in the Scanning Electron Microscope (SEM). As shown in Figs. 6a and 6b, using this technique it is possible to see how the coating has wetted the surface and flowed into cracks in the rust. Although low viscosity systems gave better penetration into fine cracks in the rust, it was found that the higher viscosity coatings gave surprisingly good penetration and a much thicker film build and hence better protection in exposure tests.

EXCELLENT PERFORMANCE BY RED LEAD OIL-BASED PAINTS

Red lead oil-based paints give excellent performance on wire-brushed steel, but their mechanism of protection is not understood. One suggestion is that Pb may diffuse into the rust layer and insolubilize sulphate and chloride, thus rendering these aggressive ions inert [10]. To examine the viability of this mechanism, both TEM with EDX and Laser Microprobe Mass Analysis (LAMMS) were used to discover what happens to the Pb in a red lead alkyd primer after 3 years exterior exposure on rusty steel. Cross-sectioned samples were examined to determine whether Pb had diffused into the rust layer and if it is found in association with sulphate. This work is reported in detail in ref [11].

Microtomed specimens were examined in a Phillips EM 400 transmission electron microscope operating at 120 kV, fitted with a field emission gun. Fig. 7 shows a typical section across the rust/paint interface. EDX analyses were carried out in the rust layer and both Pb and S were identified: however, they were differently distributed and the ratio of Pb:S was about 0.2 and not 1.0 as would be required if $PbSO_4$ had been formed.

LAMMS uses a pulsed laser beam to evaporate and ionize a micro-volume of sample (5µm diameter), which is then

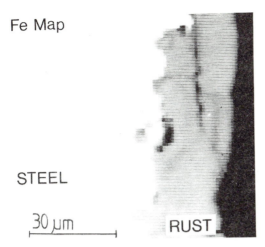

Figure 3a: Section of rusty steel. WDXA-Fe map.

Figure 3b: Section of rusty steel. WDXA-S map.

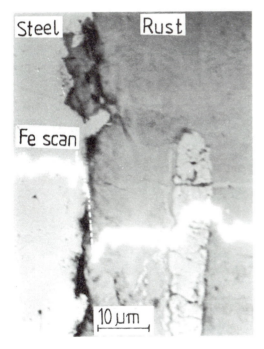

Figure 4: Section of rusty steel, showing Fe line scan. SEM backscattered image.

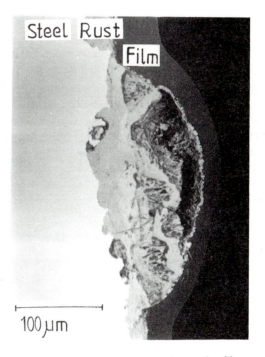

Figure 5: Rust blister growing under film.

analysed in a time-of-flight mass spectrometer. It is easy to distinguish between Pb and S with this technique, as shown by the positive ion mass spectrum of Fig. 8. This is a typical LAMMS analysis taken from the rust layer of a polished cross-section of red lead coated rusty steel (Fig. 9). Histograms plotting the distribution of Pb and S found in the rust layer showed no correlation in the distribution of these species (Figs. 10a and b). From these results it was concluded that red lead oil-based paints do not protect rusty steel by the insolubilization of aggressive sulphate ions.

CONCLUSIONS

A range of complementary analytical techniques have been used to investigate the problem of coating poorly prepared steel substrates.

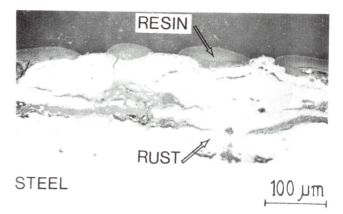

Figure 6a: Chlor-rubber coating on rusty steel.SEM backscattered image.

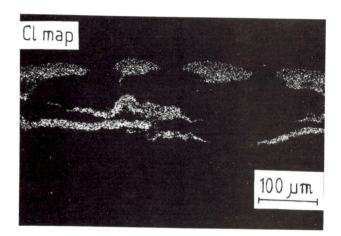

Figure 6b: Cl map of above area-EDXA.

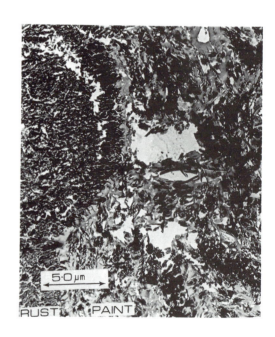

Figure 7: Red lead alkyd on rusty steel. Transmission electron micrograph.

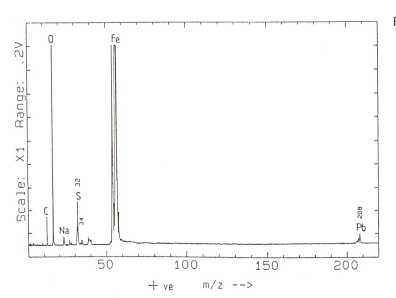

Figure 8: LAMMS spectrum from rust layer.

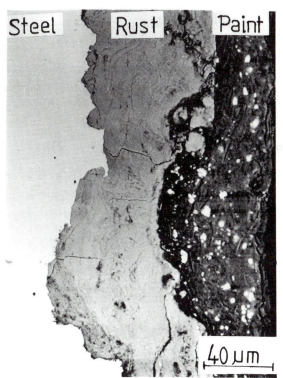

Figure 9: Red lead alkyd on rusty steel. Backscattered SEM image.

Figure 10: Histograms of LAMMS analyses.

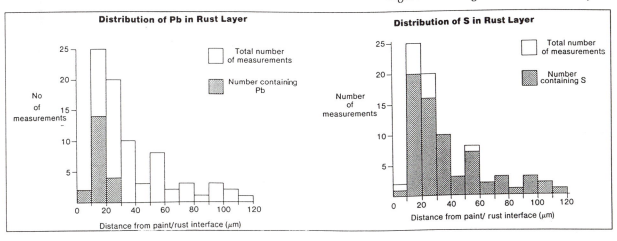

Characterisation of Rust

SEM studies have shown that rusts produced by natural weathering are much more solid and compact than the porous, spongy rust grown in salt spray cabinets. Deep pitted regions were found in the surface of atmospherically rusted steel: these are the anodic sites and are the areas of active corrosion. It was confirmed using Electron probe X-ray microanalysis that sulphate contaminants are located in 'nests' at the anodic sites. Because they are embedded in compact rust, contaminants are not easily removed by wire brushing.

Coating Failure Mechanisms

The biggest problem in painting rusty steel is the presence of aggressive ions at the rust/metal interface. Therefore surface tolerant coatings are often designed to have very good barrier properties to prevent the ingress of water and oxygen. In accelerated weathering tests, the failure mechanism was due to growth of magnetite blisters at anodic sites, causing local bulging and eventual rupture of the coating. Hence strong, flexible coatings performed better than weak, brittle ones.

Excellent Performance by Red Lead Oil-based Paints

Both TEM with EDX and LAMMS were used to discover whether Pb from red lead alkyd paints coated on to rusty steel can diffuse into the rust layer and insolubilize aggressive sulphate ions. Although Pb and S were both detected in the rust layer, there was found to be no correlation between the distributions of these two species, indicating that sulphate insolubilisation is not the mechanism of protection.

ACKNOWLEDGEMENTS

The author would like to acknowledge the help of the following people from ICI, Chemicals and Polymers, Runcorn. Marsha Lee and Colin Gould are thanked for carrying out the LAMMS and TEM analyses respectively. Thanks are also due to Jeff Booth for assistance with the SEM work. The help of Chris Salter (Oxford University Department of Metallurgy and Science of Materials) is acknowledged in carrying out the EPMA maps.

REFERENCES

1. P.K. Agrawal, Paint India 39, 1989, p.59.

2. T.B. Lemmon and M.J. Mitchell, Industrial Corrosion, 1987, p.C4.

3. K.A. Chandler, Brit. Corros. J. 1, 1966, p.264.

4. C. Calabrese and J. R. Allen, Corrosion 34, 1978, p.331.

5. M. Morcillo, S. Feliu, J. C. Galvan and J. M. Bastidas, J. Protective Coating and Linings 4, 1987, p.38.

6. J. E. O. Mayne, J. Appl. Chem. 9, 1959, p.673.

7. N. L. Thomas, J Protective Coatings & Linings 6, 1989, p.63.

8. U. R. Evans, Nature 206, 1965, p.980.

9. D. Dasgupta and T. K. Ross, Brit. Corros. J. 6, 1971, p.241.

10. G. Lincke and W. D. Mahn, Proc 12th Fatipec Congr. Garmisch - Partenkirchen, 1974, p.563.

11. N.L. Thomas, 'Advances in Corrosion Protection by Organic Coatings', Electrochem Soc.,1989, p.451.

6 OBSERVATIONS ON SILICATES IN SCALE ON A LOW-ALLOY STEEL

J. Smuts and H. J. De Klerk

Research Division, ISCOR Ltd, Box 450, Pretoria 0001, South Africa.

ABSTRACT

The oxidation of low alloy steels during the hot rolling process is one of the serious constraints that often prevent the use of high temperatures in order to fully dissolve micro-alloying element phases (such as niobium carbides) or to obtain higher production rates associated with lower rolling forces. One of the important species in the oxidation process is silicon which is often associated with internal oxidation and the formation of a silicate layer on the metal/scale interface during high temperature exposure.

The development of internal oxides with silicon, the formation of a sub-scale, and precipitation reactions involving silicates and iron oxides within the scale are illustrated. Special attention is given to the various iron silicate species that can develop in the scale on slabs of BS 4360 Gr.50E steel during exposure in the reheating cycle prior to hot rolling. Examples of scale morphologies are shown and the influence of silicates on descaling during hot rolling is indicated.

INTRODUCTION

The presence of Si in relatively low concentrations in steel can have a significant influence on the properties of the scale formed on the steel before and during hot rolling. This influence was realised in particular in investigations concerned with descaling of BS 4360 Gr 50E steel which contains, amongst other elements, 0,4% Si, 0,10% C and 0,22% Ni. Incomplete descaling of the steel during hot rolling leaves patches of rolled-in scale which have to be removed by scarfing operations from the finished product. Adverse descaling behaviour of some steel slabs asked for critical examination of scale structure and this showed that the presence and morphology of silicates in the scale could be responsible for incomplete descaling. Fascinating aspects were observed in a number of scale specimens prepared from scale samples from reheated slabs and also from scale produced in laboratory oxidation simulations on the steel. The presence of the iron silicate iscorite (Fe_7SiO_{10})[1] was observed in parts of the scale and provided new insight into the formation and morphology of this relatively unknown phase.

EXPERIMENTAL

Specimens of scale and steel oxidised in air at temperatures up to 1260°C were mounted in epoxy resin, sectioned and polished for metallographic examination with light microscopy and scanning electron microscopy (SEM). All observations were made on unetched specimens. In SEM, the BSE(back-scattered electron) mode was mainly used. The SEM instrument is equipped with a built-in light microscope which was extremely valuable for this type of investigation.

RESULTS OF OBSERVATIONS

General Observations

While it is easy to distinguish magnetite and wustite from hematite and fayalite with incident light microscopy, it is not so easy to distinguish between magnetite and wustite. Primary wustite has a slightly lower reflectivity than magnetite and appears slightly bluish in reflected light while magnetite reflects pinkish. Secondary wustite reflects stronger than magnetite and has a yellowish appearance. The reflectivity of iscorite is lower than that of magnetite or wustite and considerably higher than that of fayalite so that it is not difficult to distinguish it from iron oxides or fayalite.

In the SEM/BSE images the "reflectivity" grading for oxide phases, in descending order, is as follows: wustite (both "primary" and "secondary" wustite); magnetite and iscorite; hematite; fayalite. In this group it is not easy to distinguish between the two types of wustite or between magnetite and iscorite. Contrast between magnetite and iscorite can vary as a result of composition variations caused by elements such as Al, and can even be inverted.

Secondary wustite is the wustite phase which forms together with magnetite during the first stage of decomposition of primary wustite. It is enriched in Fe and has a composition approximating FeO [2].

Oxidation at the Scale/Metal Interface

The BSE images in Fig. 1 show a number of aspects in the subscale regions of steel oxidised in air in the laboratory. All the facets represent relatively advanced stages of oxidation at temperatures in excess of 1200°C and in all cases the subscale is found underneath relatively thick layers of iron oxide. The images illustrate, amongst others, facets of internal oxidation, iron silicate accumulation and the formation of metallic and iron oxide islands in the silicate. Internal oxide in the deeper zones near the oxidation front is single-phase silicate, probably SiO_2[4], but in the outer regions of the subscale closer to the massive oxide, the internal oxide particles consist of two phases namely wustite and fayalite.

General Scale Structure

The micrographs in Fig. 2 represent the main features of the structure of scale on the steel slabs. In broad terms, these are the following: (a) The scale consists of recognisable, sometimes very distinct, layers in various stages of separation. (b) The massive outer layer contains the familiar sequence of hematite, magnetite and a main body of wustite. (c) Inner layers contain wustite grains of various shapes ranging from small rounded on the metal side, through polygonal, to large columnar towards the outer surfaces. (d) Oxide grains in the inner layers are separated in many places by sheets or pools of iron silicate (fayalite). Intergranular silicate is also present in places in the massive outer scale but only as very thin layers. (e) Spherical or oval-shaped silicate inclusions are present in many oxide grains in inner layers. (f) Ni-rich metallic grains are concentrated in regions adjacent to the subscale.

Silicate Morphology

Detail of silicate and scale structures in a few selected regions is shown in Figs. 3 to 6. The following features are shown in these images:

(a) The silicate in the larger pools has a eutectic structure (Fig. 3). The eutectic is prominent in the inner regions of the scale where iron oxide grains are rounded and where silicate liquid existed only in equilibrium with wustite during scale growth.

(b) The crystallisation of the iron silicate iscorite can be seen in Figs. 5 and 6. Iscorite can be seen as globular inclusions in partly-decomposed wustite grains (Figs. 5a, 6), as needle-like crystals in intergranular silicate (Fig. 5) and as needle-like crystals in larger oval-shaped silicate within iron oxide grains (Fig. 6c). Globular iscorite is frequently associated with voids and can have a layered appearance depending on the plane of sectioning (Figs. 6b,c). Iscorite is always attached to iron oxide and needle crystals attached to the same grain have the same or related orientation. Iscorite is mainly found in wustite regions of delaminated scale with only rare occurrences in primary magnetite. No occurrence of iscorite was observed in undecomposed (primary) wustite and in scale/metal interface regions of adherent scale.

(c) Decomposition of primary wustite can be observed in all the images. A thin layer of magnetite can be seen in many places between silicate and secondary wustite. In regions where iscorite is formed these magnetite layers are thicker, indicating a higher oxygen content and more wustite decomposition. Patches of primary wustite (Fig. 4) were only observed in interior regions of iron oxide grains and not at silicate interfaces where decomposition of primary wustite has always occurred during cooling of the scale.

DISCUSSION

The formation and structure of iron oxide scale have been investigated and reviewed extensively (see e.g. ref[3]), and need no introduction. In the work reported here, only oxidation at temperatures above 1200°C is concerned and the formation of liquid silicate via internal oxidation is relevant. The results are qualitatively in agreement with previous observations (see e.g. the review of Van Vlack [4]). Small internal oxide particles near the internal oxidation front have a high Si-content and consist mostly of SiO_2. Dual phase inclusions consisting of wustite and silicate (probably fayalite) are present near scale/subscale interfaces (Fig. 1). Combination and accumulation of silicate lead to accelerated grain boundary oxidation and metallic grains are isolated from the metal substrate by liquid silicate. Formation of liquid appears to be a self-catalysing process through rapid transport of oxygen to the interface. At the high temperatures concerned, no protective barrier oxide layer of SiO_2 can be formed. The silicate which is formed acts as an easy path for oxygen diffusion and not as protective barrier.

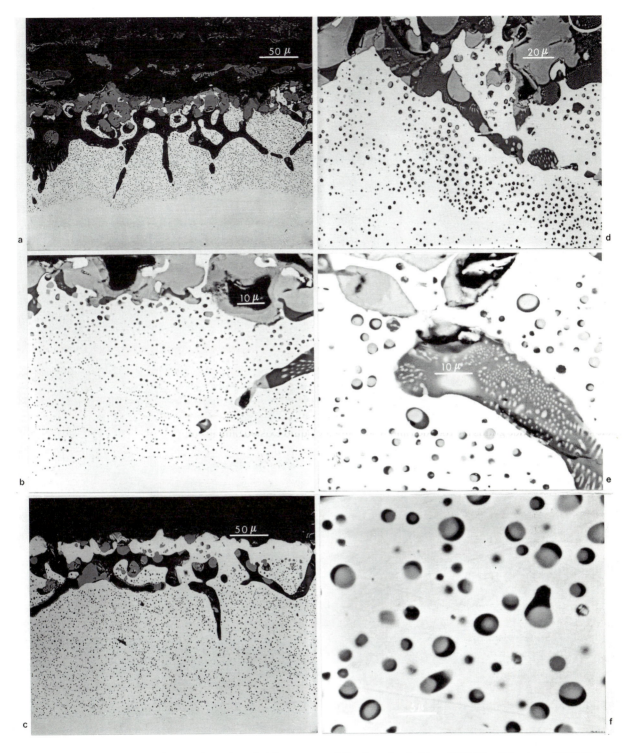

Figure 1: SEM/BSE images of subscale regions in laboratory-oxidised steel. a,b: 2 h at 1200°C; c: 7.5 min. at 1230°C; d,e,f: 3.5 h at 1230°C.

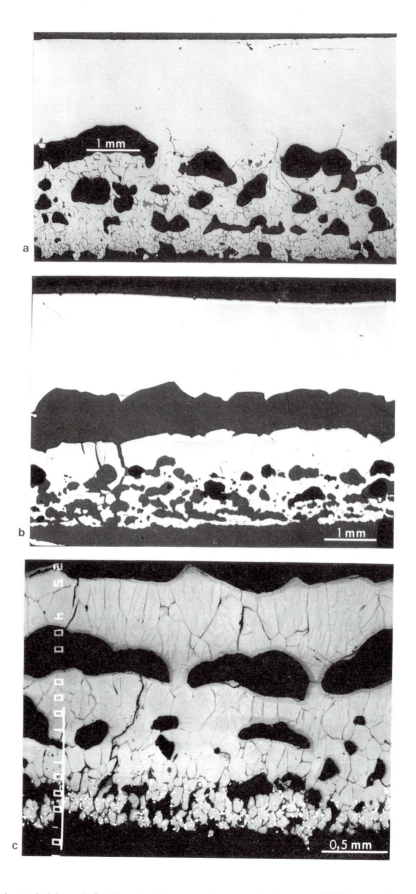

Figure 2: Scale on reheated slabs. a,b: Light optical images of outer scale layers; c: BSE image of intermediate layer.

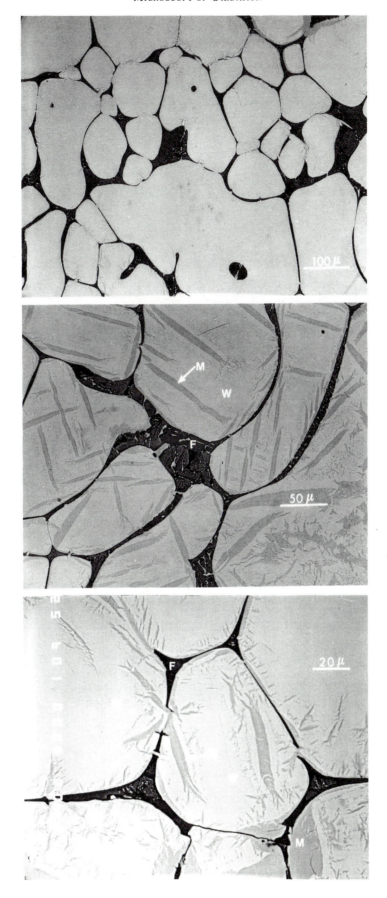

Figure 3: BSE images of inner scale on slabs showing fayalite(F), wustite(W) and magnetite(M).

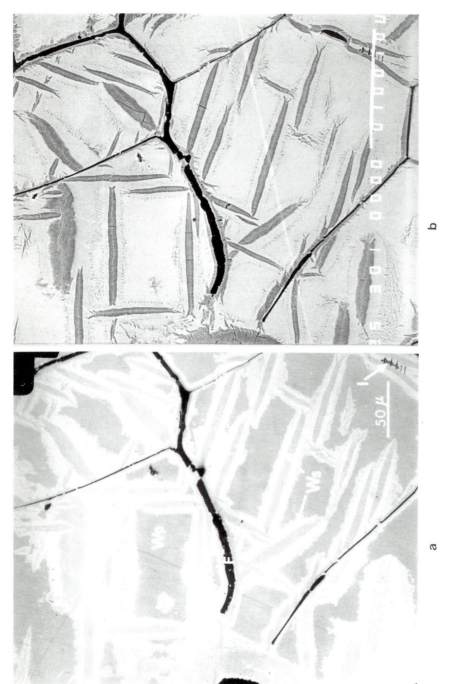

Figure 4: Light optical (a), and BSE (b) images of the same region in scale with primary wustite (Wp), secondary wustite (Ws), magnetite (M) and intergranular fayalite (F). Small crystals of iscorite (I) can be seen in one corner.

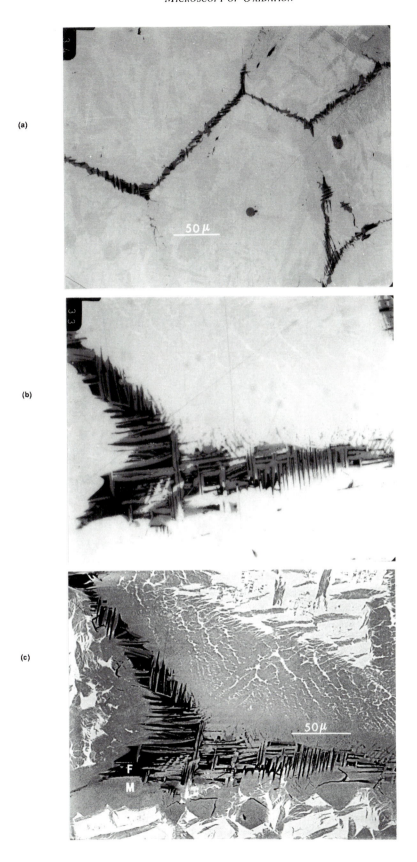

Figure 5: Detail of scale with iscorite in intergranular silicate. Globular iscorite can also be seen in a. a,b: Light optical images; c: BSE image of b.

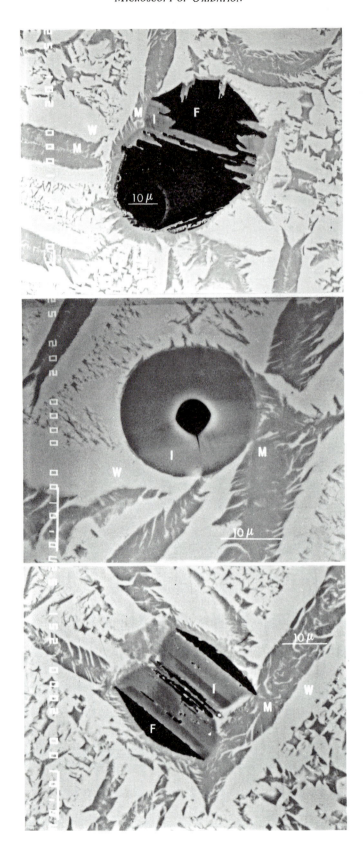

Figure 6: BSE images of scale to show inclusions of iscorite (I) and fayalite (F) in wustite grains which are partly decomposed to secondary wustite (W) and magnetite (M).

Studies on the formation and morphology of the mixed valence iron silicate iscorite(Fe_7SiO_{10})[1,5,6], indicate that the phase is preferably formed at temperatures in excess of 1150°C from liquid iron silicate and iron oxide. All the constituents can be supplied by fayalite and iron oxide of composition $3FeO.Fe_2O_3$ which will consist of a mixture of wustite and magnetite below 1400°C [7]. Iron oxide of this composition is only present in the oxygen-rich outer regions of scale or in layers which have separated from the steel and in which the wustite is oxygen-rich. Iscorite crystallises from liquid silicate but is nucleated on wustite and the crystallised phase maintains predictable and consistent orientation relationships with the wustite and associated magnetite. The epitaxy is associated with the similar packing of oxygen atoms in the structures of iscorite and the cubic iron oxides. The orientation relationship is such that iscorite needles or laths will be perpendicular to wustite (110) planes. In the regions in the scale in which iscorite is found, each iron oxide grain is a single crystal and the iscorite crystals which are nucleated on a particular grain have the same or related orientation.

The exact requirements for the crystallisation of iscorite are not yet fully understood. The phase occurs haphazardly in some regions of matrices of apparently similar composition and structure. It is speculated that variations in conditions related to composition changes of wustite and liquid silicate could be the critical factor. The composition of wustite on which iscorite will nucleate is on the boundary of the wustite field and the formation of magnetite nuclei at the wustite/silicate interface could initiate crystallisation. Although initial crystallisation of iscorite appears to occur rapidly, crystal growth can only proceed at a low rate since it depends on a process of dissolution of iron oxide in the remaining liquid silicate. Evidence of this process can be seen in Fig. 5 where the original silicate/wustite interfaces have clearly been disturbed by the formation of iscorite. It is particularly noticeable where liquid silicate was consumed by the formation of iscorite or the interface area became limited.

The descaling behaviour of steel which has been reheated to about 1250°C is significantly influenced by the presence of iron silicate especially in scale which has developed a well defined layered structure. The relatively massive outer scale layers will not be deformed plastically and can become separated from inner scale layers at the layer boundaries which often consist of large void spaces (see Fig. 2). Inner scale layers which contain most of the iron silicate, can deform plastically since fayalitic iron silicate is liquid at temperatures above 1170°C [5]. As a result some of the inner scale will not become separated from the steel substrate during hot rolling and will be rolled into the steel surface. This tendency is aggravated by the presence of metallic residues which are found in scale on steel which contains Ni.

The presence of iscorite in intergranular iron silicate is expected to decrease the fluidity of the silicate and therefore the deformability of the scale. In practice, however, iscorite is not found in the critical interface regions where scale adherence is relevant. It is mostly found in regions where iron oxide grains have straight edges, where silicate is less abundant and present as thin intergranular layers, and where the scale has a more massive appearance.

ACKNOWLEDGEMENTS

This paper is published by permission of ISCOR Ltd.

REFERENCES

1. J. Smuts, J.G.D. Steyn and J.C.A. Boeyens, Acta Cryst. A29, 1969, p.1251.

2. J. Smuts and P.R. De Villiers, J. Iron and Steel Inst. 204, 1966, p.787.

3. A.G. Goursat and W.W. Smeltzer, High Temperature Materials Coatings and Surface Interactions, Ed. J.B. Newkirk, Freund Publishing House, Tel Aviv, p.49.

4. L.H.Van Vlack, Int. Metals Rev. 22, 1977, p.207.

5. A. Muan and E.F. Osborn, Phase Equilibria Among Oxides in Steelmaking, Addison-Wesley publishing Co., 1965, p.53.

6. J. Smuts, On the Formation and Morphology of the Iron Silicate Iscorite. To be published.

7. L.S. Darken and R.W. Gurry, J.Am.Chem.Soc. 68, 1946, p.798.

7 EFFECT OF MICROSTRUCTURE ON THE OXIDATION BEHAVIOUR OF A LOW ALLOY FERRITIC STEEL

R.K. Singh Raman, A.S. Khanna and J.B. Gnanamoorthy

Metallurgy Division, Indira Gandhi Centre for Atomic Research, Kalpakkam - 603102, India.

ABSTRACT

The influence of various metallurgical parameters such as grain size, cold work and heat-treatment on the oxidation behaviour of 2 1/4Cr-1Mo steel has been studied. Optical microscopy and other electron optical techniques such as SEM/EDX, EPMA and XRD have been used to study the scale morphologies and composition of the scale as well as the scale/metal interface. Preferential oxidation along grain boundaries, changes in the rate of oxidation as a result of prior cold work, and the role of secondary phases due to various heat-treatments on the oxidation behaviour have been investigated.

INTRODUCTION

Low alloy ferritic steels such as 2 1/4Cr-1Mo and 9Cr-1Mo are well known steam generator materials. The oxidation behaviour of these materials depends on the formation of a mixed oxide of iron and chromium as there is not sufficient chromium to form a protective chromia layer. Hence any metallurgical treatment which results in changing the effective Cr concentration of the alloy will affect the oxidation behaviour. Prior cold work is well kown to affect the oxidation behaviour[1,2]. Cold work produces short circuit diffusion paths for Cr to reach the surface. Carbides formed during heat treatment fix the available chromium in the alloy and hence reduce the oxidation resistance. Decreasing the grain size usually results in faster chromium diffusion through the larger grain boundary area and hence helps in improving the oxidation resistance [3,4].

In this paper, an attempt has been made to assess the influence of a few metallurgical factors on the oxidation behaviour of a low alloy ferritic steel. The oxidation was carried out in air or oxygen using accelerated tests i.e. at temperatures higher than those at which these materials are normally used. The effect of prior cold work, tempering treatments and variation of alloy grain size were investigated.

EFFECT OF COLD WORK ON THE OXIDATION BEHAVIOUR OF FERRITIC STEELS

The effect of cold work on the oxidation of 2 1/4Cr-1Mo steel in air has been reported elsewhere [5]. Specimens with cold work of 37, 46, 63, 76 and 89% were prepared from 6mm sheet by cold rolling. Oxidation was carried out at 773, 873, 973, 1073, 1173 and 1223K. No significant effect of cold work was observed up to 973K. This could be due to very slow diffusion of Cr in the iron matrix up to this temperature [6]. Oxidation above 973K showed marked effect of the prior cold work on the total weight gain (Fig. 1). At 1073K, the lower weight gain of the cold worked specimens was due to the lower parabolic rate constant. At 1173K no breakaway oxidation was observed on the specimens with prior cold work. For specimens oxidized at 1223K, the lower weight gain was due to the lower post-breakaway rate of the cold worked specimens compared to that of the annealed specimens.

Considerable difference was noticed between the cold worked and the annealed specimens when detailed microscopic analyses of the oxidized specimens were carried out at 1173K. The outer oxide layer on the annealed specimen consisted of a blocky mass with substantial oxide cracking, indicating excessive oxidation. This was in contrast to a uniform oxide layer on the cold worked samples (Fig. 2). XRD identified the outer oxide as α-Fe_2O_3 on the annealed specimen compared to Fe_3O_4 on the cold worked specimens. Figure 3 shows the scale cross-section along with the EDX analyses. One of the major differences between the annealed and the cold worked samples was the higher scale thickness of the former. Another important difference was the formation of internal oxides below the inner oxide layer in the cold worked samples. No such internal oxides were formed in the annealed sample. These internal oxides were found to be rich in chromium, as shown in the X-ray maps generated by microprobe analyses (Fig. 4). A clear difference in the compositions of the inner layers of the annealed and the cold worked samples is also apparent from this analysis. The inner layer in the annealed sample showed more Cr-counts than the inner layer on the cold worked samples. In the case of cold worked specimens,

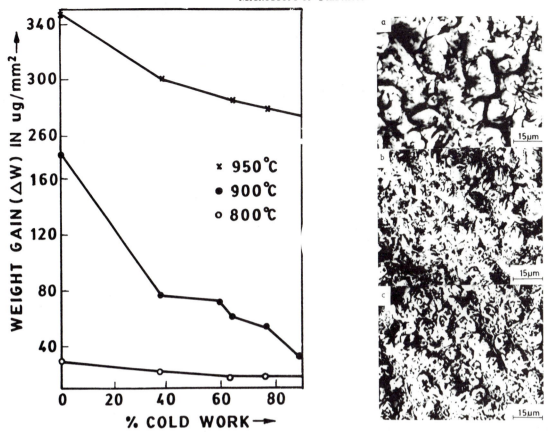

Figure 1: Total weight gain during 4 n oxidation as a functiuon of percentage of cold work.

Figure 2: SEM micrographs showing the surface topography: (a) annealed specimen, (b) 63% CW, (c) 89% CW, oxidised at 1173K for 4 h.

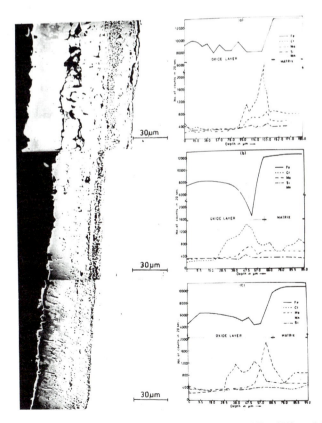

Figure 3: Scale cross-section and depth profile for (a) annealed specimen, (b) 37% CW, and (c) 46% CW sample during 4h oxidation at 1173K.

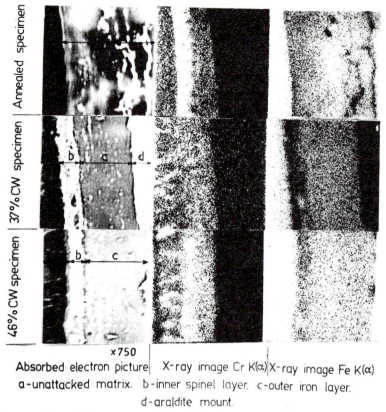

×750

Absorbed electron picture | X-ray image Cr K(α) | X-ray image Fe K(α)
a-unattacked matrix.   b-inner spinel layer.   c-outer iron layer.
d-araldite mount.

Figure 4: Electron micrographs and X-ray maps of Cr and Fe of the oxide scale formed during 4th oxidation at 1173K.

preferential diffusion of Cr takes place through short circuit paths, especially grain boundaries resulting in the formation of a spinel layer in the beginning of oxidation. This acts as a protective layer to restrict the transport of Fe and hence helps in reducing the overall oxidation rate.

INFLUENCE OF TEMPERING TREATMENTS

The influence of tempering time on the oxidation behaviour of 2 1/4Cr-1Mo steel in air has been reported elsewhere [7]. When the normalized steel was tempered at a temperature above 873K, part of the chromium present in the solid solution was used in carbide formation ($M_{23}C_6$ and M_7C_3). The extent of Cr used depends upon both time (duration) and temperature of the tempering treatment. Hence a variation of tempering parameters (temperature and time) is expected to affect the oxidation behaviour by forming scales of varying chromium content. Optical microscopy has been used to show the degree of carbide precipitation during tempering, i.e. their distribution and extent of coarsening. Figure 5 illustrates the coarsening of the carbide particles with time at 998K. As reported earlier [8,9], these carbides are chromium-based; therefore an increase in tempering treatment parameters (both time or temperature) results in fixing more Cr as carbides and hence, a decrease in the oxidation resistance of the alloy. Figure 6 depicts the oxidation results at 873K confirming that with an increase in tempering time the oxidation resistance decreases. Figure 7 shows SEM/EDX results indicating that the inner scale formed at the alloy/matrix interface of the specimens tempered for a short duration has a higher Cr content than in the samples tempered for longer durations. X-ray diffraction analysis further indicated that an additional phase, $FeCr_2O_4$ (having higher Cr content) was formed on the alloy tempered for shorter duration. This oxide phase was absent in the scale formed on the samples tempered for longer duration. Lower availability of Cr as a result of longer treatment favours the formation of a mixed oxide of Fe and Cr, $(Fe,Cr)_2O_3$, instead of a spinel.

INFLUENCE OF GRAIN SIZE

The effect of variations in grain size of the base alloy on its oxidation behaviour is well known [3,4]. Several high temperature iron base alloys with chromium as a major alloying element show better oxidation resistance when the grain size is small. The beneficial effect is attributed to the faster transport of the protective oxide-forming element through the large grain boundary area associated with the smaller grain size. This in turn helps in the formation of the protective chromia layer at a very early stage of oxidation thereby reducing the overall oxidation rate of the alloy. However, in the case of the low alloy 2 1/4Cr-1Mo ferritic steel, the results were the opposite: decreasing the grain size resulted in the increase in the oxidation rate. This could be due to either or both of the following:

(i) because of the lower Cr content, inspite of having smaller grain size, it is difficult to form a protective chromia scale,

(ii) part of chromium gets internally oxidised at grain boundaries restricting the chromium diffusion outward.

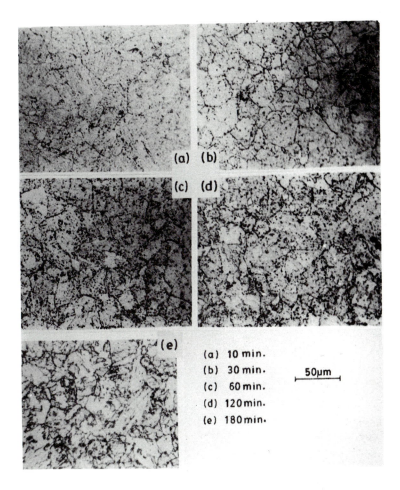

Figure 5: Microstructures of 2 1/4Cr-1Mo steel after tempering at 998K for different exposures.

The influence of the variation of grain size on the oxidation behaviour of 2 1/4Cr-1Mo has been studied earlier [10]. Specimens with different grain sizes were prepared by varying annealing treatments. Grain sizes were measured by the linear intercept method (ASTM E 112-85) using optical micrography. Oxidation tests were carried out at 873, 973 and 1073K in air. For correlation of oxidation behaviour with grain size, the inverse of grain diameter (d^{-1}) was plotted against weight gain (Fig. 8). The results show a linear increase in weight gain in a given duration (6h) with the inverse of grain diameter indicating an increase in oxidation with decrease in grain size. Using a least square fit, the following relationship was found to exist between weight gain and grain size:

$$W = W_0 + n\, d^{-1}$$

A very good correlation exsists between weight gain and grain size at the three temperatures of oxidation(0.99, 0.95 and 0.97 at 873, 973 and 1073K respectively).

CONCLUSION

The microstructure of the metal/alloy is very important in deciding its oxidation behaviour. Metallurgical factors such as cold-working, heat-treatments or variation in grain-size cause an effective change in the availability of certain oxide-forming elements. This in turn affects the oxidation behaviour of the alloy. The present results indicate that large grained, cold worked 2 1/4Cr-1Mo steels which have not been given a tempering treatment have the optimum oxidation resistance.

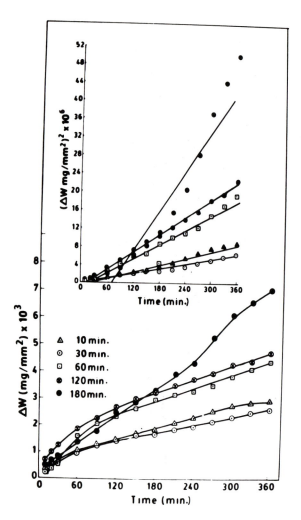

Figure 6: Weight gain data for the oxidation of 2 1/4 Cr-1Mo steel at 873K for specimens, tempered for different times.

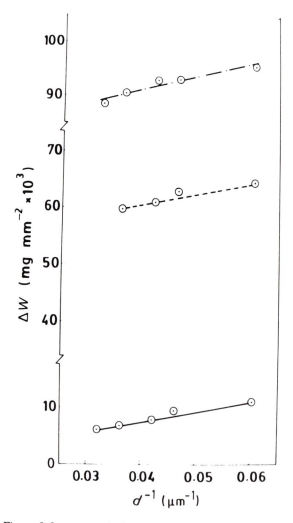

Figure 8: Inverse grain diameter as a function of weight gain at various temperature of oxidation – 873K, — 973K, --1073K.

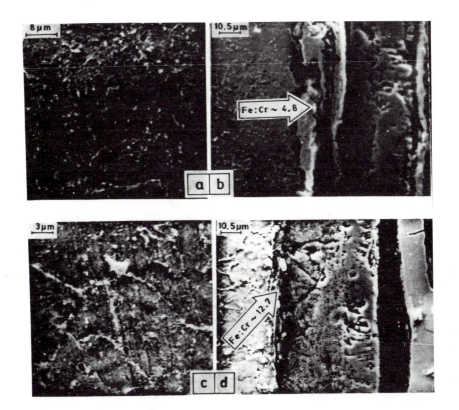

Figure 7: Surface topography (a and c) and scale cross-section (b and d) during oxidation of a sample tempered at 998K for 30 min (a and b) and (c and d) - tempered for 180 min.

REFERENCES

1. W.E. Ruther and S. Greenberg, J. Electrochem. Soc. 111, 1964, p.116.

2. M. Warzee et al., J. Electrochem Soc. 112, 1965, p.670.

3. S. Leistikow, I. Wolf and H.J. Grabke, Werkst. und Korros. 38, 1987, p.556.

4. Y. Shida, N. Ohtsuka, J. Muriama, N. Fujino and H. Fujikawa, Trans. JIM 3, 1983, p.163.

5. A.S. Khanna and J.B. Gnanamoorthy, Oxid. Met. 23, 1985, p.17.

6. K.F. Smith, Met. Sci. 10, 1976, p.408.

7. R.K. Singh Raman, A.S. Khanna and J.B. Gnanamorthy, Oxid. Met. 30, 1988, p.345.

8. R.J. Baker and J. Nutting, JISI 192 , 1959, p. 257.

9. J. Pilling and N. Ridley, Metall Trans. (A) 13, 1982, p.557.

10. R.K. Singh Raman, R.K. Dayal, A.S. Khanna and J.B. Gnanamoorthy, J. Mat. Sci. Lett. 8, 1989, p.277.

8 BORON INHIBITED STEAM OXIDATION OF Fe-Cr ALLOYS STUDIED USING EELS

P. N. Rowley, R. Brydson*, J. Little and S. R. J. Saunders†

Department of Materials Science and Metallurgy, University of Cambridge, Pembroke Street, Cambridge CB2 3QZ, UK.
*The Blackett Laboratory, Imperial College, Prince Consort Road, London SW7 2BZ, UK.
† National Physical Laboratory, Queens Road, Teddington, Middlesex TW11 0LW, UK.

ABSTRACT

Boron inhibition of the steam oxidation at 873K of an Fe-10%Cr steel and a Fe-9%Cr1% Mo alloy has been studied by TEM and EELS examination of the composition and structure of the oxide films formed. Protective behaviour appears to derive from the rapid formation of an ultrafine grained $(Cr_xB_{1-x})_2O_3$ film. The paper discusses the mechanism by which this film formed and how it provided continuing oxidation resistance.

INTRODUCTION

The ability of boron additions to significantly improve the oxidation resistance of Fe-Cr alloys in a variety of conditions has been well documented in recent years [1]. In order to study the composition and structure of the oxide films so formed we have employed the technique of electron energy loss spectroscopy (EELS) conducted in a transmission electron microscope (TEM). The use of an electron microscope capable of analysis from sample areas of sub-micron dimension is essential due to the heterogeneity inherent in such oxide films and EELS permits sufficiently accurate chemical quantification especially of light elements, $Z > 2$ (e.g. boron). Analysis of the electron loss near-edge structure (ELNES) associated with a particular core loss edge also provides an insight into the local environment of the atomic species under consideration when combined with the results of theoretical modelling using multiple scattering ICXANES calculations [2].

EXPERIMENTAL

The samples chosen were an industrial Fe-9%Cr-1%Mo alloy (additionally containing 0.5% Mn) and a binary Fe-10%Cr alloy. Rectangular specimens were ground and polished to a 1µm finish and then degreased. Subsequent oxidation was conducted in steam (linear flow rate 0.18 ms⁻¹) at 873 K for a variety of time periods ranging between 1 minute and 24 hours. Both control (untreated) and boron doped samples were exposed. Boron doping was achieved in one of three ways:

(1) a constant injection of boric acid solution (1.8×10^{-3} moles of boron atoms per hour) into the steam flow; (2) Vapour deposition of crystalline boric acid onto the alloy surface prior to oxidation; (3) A combination of methods (2) and (1) together with an intermediate pre-oxidation step in oxygen prior to subsequent steam oxidation.

The oxide films so formed were stripped using a methanol/bromine solution and mounted on holey carbon support grids for TEM analysis. Electron microscopy, EELS and energy dispersive X-ray (EDX) measurements were performed on a JEOL 2000FX operating at 200 keV and fitted with a Gatan magnetic prism spectrometer and serial recording system together with a LINK systems X-ray detector. Further high resolution EELS measurements were made on a dedicated EELS scanning transmission electron microscope (STEM) possessing a parallel recording system coupled with an energy resolution of *ca.* 0.5 eV [3].

RESULTS

Morphology

Figure 1 shows typical TEM micrographs with inset diffraction patterns from the control and boron doped oxides respectively. These show a number of differences, most notably in the oxide grain size. The control oxide exhibits an average grain size of *ca.* 250 nm (with considerable evidence of void or blister formation at grain boundaries) whilst the boron doped oxide possesses a more microcrystalline structure with an average grain size of *ca.* 10-20 nm. Thicker

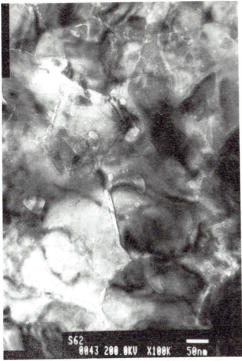

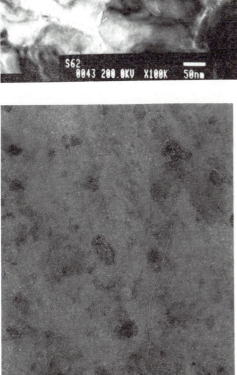
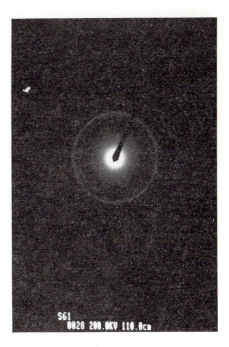

Figure 1: Typical TEM micrographs and inset diffraction patterns from (Top) a control oxide and (Bottom) a boron doped oxide.

nodular breakthrough is evident at isolated positions on the thin film. EELS analysis indicates that the limiting thickness for the protective areas of the boron doped oxide films is of the order of 500 A.

Microanalysis

Control Samples

EELS and EDX analyses of the oxide formed after 3 minutes on the untreated alloys were consistent with the formation of a duplex oxide composed of Fe_2O_3 and $FeCr_2O_4$. According to this model, a graph of O/Cr (the oxygen to chromium

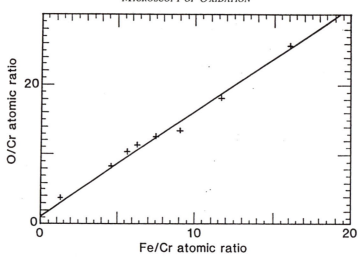

Figure 2: A graph of the oxygen to chromium elemental ratio versus the iron to chromium elemental ratio for a number of measurements on control oxide films. The results are consistent with the oxide being comprised of Fe_2O_3 and $FeCr_2O_4$.

elemental ratio) versus Fe/Cr (the iron to chromium elemental ratio) should then give a slope of 1.5 and an intercept of 1.25. Our EELS results are shown in Fig. 2 where we obtain a slope of 1.45±0.06 and an intercept of 1.7±1.3. The relative amounts of the two oxides varied considerably over the sample areas studied, thinner areas generally containing more of the spinel M_3O_4 (M=Fe,Cr).

Boron Doped Samples
EELS and EDX measurements on the oxide films produced after short term oxidation in the presence of boron doping revealed a large increase in the Cr (and Mn in the case of the commercial alloy) content over that observed in the control oxide. Boron concentrations determined by EELS were considerable and varied over the sample area under study. However, the oxygen to total cation (transition metal plus boron) elemental ratio, O/Total, was approximately constant at 1.5. This implies the formation of an $(M_xB_{1-x})_2O_3$ compound (M = transition metal). For a given doping procedure, the relative amounts of boron and iron were observed to decrease, while the relative amount of chromium (and manganese in the case of the commercial alloy) was observed to increase with increasing oxidation time. Auger analysis indicated that boron was present throughout the oxide film, falling to *ca.* 40% of the surface concentration at the metal/oxide interface. Heavy boron pre-doping led to an oxide stoichiometry which was consistent with a model comprising both $(M_xB_{1-x})_2O_3$ and M_yB_z, where M=Cr and Mn (if present), i.e. the formation of metal borides (or elemental boron) in addition to the oxide stoichiometry found previously. Here, thin microcrystalline regions were found to be iron-deficient, but thicker, nodular areas did contain significant amounts of Fe.

Near-edge Structures

Analysis of the ELNES may be summarized as follows:

(1) Measurement of the L_3/L_2 white line ratios for the transition metal L-edges in the boron doped oxide films suggested that all transition metal cations were present in the +III oxidation state [4]. This agrees with the postulated oxide stoichiometry.

(2) The boron K-edge from the doped oxide films possessed a structure which was indicative of mixed trigonal and tetrahedral co-ordination. A typical boron K-edge measured with a resolution of 0.5 eV is shown in Fig. 3, the sharp π^* peak at 193±0.6 eV and the broader s* peak at *ca.* 203 eV are characteristic of a trigonal BO_3 group [1] and arise due to a splitting of the final state p-type molecular orbitals (MO's) by the trigonal symmetry. The additional peak present at *ca.* 199 eV is due to the presence of tetrahedral BO_4 groups [5]. This 'mixed co-ordination fingerprint' arises from the presence of at least two differing phases, which may be inferred from the fact that the relative intensities of the two s* peaks was observed to vary with the position of the STEM probe (diameter *ca.* 50Å).

After a suitable normalisation procedure it is possible to compare the boron K-edge π^* peak areas from the oxide film and from a sample containing 100% trigonally co-ordinated boron (eg. Fe_3BO_5) and so infer something about the relative proportions of trigonal and tetrahedral boron sites in the oxide. Although, as stated, this ratio varies across the sample the

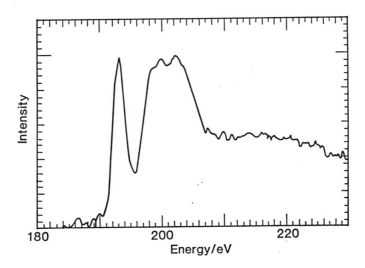

Figure 3: Background subtracted boron K-edge from a boron doped oxide film measured using EELS. The twin peaked structure above 199 eV is indicative of mixed trigonal and tetrahedral boron co-ordination.

average value so obtained was approximately 3:1 trigonal to tetrahedral site occupancy. Heavily Fe-doped samples exhibited a small pre-peak at *ca.* 189 eV in addition to the borate K-edge structure which was consistent with the presence of a metal boride in concordance with the results of the quantitative analysis.

(3) The oxygen K-edge from the boron doped oxide films suggests that in the initial stages of oxidation some form of metal (predominantly iron) borate is formed on the surface of the alloy. A typical oxygen K-edge from a control oxide is shown in figure 4A, curve **A**. Figure 4A, curve **B** shows the results of ICXANES calculations assuming a rhombohedral a-Fe_2O_3 structure. The initial peak present at *ca.* 531 eV is observed in many transition metal oxides and arises from oxygen 2p levels hybridising with unoccupied transition metal 3d levels [6]. Figure 4B, curve **A** shows the oxygen K-edge from the boron doped oxide produced after short term oxidation, whilst curve **B** shows that from vonsenite (Fe_3BO_5). The ELNES of these two oxygen K-edges are very similar and differ from that of the control oxide mainly in the height of the initial peak, although the second peak centered at *ca.* 542 eV is slightly narrower in the control oxide. This diminution of the first peak results from a decrease in the unoccupied d-like density of states (DOS) available for hybridisation with oxygen 2p levels and this must be associated with the presence of boron. The results of our ICXANES calculations for the oxygen K-edge in Fe_3BO_5, shown in Figure 4B, curve **C**, also predict this change.

CONCLUSIONS

The results suggest both the outline of a possible mechanism by which boron affords protection to low chromium steels in high temperature steam and the optimum doping procedure. A detailed discussion is provided elsewhere [7]. However, the direct cause of protective behaviour seems to be the rapid formation of an ultrafine-grained $(Cr_xB_{1-x})_2O_3$ film which acommodates growth stresses and prevents further oxidation of the scale. Briefly, this is achieved because the boric acid/steam environment appears to react preferentially with Fe/Fe_2O_3 in the initial stages of oxidation, forming $FeBO_3$ and Fe_3BO_6 (which posesses trigonally and tetrahedrally co-ordinated boron respectively). This increases the effective surface concentration of Cr and allows the selective formation of chromia nuclei (from thermodynamic considerations) which then undergo lateral growth to form a continuous film. Further boron dopant is segregated to grain boundaries and restricts subsequent grain growth. Continued oxidation resistance results from a reduction in the rates of cation diffusion through the scale.

ACKNOWLEDGEMENTS

The authors acknowledge with gratitude financial support from the S.E.R.C. and N.P.L. (for P.R.) and the Royal Society (for a Pickering research fellowship to R.B.). We also wish to thank the National Physical Laboratory, Zeiss Oberkochen Ltd., U.K. the Earth Sciences department, Cambridge and the British Museum for continued assistance.

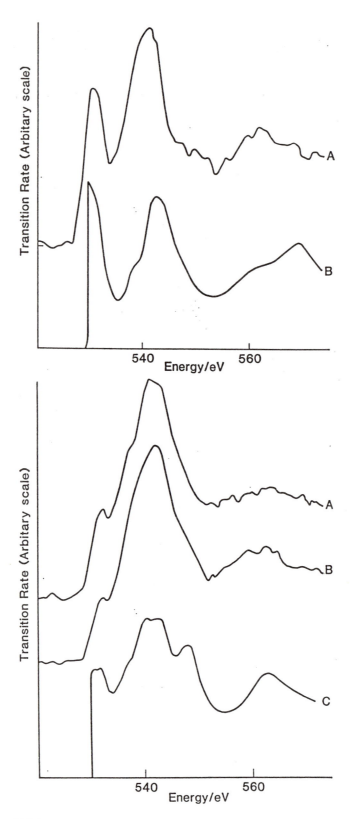

Figure 4 (a): A - Experimental EELS oxygen K-edge from control oxide after deconvolution and background subtraction. B - The results of ICXANES calculations for the oxygen K-edge in α-Fe_2O_3.

Figure 4 (b): A - Experimental EELS oxygen K-edge from a short term boron doped oxide film after deconvolution and background subtraction. B - Experimental EELS oxygen K-edge from vonsenite, Fe_3BO_5, after deconvolution and background subtraction. C - The results of ICXANES calculations for the oxygen K-edge in Fe_3BO_5.

REFERENCES

1. P. N. Rowley and J. A. Little, 'NACE Corrosion Research Symposium', Proc. Conf., New Orleans, La., April 17-19th, 1989, p.21.

2. D. D.Vvedensky, D. K.Saldin and J. B. Pendry, Computer Phys. Commun., 40, 1986, p.421.

3. W. Engel, H. Sauer, R.Brydson, B. G. Williams, E.Zeitler and J. M. Thomas, J. Chem. Soc. Faraday Trans. 1., 84, 1988, p.617.

4. P. N. Rowley, R. Brydson, J. Little and S. R. J.Saunders, Proc. EMAG-Micro '89 Conf. Ser. No. 98, Vol.2,1990, p.41, Eds. P. J. Goodhew and H. Y. Elder, London: IOP; P. N. Rowley, R. Brydson, J.Little and S. R. J. Saunders, 1990, Phil. Mag. B, 62 (2), p. 229.

5. R. Brydson, D. D.Vvedensky, W. Engel, H. Sauer, B. G.Williams, E. Zeitler and J. M. Thomas, J. Phys. Chem., 92, p.962.

6. F. M. F. de Groot, M. Grioni, J. C. Fuggle,J. Ghijsen, G.A. Sawatzky and H. Petersen, Phys. Rev. B, 40, 1989, p.5715.

7. P. N. Rowley, Ph.D. Thesis, University of Cambridge, 1990.

9 ANALYTICAL ELECTRON MICROSCOPY OF 9wt%Cr - 1wt%Mo STEELS

I.M. Reaney and G.W. Lorimer*

Department of Physics, University of Essex, Wivenhoe Park, Colchester CO4 3SQ, UK.
*Manchester Materials Science Centre, University of Manchester/U.M.I.S.T., Grosvenor Street, Manchester M1 7HS, UK.

ABSTRACT

The oxidation of two 9wt%Cr-1wt%Mo steels in CO_2 have been examined using analytical electron microscopy. The two steels have significantly different rates of oxidation and also have slight differences in composition. In particular, Steel B has a higher Si concentration than Steel A. The microstructure of the oxide scale on the two steels was similar but the distribution and type of precipitates beneath the oxide/metal interface was found to be significantly altered by the small changes in composition.

INTRODUCTION

The protective oxide which forms on the surface of 9wt%Cr-1wt%Mo steels oxidised at 560°C in CO_2 has a duplex structure of outer magnetite and an inner layer of Fe-Cr spinel [2-4]. The interface between the two layers lies at the site of the original metal/gas interface. The outer scale of magnetite grows at the scale/gas interface by the outward diffusion of Fe-ions and the inner scale grows at the metal/oxide interface by the inward diffusion of CO_2 [2-4]. During the oxidation process, carbon becomes deposited within the metal below the oxide/metal interface and this is accompanied by the formation of several types of carbides below the oxide/metal interface.

It has been shown that small additions of Si to a 9wt%Cr-1wt%Mo steel results in a dramatic reduction in the overall rate of protective oxidation [6] and in an increase in the time to formation of 'breakaway' oxide. (Breakaway oxide has a linear oxidation rate and forms after long term exposure to the CO_2 atmosphere at temperatures above 560°C.) In this paper, the results of an examination of the oxidation of two 9wt%Cr-1wt%Mo steels which have different Si concentrations are reported.

EXPERIMENTAL

The composition of the two steels examined is given below in Table 1.

Table 1: Composition (wt%) of the two steels examined within this paper (balance Fe). The steels were oxidised at 560°C and 41atm in a 99%CO_2/1%CO atmosphere for up to 1000h. The oxidation conditions recreate those within the AGR.

spec.	C	Si	S	P	Mn	Cr	Mo	Co	Ni	Ti	V	Cu	W
Steel A	0.29	0.44	0.01	0.01	0.51	9.09	0.87	-	0.18	0.02	0.02	0.09	-
Steel B	0.59	0.65	0.01	0.01	0.05	9.60	0.90	-	0.30	0.02	0.08	0.04	-

Two-Stage Extraction Replicas

The carbide type and distribution below the oxide/metal interface was monitored using two-stage extraction replicas. Oxidised bulk specimens were mounted in Bakelite, polished to 1 micron and electrolytically etched in 10vol%HCl at 10V. A small strip of acetate sheet was dipped in acetone, placed over the specimen and allowed to dry. The plastic sheet was stripped from the sample removing carbide precipitates in the process. The sheet was coated in carbon and cut into 3mm squares. Acetone was again used to dissolve the plastic sheet which left a free floating carbon film that contained precipitates from the oxidised sample.

Ion-beam Thinned Transverse Sections

Thin electron transparent sections through the oxide/metal interface were produced using the technique developed by Newcomb et al. [4]. Half cylinders (20mm long, 2.5mm diam.) were cut from the steels and oxidised. A layer of nickel approximately 50μm thick was plated onto each oxidised half cylinder using a Watts bath (nickel chloride, boric acid and nickel sulphamate). The two Ni-plated half cylinders were stuck together using an epoxy resin and the whole sample was Ni-plated (0.3-0.4mm). Slices, perpendicular to the length of the reformed cylinder, were removed on a spark machine and ground to a thickness of approximately 120μm on SiC grit. The sample was indented using a dimpler to a depth of 80μm. The sample was ion-beam thinned using an Ion-tech 'fast atom' beam thinner.

Electron Microscopy

The samples were examined with a Philips EM430 analytical electron microscope operated at an accelerating voltage of 300kV which was equipped with an EDAX energy dispersive X-ray detector (EDX). Quantitative analyses were obtained by the ratio technique of Cliff and Lorimer [7]. A Philips 525 scanning electron microscope was used to examine bulk samples in cross-section.

RESULTS AND DISCUSSION

Figure 1 shows the weight gain verses time curves for steels A and B which have been oxidised in CO_2 at 560°C and 41atm. It can be seen from these two curves that the weight gain associated with the oxidation of Steel A is approximately five times greater than that associated with Steel B after exposure for 50 000h. However, the oxidation time examined within this paper is 1000h; over this period the oxidation rates are similar and the weight gains of the two steels differ by a factor of two. Figures 2a and b are scanning electron micrographs which show typical duplex oxides that have formed on Steels A and B after exposure to the CO_2 atmosphere for 500h. A similar microstructure is observed for both steels which consists of an inner and outer oxide as well as carbides (arrowed) that have grown along the ferrite grain boundaries (GB).

Figures 3a and b are transmission electron micrographs showing the interface between the outer magnetite and the inner spinel which have grown on Steels A and B after oxidation for 72h and 500h, respectively. The microstructures, in each case, consisted of large columnar grains of outer magnetite and an aggregate of fine oxide grains and microchannels within the inner spinel. EDX analyses of the regions shown in Figs. 3a and b did not reveal any areas with a high Si-concentration. However, a large variation in the Cr-concentration within the inner oxide was observed. It has been suggested that this arises as a result of the oxidation (internal and external) of Cr-rich carbides in conjunction with the oxidation (external) of Cr-depleted ferrite [8].

Figures 4a and b are extraction replicas which show typical distributions of carbides across the oxide/metal interfaces of Steels A and B oxidised in CO_2 gas at 560°C for 1000h. A similar distribution of carbides is observed in the two steels. However, invariably a smaller number of carbides were extracted from Steel B than from Steel A. Electron diffraction patterns and EDX analyses from individual precipitates revealed that the majority of carbides from both steels were either acicular M_2C or granular $M_{23}C_6$. However, some granular M_7C_3 carbides were observed below the oxide/metal interface of Steel B, Fig. 5. A high density of granular M_7C_3 and $M_{23}C_6$ precipitates were observed immediately below the oxide/metal interface whereas acicular M_2C were observed beneath the M_7C_3 and $M_{23}C_6$. The ratio of the $M_2C/M_{23}C_6$ precipitates observed in Steel A was greater than that observed in Steel B. EDX analyses revealed that the concentration of Fe decreased in both M_2C and $M_{23}C_6$ as the distance from the interface increased. Banks et al. [5] suggested that a high carbon activity occurs at the oxide/metal interface this created a high 'carbide-forming' potential. Therefore, the carbides nucleate and grow rapidly in this region but never achieve equilibrium with their surroundings.

Windowless EDX analysis and electron energy loss spectrometry revealed that some intermetallic precipitates were present within the carburised zone which had the composition 43wt%Fe, 37wt%Mo, 11 wt%Cr and 8wt%Si, Figs. 6a, b and c. These precipitates were observed to grow in a narrow band approximately 100μm below the oxide/metal interface. Electron diffraction patterns obtained from these precipitates exhibited streaking in the high order Laue zones and only one pole was observed in reciprocal space which exhibited almost perfect 5-fold symmetry, Fig. 6d [9,10]. The Mo,Fe-rich phase was only observed after oxidation for 1000h in Steel A (0.4%Si) whereas it was observed from 72h in Steel B (0.7%Si). In addition, the concentration of Si within the precipitate in Steel B (8wt%Si) was greater than that found within the precipitate in Steel A (4wt%Si).

The distribution and occurrence of the phases observed during the oxidation of Steels A and B are schematically illustrated in Figs. 7a and b.

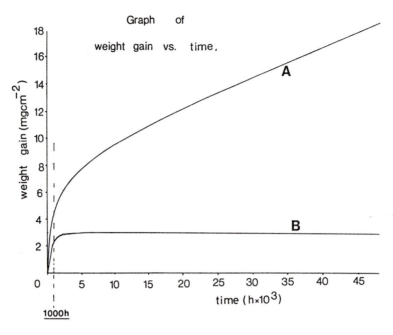

Figure 1: Graph of weight gain per unit area versus time for Steels A and B.

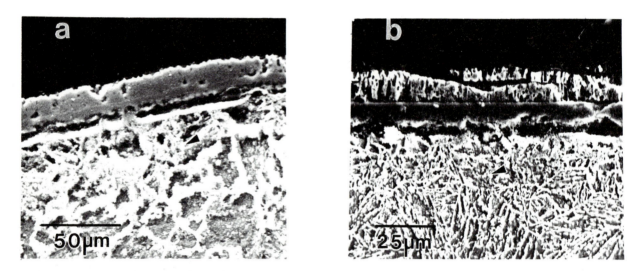

Figure 2: Scanning electron micrographs showing the oxide scales on (a) Steel A and (b) Steel B in transverse section after oxidation for 500h. Carbides below the oxide/metal interface are arrowed.

SUMMARY

The microstructures of the oxide scales formed on Steels A and B, which contained 0.44 and 0.65%Si respectively, are similar despite the difference in oxidation rates between the two steels. For periods of oxidation up to 1000h, no evidence was obtained for the presence of a Si-rich layer which could reduce the rate of oxidation. However, our observations do not eliminate the possibility that a low Si concentration dispersed throughout the oxide may 'block' or 'poison' the diffusion paths for Fe-ions. This would result in a decrease in the overall rate of oxidation.

The most significant effect we have observed is that an increase in the Si-content appears to decrease the total number of carbides and changes the carbide sequence within the carburised zone. It appears that an increase in the Si concentration not only reduces the total volume reaction of carbide but decreases the $M_2C/M_{23}C_6$ ratio. The increase in Si concentration

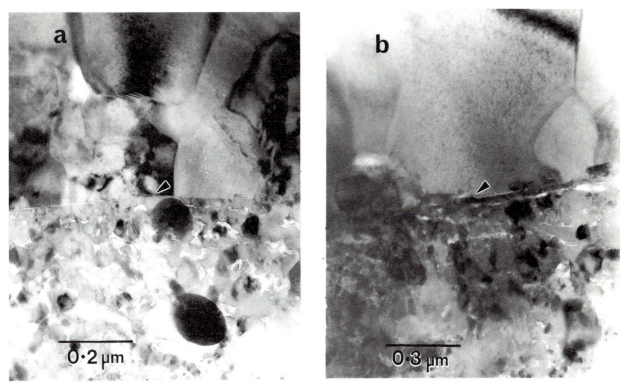

Figure 3: Transmission electron micrographs which show the interface (arrowed) between the inner and outer oxide on (a) Steel A (72h) and (b) Steel B (500h). A similar microstructure is observed in each image.

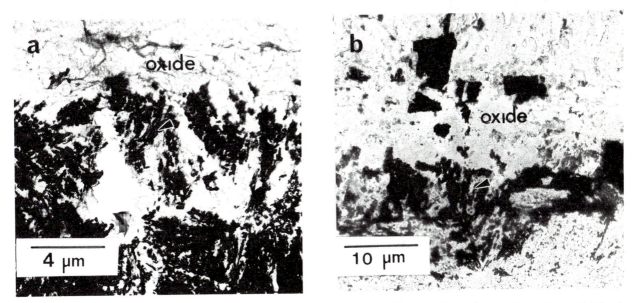

Figure 4: Extraction replicas showing the distribution of carbides (arrowed) across the oxide/metal interface of (a) Steel A and (b) Steel B after oxidation for 500h. Fewer carbides were extracted from Steel B than Steel A.

also appears to promote the formation of M_7C_3 and the intermetallic Mo,Fe-rich phase. However, the two steels contain a large number of other alloying additions. In particular, Steel A contains significantly more Mn than Steel B and it is unlikely that a simple mechanism could be proposed which would fully explain the above phenomena.

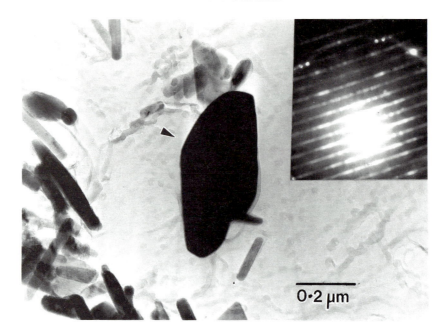

Figure 5: Transmission electron micrograph which shows an M_7C_3 precipitate (arrowed) that has formed beneath the oxide/metal interface of Steel B after oxidation for 1000h. Insert: streaked zone axis electron diffraction pattern characteristic of M_7C_3.

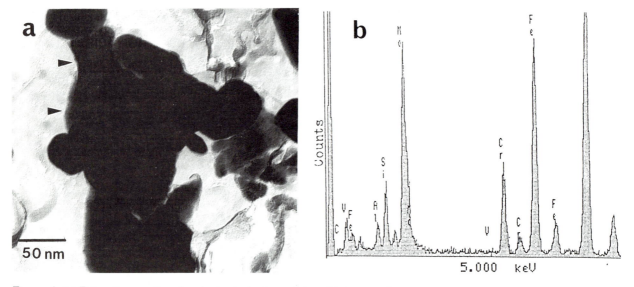

Figure 6: (a) Extraction replica showing a faulted precipitate(s) (arrowed) growing adjacent to an $M_{23}C_6$ carbide. (b) Windowless EDX spectrum which does not reveal the presence of C or O.

ACKNOWLEDGEMENTS

The work was supported by Nuclear Electric Plc, formerly the CEGB. The authors greatefully acknowledge many useful discussions with Drs P. J. Nolan and J. C. P. Garrett of Nuclear Electric Plc, formerly with CEGB, Wythenshawe Laboratories.

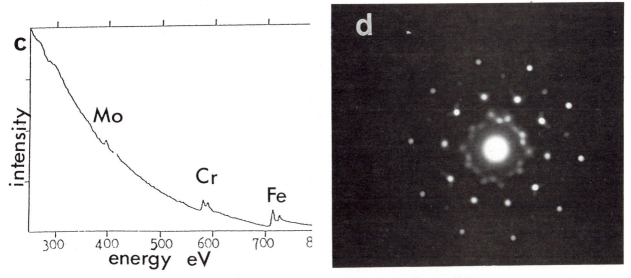

Figure 6: (c) Electron energy loss spectrum which does not reveal the presence of either C or O. (d) Electron diffraction pattern which exhibits almost perfect 5-fold symmetry.

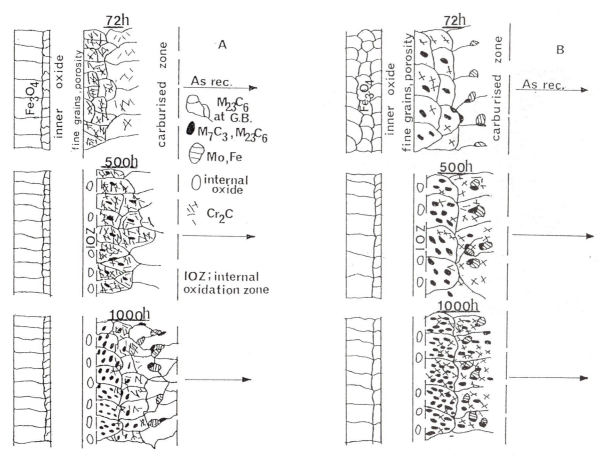

Figure 7: Schematic representation of the distribution of the phases associated with the oxidation of (a) Steel A and (b) Steel B for periods of exposure up to 1000h.

REFERENCES

1. P.C. Rowlands, J.C. Garrett and A. Whittaker, C.E.G.B. Research, No. 15, Nov. 1983, p.3.

2. S.K. Lister, P.J. Nolan and J.C.P. Garrett, 9%Cr steel Oxidation Monitoring Pt. iii, C.E.G.B. Confidential Report, 1974.

3. P.J. Castle and K.E. Smallman, Corros. Sci. 24, 2, 1984, p.99.

4. S.B. Newcomb, C.B. Boothroyd and W.M. Stobbs, J. Microscopy 140, 1985, p.195.

5. P. Banks and G.W. Lorimer, Environmental Degradation of High Temperature Materials, Inst. Metall., Ser. 3, No. 12, 1980, p.28.

6. P.C. Rowlands, J.C.P. Garrett, F.G. Hicks, S.K. Lister, B. Lloyd and J.A. Twelves, B.N.E.S. Inst. Conf. on Corr. Steels in Co, Ed. D.R. Holmes, R.B. Mill and L.M. Wyatt, 1974, p.193.

7. G. Cliff and G.W. Lorimer, J. Microscopy 103, 1975, p.203.

8. I.M. Reaney and G.W. Lorimer, Mat. Sci. and Tech. 4, 1988, p.391.

9. H.Q. Ye, D.W. Wang and K.H. Kuo, Ultramicroscopy 16, 1985b, p.273.

10. I.M. Reaney and G.W. Lorimer, Phil. Mag. Lett. 57, 1988, p.247.

10 CARBURISATION-OXIDATION INTERACTIONS IN BREAKAWAY SCALING

S.B. Newcomb and W.M. Stobbs

Department of Materials Science and Metallurgy,
Cambridge University, Pembroke Street, Cambridge CB2 3QZ, UK.

ABSTRACT

The microstructures of a number of breakaway scales formed in CO_2 of apparently different form are described in an attempt to see whether or not the process has a single mode of control. Particular attention is paid to the relationship between the way carburisation occurs and the form of the scale as a function of both the oxidation temperature and alloy carbon content.

INTRODUCTION

The change in oxidation behaviour from protective to breakaway scaling which is shown by some Fe-Cr alloys is as dramatic as it is generally unpredictable. Breakaway scales are characteristically formed at an approximately linear rate and are typically porous as well as having a tendency to spall, factors which have imposed operational constraints on commercial power plants. Considerable effort has thus been put into the development of mechanistic models [see e.g. 1] to support the phenomenological methods used to predict the lifetimes of Fe-Cr components. Most current models stem from the work of Gibbs [2] who suggested that breakaway requires the initially protective inner layer scale to develop an open network of carbon to allow the direct gaseous attack of the underlying metal. On the basis that the time to breakaway falls as the carbon content of an Fe-9Cr alloy is increased, Rowlands et al. [3] have suggested that such a network can form only once the metal beneath the scale can take up no further carbon. However, it would also appear that there can be a change in the form of the carburisation and in the balance of the scaling reactions as a function of the oxidation temperature: between 480°C and 520°C carbon is partitioned in an approximately 1:1 ratio between the alloy and the oxide whereas at 560°C this ratio is 3:1 [4].

While we are clear that carbon must play an important rôle in breakaway we have been surprised by the way in which a universal model is currently applied for the process over a whole range of oxidation temperatures and alloy carbon contents. The extraordinary heterogeneity of breakaway oxides [5] coupled to the fact that carbon is sometimes not deposited in the protective scale at all [6] has fuelled our concern, as has the way in which breakaway can occur either at the beginning of an oxidation test or with a delayed onset. Our aim here has thus been to determine whether or not there is now unequivocal evidence that there is more than one way in which breakaway can be initiated. With this in mind we draw on some of our TEM results on the differences in the scale morphologies seen for a range of Fe-Cr alloys which exhibit breakaway in order to highlight the variety of ways in which carbon can become incorporated into both the metal and the oxides.

RESULTS AND DISCUSSION

We first describe the breakaway scale morphologies seen for two high Si content Fe-11Cr alloys oxidised at 600°C and 650°C and then compare the data described with that obtained for Fe-9Cr alloys, as oxidised at 600°C and 480°C.

Fe-11.3Cr-0.41Si: 600°C

The oxidation behaviour of Fe-11.3Cr-0.41Si-0.2C shows a remarkable change from healing kinetics to breakaway scaling after approximately 5000 hours exposure to 3%CO-CO_2 at 4.1MNm^{-2} and 600°C. We have investigated the temporal development of the scales formed on this alloy prior to breakaway [7] and found that carbon is deposited in the protective oxides after 3088 hours but not after a much shorter oxidation period (294 hours). While this observation would appear to be broadly consistent with Gibbs' ideas [2], it is interesting that carbon was not found in a 1-2.5μm band of the oxide at the metal/oxide interface, where the chromium concentration was also high (74-78wt%). It thus seems difficult to argue that the change from healing to breakaway scaling is caused solely by the greater diffusivity of the carbon deposits when there is a zone still present which should act as a good diffusion barrier to reduce the overall rate of oxidation. There would have to be a gross change in the overall balance of reactions which are occurring at the base of the scale to favour the deposition of carbon within the Cr-rich oxide. On the other hand it should be noted that carbon was not seen in the scale formed on Fe-Cr-1.10Si which contained sufficient Si to prevent the initiation of breakaway [7]. The possibility that the scale fails mechanically must also be considered though the fairly gradual transition to breakaway scaling which is exhibited suggests that any such spalling would have to be partial, with accelerated oxidation thus initially being promoted only locally.

Any interpretation of the role of carbon in the initiation of breakaway is further confused by the way that the microstructure of the scales changes radically once breakaway has started. After 15 920 hours exposure to 3%CO-CO$_2$ at 600°C, three distinct breakaway oxide morphologies are found for the 0.41 Si alloy. For the first of these, as shown in Fig. 1a, the spinel is fine grained (~100nm) and the oxide is intermixed with untextured graphite, the typical form of which is exemplified by the dark field micrograph in Fig. 1b. More extended regions of porous graphitic deposits were also seen, as shown in Fig.1c, and in such regions the grain size of the intermixed spinel could be as small as 20nm. Other parts of the breakaway scale contained much coarser grains of spinel (0.5-2μm), as shown in Fig. 1d, and these did not contain carbon. This correlation between the spinel grain size and the presence or otherwise of graphite would suggest that extended grain growth occurs only when graphite is not deposited but this is not our observation for other breakaway scales.

The sub-scale region of the alloy, like the oxide formed above it, was found to be significantly changed once breakaway had been initiated. Whereas during the formation of the protective scale carbon was not injected into the alloy beneath the scale (a behaviour which is unlike that seen for the the protective oxidation of a binary Fe-9Cr not containing Si [6]) we now find a very high volume fraction of M$_{23}$C$_6$ precipitates in a 250μm band beneath the scale. The carbides, which were up to 2μm in length, had a faulted lenticular morphology (Fig. 1e) and typically had an Fe:Cr ratio as high as ~70:30.

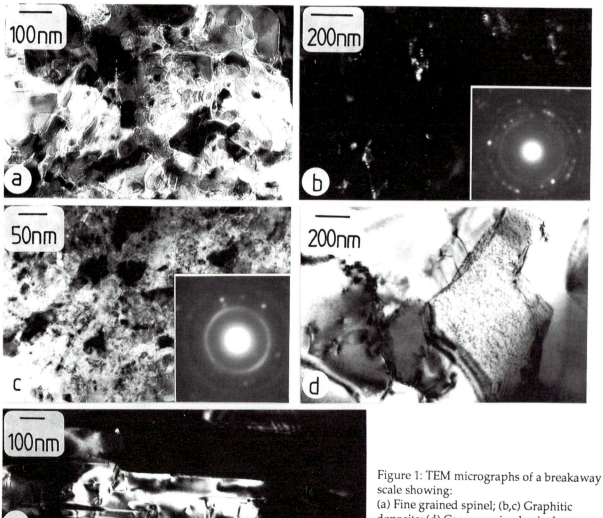

Figure 1: TEM micrographs of a breakaway scale showing:
(a) Fine grained spinel; (b,c) Graphitic deposits; (d) Coarse grained spinel;
(e) Faulted M$_{23}$C$_6$ beneath the scale.

The growth of the $M_{23}C_6$ has apparently involved the assimilation of Fe in the absence of a good Cr supply but at a sufficiently low C 'potential' that M_2C is not formed in the way it is during breakaway on Fe-9Cr when this occurs through a redox reaction with the formation of wustite [8].

Fe-11.6Cr-0.29Si: 650°C

The breakaway scaling characteristics of Fe-Cr-0.41Si, as described above, were compared with that of an Fe-11.58Cr-0.91Mo-0.03V-0.079Ni-0.29Si-0.2C alloy oxidised in $3\%CO/810$vpm H_2O in CO_2 at $4.1MNm^{-2}$ for 4563 hours. The slightly lower Si content of this alloy clearly makes it less protective than the 0.41Si alloy, but even so a tapered specimen was oxidised in an attempt to ensure that the metal beneath the breakaway scale became saturated with carbon during scaling so as to see the effects of the high carbon activity on the microstructural development of the breakaway scale caused by using a higher temperature (650°C) than for the 0.41Si alloy (600°C).

Much of the breakaway scale was morphologically very similar to that formed on the 0.41Si alloy at 600°C. However, we now found some of the coarser grained (~1mm) spinel to be interspersed with carbon in a way which was generally indicative of a higher carbon content scale than was previously examined. A low magnification micrograph of one such region is shown in Fig. 2a and the carbon deposit is shown at higher magnification in Fig. 2b. Most of the carbon was found to be amorphous although non-intercalated graphite may be seen adjacent to the spinel oxide. Further changes in the breakaway scale were found nearer to the metal/oxide interface where approximately 20µm from it wustite was found. The coarse grained FeO which was found is shown in Fig. 2c, the non-stoichiometry of the oxide being demonstrated by the (100) inset diffraction pattern, and the morphology seen may be compared to that of wustite which is characteristic of the layered breakaway scale formed on Fe-9Cr at 600°C [8]. For this latter alloy we argued that CO diffuses both up and down in the scale, so that its concentration peaks at the point where FeO is formed, once the alloy becomes sufficiently carburised to form M_2C, the oxidation of these carbides providing the secondary source of carbon. The Fe-11Cr alloy

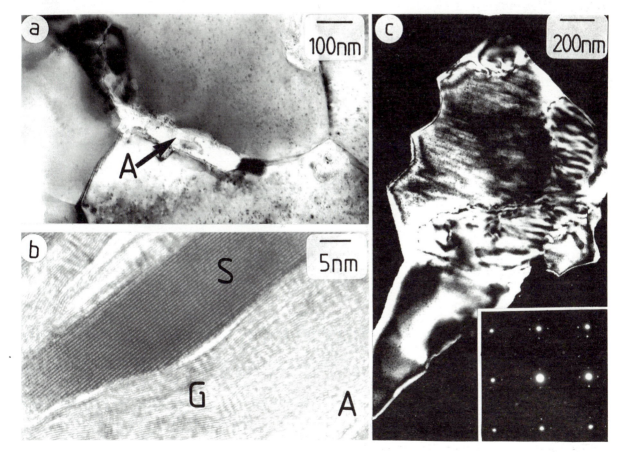

Figure 2: TEM micrographs showing (a) C deposit (as at A) and coarse grained spinel; (b) a deposit containing both graphite(G) and amorphous(A) carbon adjacent to a spinel(S) grain; and (c) coarse grained FeO.

examined here, however, was again found to contain $M_{23}C_6$ precipitates rather than M_2C despite its high (0.2) carbon content prior to oxidation. This indicates that very high CO concentrations can be attained within a breakaway scale without carbon supersaturation of the alloy, though the absence of any internal oxidation zone suggests that carbon can quite readily be injected back into the scale as the oxidation front advances.

Fe-9Cr-0.1C: 600°C

In comparing the scales formed after five different oxidation periods on an Fe-9Cr-0.10C binary alloy, as oxidised at 600°C in $3\%CO/50ppmCH_4/250ppmH_2O/100ppmH_2$ in CO_2 for up to 3807 hours, we saw gross variations in the thicknesses of the scales formed. Differences were also found in the depth of the sub-scale carburisation on any one specimen in addition to apparent inconsistencies in the progressive development of both the scale and carburisation layers despite all the surfaces of the alloy having been given the same pre-oxidation treatment (pickling in 15v/o HNO_3 at 60°C for 20 minutes). For example, while the scale formed after 619 hours oxidation was generally about 175μm in thickness the depth of carburisation in the alloy beneath the scale was extremely variable (50-250μm) even if it was of course generally much more developed at the specimen corners. By comparison, the ~800μm scale formed after 3807 hours showed alloy carburisation to a depth of only 125μm in some areas and 600μm in others; we are thus seeing *less* carburisation, in at least some regions, after 3807 hours than after 619 hours oxidation.

All the different breakaway scales which formed were found to be generally layered. A schematic summary diagram of the scale formed after 619 hours is shown in Fig. 3a and the microstructures of the different layers are not too dissimilar from the layered breakaway scale examined previously [8], a high volume fraction of carbon again being seen in the scale. Isolated, but relatively coarse grains of α-Fe, were also found within the scale, as shown in Fig. 3b. Iron was not, however, observed after longer periods of oxidation suggesting that it is progressively oxidised to M_3O_4 and was present as islands of unoxidised alloy rather than as a product of redox scaling. Equally, wustite was not seen in any of the scales despite the formation of M_2C within the sub-scale metal, in a way which we have previously taken to be indicative of a sufficiently high carbon activity to lead to FeO formation. While the formation of M_2C in the 0.1C alloy examined here is consistent with its observation in the 0.1C alloy described previously [8], the retention of $M_{23}C_6$ in the Fe-11Cr alloy, which contains 0.2C, is all the more surprising. The differences in the mode of carburisation of the alloys examined further highlights the various ways in which wustite formation can be initiated, since we are now seeing high carbon activities in the sub-scale metal without FeO formation (whereas the Fe-11Cr alloy formed $M_{23}C_6$ precipitates beneath a scale containing

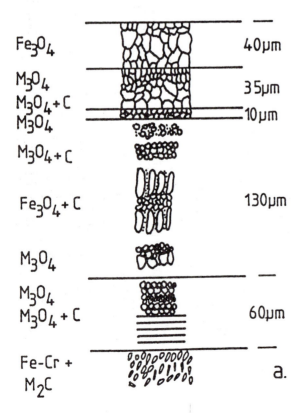

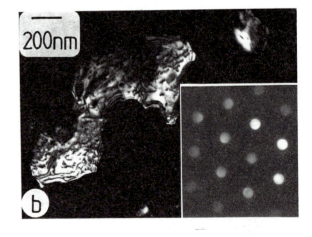

Figure 3: (a) A schematic summary diagram of the breakaway scale formed on Fe-9Cr at 600°C. A dark field micrograph of unoxidised Fe is shown in (b) where the inset diffraction pattern shows a (110)Fe normal.

FeO). There are yet further differences in the oxidation behaviour in relation to whether or not an internal oxidation zone is formed. In the alloy examined here, for example, we find a well developed 3μm band of carbide free oxide within the metal, suggesting that carbon may not be being injected into the scale from the alloy. At first sight this would naturally explain the non-formation of FeO except that an internal oxidation zone was seen in the Fe-9Cr alloy which did form FeO [8].

Fe-9Cr:480°C

We have also examined a breakaway scale formed on the binary Fe-9Cr alloy at 480°C after 54617 hours exposure to 1%CO-CO$_2$ and find that the alloy is, after this treatment, *less* carburised than during breakaway scaling at 600°C.

Most of the oxides seen in the breakaway scale had very similar morphologies to those described above. For example, an outer layer of magnetite was seen and there were inner layers of M$_3$O$_4$ containing various amounts of carbon though we noted that the columnar oxide grains could contain high volume fractions of carbon (see Fig.4a) where the deposits were found to be textured (see inset diffraction pattern). The bottom 60μm of the scale was sub-layered and was found to consist of alternating layers of 200nm M$_3$O$_4$ grains, which contained no carbon, and much finer grains (50nm) with intermixed carbon. A low magnification micrograph of a typical area from this zone is shown in Fig. 4b and the different layers at higher magnification in Figs. 4c and 4d. However, at the base of the scale no internal oxidation layer was observed, in contrast to our observations of oxide formation at 600°C, and in the specimen examined fine acicular M$_2$C

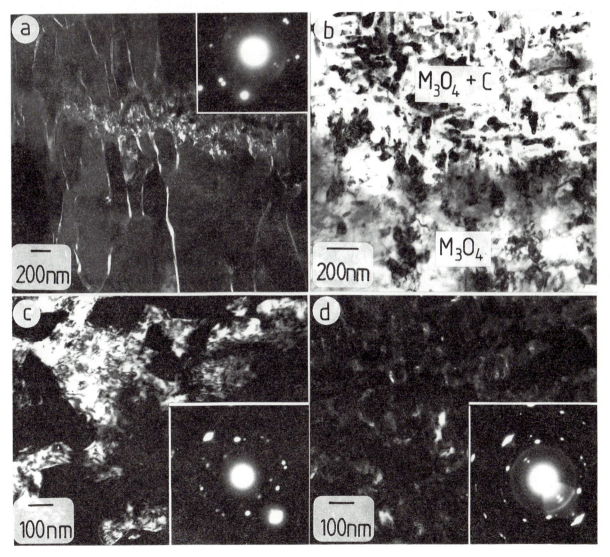

Figure 4: TEM micrographs of (a) textured graphite at coarse grained spinel grain boundaries and (b) the microstructure of the layered scale near the metal/oxide interface. The different non-carbon and carbon containing layers are shown in more detail in (c) and (d) respectively.

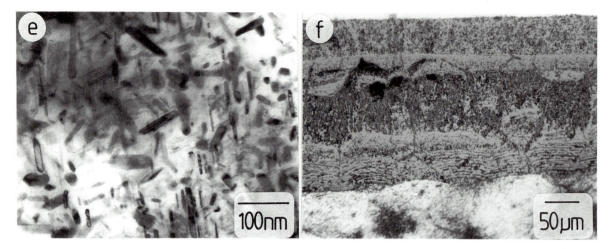

Figure 4: (e) Bright field micrograph of M_2C carbides formed beneath the breakaway scale and (f) light optical micrograph of the breakaway scale showing inhomogeneous carbide formation beneath the oxide.

was seen to a depth of ~150μm from the alloy/oxide interface (Fig. 4e). At the same time optical micrographs showed that, despite the general pearlitic microstructure of the alloy prior to oxidation, carbide precipitation beneath the breakaway scale did not always occur. The optical micrograph shown in Fig. 4f is typical of a region of the metal/oxide interface where carbide precipitation has occurred in some areas but not in others. There are, however, clear indications that carburisation has been enhanced around pre-existing pearlite colonies.

It is perhaps surprising that M_2C formation can be so localised without apparently affecting the morphologies of the layered breakaway scale forming above. The non-carbon containing layers seen at the base of the scale are equally indicative of the way in which carbon does not diffuse generally into the 9Cr alloy so that oxidation, in these circumstances, would appear to be controlled by diffusional processes within the scale.

CONCLUSIONS

Our comparisons are to some extent limited by the different origins of the Fe-Cr alloys examined and we are also concerned that test specimen pre-treatments remain insufficiently standardised for the unequivocal comparison of data from different laboratories. However the way our TEM observations have highlighted some of the many different forms which breakaway scales can adopt further requires that the fundamental mechanisms associated with breakaway be plural. Prior to this survey we had thought that there were basically two different mechanisms, for one of which carbon incorporation in the scale simply allows enhanced gaseous ingress in the manner described by Gibbs while for the other there is a chemically initiated process in which a redox reaction involving internal wustite formation is promoted by saturation of the subscale regions with carbon. It is now clear that not only can there be mixed forms of these processes but also both that wustite can be formed without the equilibrium saturation of the matrix by carbon and that the first process can be effective without the full carburisation of the spinel grain boundaries. While differences in the types of carbides which are formed, as well as in whether or not internal oxidation occurs, have been observed, it remains that it would appear to be carbon deposition within the scales which is most important in maintaining a high rate of oxidation. Equally this is not a prerequisite for breakaway.

ACKNOWLEDGEMENTS

We thank Professor D. Hull FRS for the provision of laboratory facilities as well as the SERC and National Power (Leatherhead) for financial support.

REFERENCES

1. Corrosion of Steels in CO_2, (Ed. D.R.Holmes et al.) BNES, London, 1974.

2. G.B.Gibbs, Oxid. Met. 7, 1973, p.173.

3. P.C. Rowlands, N. Thorne and M.J. Holt, CEGB Report RD/L/N137/79, 1979.

4. P.C. Rowlands, M.J. Holt and P.L. Harrison, CEGB Report TPRD/L/2461/N83, 1983.

5. P.L. Harrison, CEGB Report Rd/L/R1933, 1976.

6. W.M. Stobbs, S.B. Newcomb and E. Metcalfe, Phil. Trans Roy. Soc. (Lond.) A319, 1986, p.219.

7. S.B. Newcomb and W.M. Stobbs, these Proceedings, p.84.

8. S.B. Newcomb and W.M. Stobbs, Oxid. Met. 26, 1986, p.431.

11 HEALING LAYER FORMATION IN Fe-Cr-Si ALLOYS

S.B. Newcomb and W.M. Stobbs

Department of Materials Science and Metallurgy,
Cambridge University, Pembroke Street, Cambridge CB2 3QZ, UK.

ABSTRACT

The inner layer scales formed at 600°C in 3%CO-CO_2 on two Fe-11Cr-Si alloys with differing Si contents are examined using cross-sectional TEM. Different types of oxide were found exhibiting chemistry changes with time. While the processes seen are dominated by the formation of SiO_2 at the metal-oxide interface the way the scaling is also critically affected by carbon not diffusing into the sub-scale alloy is discussed.

INTRODUCTION

The oxidation properties of many materials can be significantly improved by the presence of even trace amounts of an element which forms a very stable oxide at a low rate. The beneficial effects of silicon in a number of metals such as iron [e.g. 1] as well as Fe-Cr [e.g. 2] and Fe-Ni-Cr alloys [e.g. 3] have, for example, been demonstrated and are of interest to the power generation industry for which the long term stability of Fe-Cr alloys is required. Robertson and Manning [4] have recently correlated the oxidation kinetics of a range of Fe-Cr alloys with their silicon and chromium contents in order to determine the critical amount of silicon needed to produce an acceptable fall in the oxidation rate ('healing') at different temperatures, though weight gains decrease generally as the alloy silicon content is increased. Here our aim has been primarily to determine whether or not there are differences in the effects of Si on the microstructures of the scales formed on two Fe-11Cr-Si alloys containing different amounts of silicon which show differing but characteristic healing kinetics. We have also investigated whether or not the structure and chemistry of the scales change during the very low weight gain period of the 'healing' kinetics since it is known that the lower Si content alloy goes in to breakaway kinetics after an extended period of low weight gain.

EXPERIMENTAL DETAILS

The two Fe-Cr-Si alloys investigated contained 0.41 and 1.10 wt% Si respectively and had a nominal composition of Fe-11.3Cr-1.0Mo-0.57Mn-0.2C. Coupons measuring 12*12*5mm were oxidised at 600°C in 3%CO-CO_2 containing 810vppm H_2O for up to 16 000 hours. The data presented here were obtained by examining cross-sectional TEM foils prepared using standard methods [5]. TEM investigation of the alloys prior to oxidation showed that both had a ~1μm subgrained ferritic structure containing $M_{23}C_6$ carbides. These carbides typically contained ~29-45wt% Cr and the surrounding matrix 6.7-10.7wt% Cr, a significantly coarser compositional variation than is present in other Fe-Cr alloys which can exhibit different modes of scaling cotemporally [6,7]. It is in general well worth remembering that if there is significant carbide deposition in an Fe-Cr alloy, whether as a function of the initial carbon content of the alloy prior to oxidation or as a result of carbon ingress during oxidation, then the alloy Cr content can be reduced to the extent that the oxidation mechanism can change locally.

RESULTS AND DISCUSSION

Optical microscopy showed that the thicknesses of the scales formed on both the two Fe-Cr-Si alloys were locally very variable. We have thus examined a number of different TEM specimens for each of the oxidation treatments in order to determine whether or not there are differences in the local scale microstructure which can be associated with the scale thickness variations. We describe the microstructures of only the inner layers given the general tendency for the outer magnetite layer to spall as cross-sectional TEM specimens are made.

Fe-Cr-0.41Si: 294 hours oxidation

Two characteristically different types of inner scale, were found to be formed on the 0.41Si alloy after 294 hours oxidation. One of these was much thicker than the other.

We will describe first the more developed inner layer oxide which was approximately 14.7μm in thickness. A bright field TEM micrograph of the metal/oxide interface region is shown in Fig. 1a which demonstrates the fine grained morphology (20-50nm) of the inner layer oxide. Elongated voids were found at the metal/oxide interface (arrowed) but no SiO_2, EDX analyses of the area indicating the presence of only trace amounts of Si (<0.25wt%). This is initially surprising, remembering the characteristic healing kinetics for this alloy, but consistent with the data of Coad et al. [8] who found that SiO_2 is not formed on a 20/25 Nb alloy during the first ~1000 hours of oxidation. Beneath the scale, $M_{23}C_6$ precipitates were seen in the alloy, as at A in Fig. 1a, which is approximately 0.5μm from the metal/oxide interface. The chromium content of this carbide (7.8wt%) was considerably lower than that of the $M_{23}C_6$ seen in bulk while the chromium content of the ferrite at the metal/oxide interface was also low (1.4wt%). There was found to be a graded depletion of chromium both in the carbides and in the ferritic matrix to a depth of about 2μm beneath the inward growing scale. The way in which the chromium in the carbides is being depleted while the alloy is showing healing characteristics is different from the way a similar alloy, which did not show healing kinetics as oxidised in 1%CO-CO_2, exhibited carbide precipitation beneath the scale [9]. The bulk of the oxide seen in Fig. 1a was M_3O_4 though interspersed within it were found to be small amounts of M_2O_3. While the majority of the inner layer scale had the unvoided morphology of the oxide seen in Fig. 1a, other regions had a more 'open' microstructure, as shown in Fig. 1b. The Fe/Cr composition profile for the 14.7μm inner layer scale described is shown in Fig. 2 (where the chemistries of all the inner layer scales examined are summarised). Here we can see that the Fe concentration of the scale is highest at the inner/outer layer interface but thereafter falls over a distance of ~7μm beyond which it remains approximately constant at a concentration of ~46wt%, the formation of M_2O_3 in the metal/oxide region of the scale apparently not resulting in a significant increase in the Cr content of the scale.

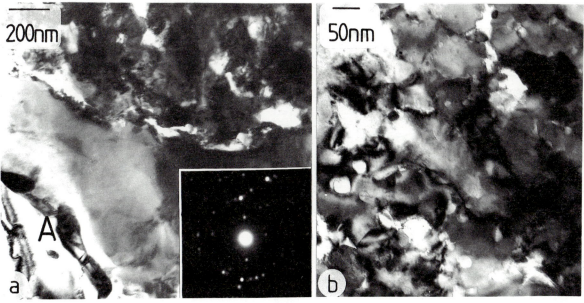

Figure 1: (a) The metal/oxide interface region of the 14.7mm scale formed on Fe-Cr-0.41Si after 294 hours oxidation. Cr depleted $M_{23}C_6$ carbides (at A) were seen close to the interface. The inset diffraction pattern shows the presence of both M_3O_4 and M_2O_3 in the scale. (b) A micrograph showing the morphology of a porous region of the M_3O_4 scale.

The second inner layer oxide which was characterised for this specimen as oxidised in the way described was only about 2.1μm in thickness and while it was structurally related to the 14.7μm scale described above it showed entirely different compositional trends. Significantly, however, M_2O_3 was not found in this scale, the oxide consisting entirely of compact and fine grained (~50nm) M_3O_4 and neither carbon nor graphite were found within it. By comparison with the 14.7μm scale described above, we now find that the M_3O_4 at the metal/oxide interface is enriched in Fe (~74wt%) and that there is a relatively sharp fall in the Fe concentration, where ~0.8μm from this interface the M_3O_4 contains ~54wt%Fe (see Fig. 2b). Thereafter the Fe concentration continues to fall though with a shallower gradient so that the M_3O_4 near to the inner/outer layer interface contains less Fe (46wt%) than Cr (54wt%). Both the alloy and carbides beneath the scale are again Cr depleted, the lack of alloy carburisation clearly indicating that carbon is not being injected into the metal during scaling. Equally the absence of either carbon or carbides in both the scales examined suggests that any carbon produced during the oxidation reactions is probably converted either to CO or to CO_2. This would then change the oxidation potential within the scales and lead to the preferential oxidation of one cation or another.

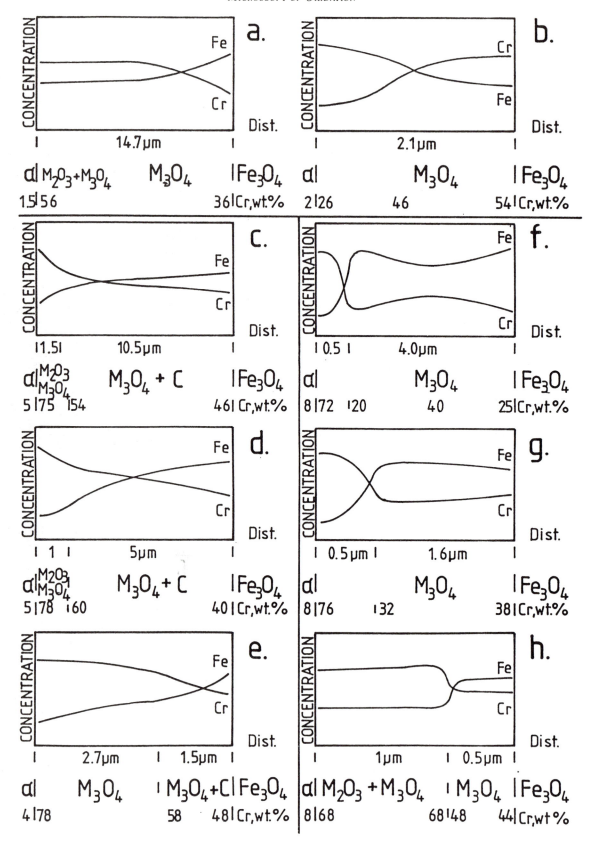

Figure 2: Fe/Cr concentration profiles for the inner layer scales formed on (a,b) Fe-Cr-0.41Si (294 hours), (c,d,e) Fe-Cr-0.41Si (3088 hours), and (f,g,h) Fe-Cr-1.10Si after 3088 hours oxidation at 600°C in 3%CO-CO_2. The different oxides found in each of the inner layer scales are also shown.

Fe-Cr-0.41Si: 3088 hours oxidation

Three different types of inner layer scales were found on the 0.41Si alloy after it had been oxidised for 3088 hours and both their morphology and chemistry as well as the composition gradients in the underlying alloy were found to be significantly changed after this longer period of oxidation. We now find, for example, SiO_2 at or near to the metal/oxide interface, Fig. 3a showing the typical morphology of the amorphous silica layer formed within the $3.5\mu m$ scale. Here the SiO_2 is ~7.5nm in thickness at the metal/oxide interface but is apparently discontinuous as it is in the 5nm layer seen at A in Fig. 3a ~100nm from this interface.

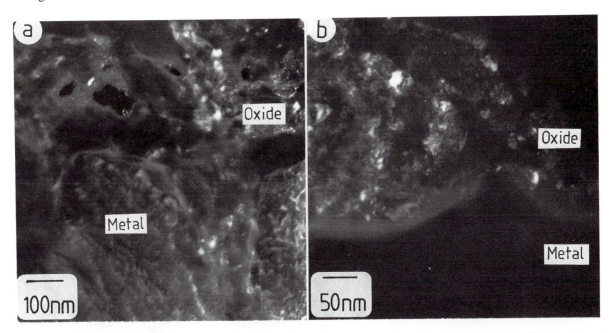

Figure 3: The metal/oxide interface seen in (a) Fe-Cr-0.41Si and (b) Fe-Cr-1.10Si after 3088 hours oxidation showing the different morphologies of SiO_2.

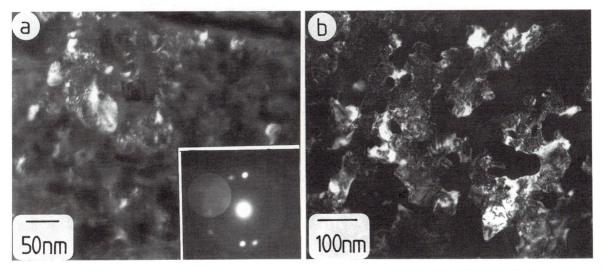

Figure 4: Dark field micrographs showing the morphology of (a) the Type A, and (b) the Type B oxides. Carbon was found in the Type A oxide.

Each of the inward growing scales examined was sub-layered, the oxides in each layer being both morphologically and compositionally distinct from one another. Nearest to the outward growing Fe_3O_4 there was an upper zone of M_3O_4 which was generally porous and contained carbon, unlike the scales examined after 294 hours oxidation. The typical morphology of this part of the inner scale, which we refer to as Type A oxide, is shown in Fig. 4a where the M_3O_4 has a grain size of ~25nm. The second sub-layer of the scale (Type B oxide), as formed beneath the Type A oxide and towards the metal/

oxide interface (see Fig. 4b), did not contain carbon and was non-porous and was thus more similar to the oxide seen after 294 hours oxidation. This was further emphasised by the presence of a low volume fraction of M_2O_3 within it. We should emphasise, however, that M_2O_3 was seen in two of the three inner layer scales but was absent in the thinnest (4.2μm) of the three scales. Given the general variability of the scales it was not surprising that the ratio of the thickness of the Type A to that of the Type B sub-layer oxides was also very variable. While there are clear morphological similarities between the inward growing Type B oxides formed at the metal/oxide interface after 294 and 3088 hours oxidation, their chemistries were however found to be very different (see Fig. 2). Whereas after 294 hours oxidation the composition of the scale at the metal/oxide interface could be apparently either Cr or Fe rich, but contained no more than 56wt% Cr, we now find that the Cr contents of all the Type B oxides are high (74-78wt%), even in the absence of M_2O_3 formation which would normally be associated with chromium enrichment [10]. When M_2O_3 is present, however, less of the Type B oxide is formed (1-1.5μm) than when it is absent (2.7μm). The compositional trends for the Type A oxides formed above the zone described were surprisingly similar in the three scales examined (see Figs. 2c-e), given the variability seen after 294 hours oxidation (Figs. 2a and b). Here the M_3O_4 formed immediately beneath the outward growing Fe_3O_4 contained ~40-48wt% Cr, the Cr content of the oxide gradually increasing towards the Type B oxide where its concentration varied from 54 to 60wt%. Given, however, the differences in the thicknesses of the various scales the Cr gradients are rather different from oxide to oxide, though it would appear that the overall chemistries of the inner oxides are generally being determined by the oxidant potential within the scales.

Turning to the metal immediately beneath the three inner layer scales we find that the alloy contains *no* carbides to a depth of approximately 5μm from the metal/oxide interface, as was confirmed by optical microscopy. A low magnification TEM image taken from the metal/oxide interface region beneath the 11.3μm inner layer scale which demonstrates the morphology of this denuded zone is shown in Fig. 5.

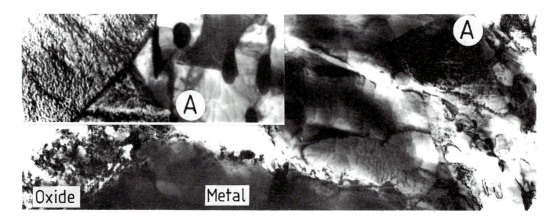

Figure 5: A low magnification bright field micrograph of the metal/oxide interface region seen in Fe-Cr-0.41Si after 3088 hours oxidation. No carbides were seen in a 5μm band immediately beneath the scale. The inset shows the interface between the zone containing no carbides and bulk.

The ferrite in the carbide denuded zone has a grain size of approximately 4μm by comparison with the 1μm sub-grains seen in the bulk of the alloy and contains approximately 5wt% Cr. Although the chromium content of the alloy here was thus somewhat higher than that seen in the metal after 294 hours oxidation (~1.4wt%), the dissolution of the $M_{23}C_6$ precipitates is consistent with the removal of chromium to the oxide forming above. The $M_{23}C_6$ carbides beneath this denuded zone showed the same general variability in their composition (20-45wt%) as was found for those in the bulk of the alloy.

Fe-Cr-1.10Si: 3088 hours oxidation

Three inner scales, which had thicknesses of 1.5, 2.1 and 4.5μm, were examined in the TEM. The 1.10Si alloy formed SiO_2 at the base of the inner layer scales in variable morphologies which were generally similar to those seen for the 0.41Si alloy though these glassy regions were generally slightly thicker than in the 0.41Si alloy. An example of a continuous layer of SiO_2 is shown in Fig. 3b where its growth to a thickness of 30nm has apparently inhibited the development of the scale above it. The alloy also contained a ~5μm carbide denuded band beneath the scales, as was seen for the 0.41Si alloy, where the Cr concentration was approximately 7.5wt%.

The three inner layer scales, like the oxides formed on the 0.41Si alloy, were all enriched in chromium at the metal/oxide interface. Here the Cr concentrations of the Type B oxides varied from 68-76wt%, though the ratios of the thicknesses of the Type A and B oxides were again found to be variable. The Type B oxides were similar to those formed on Fe-Cr-0.41Si in being non-porous and not having carbon deposited in them while M_2O_3 was again seen in some of the scales but not in others (see Fig. 2). Intriguingly the most developed high chromium content Type B oxide layer (1μm) contained M_2O_3 (though it was in the thinnest of the three scales examined) whereas M_2O_3 was absent in the thickest of the Type B oxides formed on the 0.41Si alloy. The Type A oxides, however, showed morphological differences to the inner layer scales formed on the 0.41Si alloy because none contained carbon and all were generally non-porous. It is thus surprising that the Type A oxides formed on the two Fe-Cr-Si alloys show similar compositional trends (see Fig. 2), the characteristic fall in the Cr concentration from the Type B oxide again being seen. The chemistries of the three inner layer scales examined reflected their relative thicknesses, and the Cr concentration at the M_3O_4/Fe_3O_4 interfaces were found to be 25, 38 and 44wt% with the scales decreasing in thickness as the Cr concentration increases (see Fig.2). The Cr concentration of the Type A oxide, however, can apparently remain almost constant or alternatively either increase or decrease though the relevant Cr gradients are generally steeper than in the scales seen on the 0.41Si alloy.

The formation of SiO_2 as well as of the chromium-rich spinel at the metal/oxide interface in the 0.41Si and 1.10Si alloys after 3088 hours oxidation is consistent with a reduction in the oxygen potential during the scaling process. The apparent local variability in this reduction in the oxygen potential is emphasised by the appearance of M_2O_3 in some of the scales but not in others (see Fig. 2). While we are clear that most of the SiO_2 as well as the chromium-rich oxides were formed during the period of oxidation from 294 to 3088 hours, the formation of the SiO_2 would appear to be more important in maintaining the low oxidation rate than the M_2O_3. However, despite the presence of the SiO_2 it would appear that both iron and chromium can continue to diffuse into the scales above it, as evidenced by the way the different oxides develop with time, the differences in the chemistries of the various scales further suggesting that the SiO_2 is a more effective barrier in Fe-Cr-1.10Si than in the lower Si content alloy. We find, for example, that the Cr content of the 0.41Si alloy at the metal/oxide interface is lower (~5.5wt%) than in Fe-Cr-1.10Si (~7.5wt%) as is the Cr content of the Type B oxide. We are thus seeing kinetically controlled oxidation with the relatively high flux of Fe into the inner scales formed on the 0.41Si alloy resulting in its eventual in-situ interaction with oxidant and the eventual deposition of carbon. In contrast the Fe flux for the 1.10Si alloy is apparently low enough to allow oxidant to diffuse back out of the inner scales.

CONCLUSIONS

Our TEM description of the various scales formed on these two Fe-Cr-Si alloys as a function of time has highlighted and, to a degree, clarified the different rôles played by Si and Cr in forming highly protective healing layers. Still more interesting, while remembering its grossly variable local effects, is however the progressively disruptive influence of carbon. In general the chemistry changes exhibited by the different scales with time point to the way the oxidant potential falls during scaling, but it is clear that only small changes in the balance of the competitive reactions seen, as are prone to occur through the various competitive behaviours of carbon, could lead to further and potentially unstable changes in the oxidation process. It is now known that the healed scale on the 0.41Si alloy can go into breakaway behaviour [11], and from our comparisons of the scale on this alloy with that on the higher Si alloy there would seem to be no reason to suppose that the scaling should not also breakaway for this latter material given a sufficiently protracted exposure to the specified oxidation environment.

ACKNOWLEDGEMENTS

We thank Professor D. Hull FRS for the provision of laboratory facilities as well as the SERC and National Power (Leatherhead) for financial support.

REFERENCES

1. C.W. Tuck, Corros. Sci. 5, 1965, p.631.

2. G.C. Wood, J.A. Richardson, M.G. Hobby and J. Boustead, Corros. Sci. 9, 1971, p.659.

3. A. Kumar and D.L. Douglass, Oxid. Met. 10, 1976, p.1.

4. J. Robertson and M.I. Manning, Mat. Sci and Technol. 4, 1989, p.741.

5. S.B. Newcomb, C.B. Boothroyd and W.M. Stobbs, J.Microsc. 140, 1985, p.195.

6. S.B. Newcomb and W.M. Stobbs, Mat. Sci. and Technol. 4, 1988, p.384.

7. S.B. Newcomb and W.M. Stobbs, Oxid.Met. 35, 1991, p. 69.

8. J.P. Coad, G. Tappin and J. Riviere, Surf. Sci. 117, 1982, p.629.

9. W.M. Stobbs, S.B. Newcomb and E. Metcalfe, Phil. Trans Roy. Soc.(Lond.) A319, 1986, p.219.

10. S.B. Newcomb and W.M. Stobbs. Oxid. Met. 35, 1991. p.69.

11. S.B. Newcomb and W.M. Stobbs, these Proceedings. p.77.

SECTION THREE
Growth and failure of chromia scales

12 A TEM STUDY OF THE OXIDE SCALE FORMED ON CHROMIUM METAL DURING HIGH TEMPERATURE OXIDATION

P. Fox, D. G. Lees and G. W. Lorimer

Manchester Materials Science Centre,
University of Manchester and UMIST, Grosvenor Street, Manchester M1 7HS, UK.

ABSTRACT

The high temperature oxidation of a series of nominally identical chromium samples has been examined, along with samples subjected to different hydrogen annealing treatments. The oxidation behaviour of the samples varied considerably, as did the effectiveness of hydrogen annealing. Some of the chromium samples were found to form flat adherent chromia scales, while the others formed convoluted non-adherent scales. When the samples which formed flat scales were contaminated with very low levels of sulphur or chloride salts the scale morphology changed and became convoluted and non-adherent. The microstructures of some of these scales have been examined using TEM techniques so that their microstructure and the effects of sulphur can be correlated.

INTRODUCTION

The high temperature oxidation of alumina- and chromia-forming metals has been considered for many years. A number of different theories have been developed to explain the lack of scale adhesion, the most recent being the influence of impurities within the metal. The "sulphur effect" [1], as it has become known, considers sulphur and possibly other impurity elements to be the major cause of convolution and spallation during high temperature oxidation. It is proposed [2-5] that the sulphur weakens the bond between the metal and the oxide leading to the observed microstructure, and also that "reactive elements" remove sulphur from solution within the alloy by the formation of stable sulphides thus improving oxidation resistance. Another method reported for improving oxide adherence is by annealing the metal in dry hydrogen at a high temperature before oxidation. This is thought to reduce the sulphur level within the metal and thus produce a flatter, more adherent scale, although this research has shown this technique to be more successful with some materials than others.

This research considers the oxidation of nominally pure chromium metal in the as-received state and after hydrogen annealing. Nominally pure chromium from different suppliers and from different batches from the same supplier have also been examined after differences in the effectiveness of hydrogen annealing had been noted between chromium supplied to the UKAEA at Harwell and that supplied to The Materials Science Centre.

EXPERIMENTAL

The samples were oxidised at 850°C, 900°C and 950°C, mostly in 0.1 atm high purity oxygen, although some were oxidised in unpurified laboratory air. Kinetics experiments were carried out, using manometric apparatus developed by Lees et al. [6], under both isothermal and cyclic oxidation conditions. The isothermal oxidation was for 100 hours at temperature, while the cyclic oxidation consisted of 80 cycles of 1 hour at temperature and 15 minutes cooling out of the furnace but within the controlled atmosphere.

Scanning electron microscopy (SEM) was carried out using a Philips 525 instrument operating at 20kV. The samples were examined on the surface, as polished cross-sections and as fractured cross-sections. The last technique allows the oxide grain structure to be examined directly.

Transmission electron microscopy (TEM) was carried out using a Philips EM400T operating at 120kV and a Philips EM430 operating at 300kV. The oxides were examined in cross-section as the scales were many grains thick and plan-view specimens were difficult to interpret.

Sample materials prepared at Manchester were treated in two ways. Either the material was cut to size, ground to 1200 grit and then cleaned using acetone and alcohol, or it was treated as specified but then hydrogen annealed at 1100°C for 17 hours. Normally oxidation was carried out without further treatment. Samples were also supplied by other researchers and some of these materials were subject to other hydrogen annealing treatments. Two materials were supplied by the UKAEA, the materials being exchanged after differences in the effectiveness of hydrogen annealing had been noted.

Four different "pure" chromiums have been considered to date. Three of these materials are Goodfellows chromiums (99.99+%) (Cat. No. CR000230/5) from different batches, and the fourth has the trade name Ducropur and was supplied by Metall. Plansee (99.95%).

RESULTS

Oxidation of all the as-received chromium supplied by Goodfellows revealed similar oxide morphologies, but the chromium supplied to Harwell showed a greater tendency to spall and suffered very rapid oxidation under cyclic oxidation (Fig. 1). The scales formed were of poorly adhered convoluted chromia (Cr_2O_3) with a small grain size (<0.1-5μm) (Fig. 2).

In contrast, the Ducropur chromium produced a flat adherent scale (Cr_2O_3), yet with a similarly small grain size (Fig. 3). The oxidation rates for isothermal and cyclic oxidation for this material were almost identical, indicating that the bond between this oxide and its metal substrate was very strong. Chemical analysis of the Ducropur chromium (supplied by the manufacturer) did not reveal the presence of significant amounts of materials which were likely to act as "reactive elements". It did, however, have a significantly lower sulphur level (5ppm) than that quoted for the Goodfellows material (150ppm).

To test whether the improvement in oxide adherence and reduced oxidation kinetics of the Ducropur material, compared to the Goodfellows chromium, was due to the reduced sulphur levels, a sample of the Ducropur material was brushed with flowers of sulphur before oxidation. When this sample was oxidised at 950°C the oxide formed was convoluted and spalled easily (Fig. 4). To see how little sulphur was required to produce this effect flowers of sulphur was added to ethanol, in which it is very slightly soluble, and then the solution was filtered and centrifuged. After centrifuging, liquid from the top half of the test tube was filtered again. A drop of this liquid was applied to the ground surface of a Ducropur sample while a drop of pure ethanol was applied to another as a standard. When the liquids had dried the samples were oxidised. Where the ethanol-containing sulphur had been on the surface of the sample, a convoluted oxide formed, while where the drop of pure ethanol had been, and over the rest of the sample, the oxide was flat (Fig. 5). A similar effect was produced by dimpling or ball cratering the surface of a Ducropur sample using diamond paste contaminated with sulphur. Where the sample had been polished by the contaminated diamond paste the scale formed on oxidation was convoluted while areas polished with uncontaminated diamond paste formed a flat scale. Similar effects have been noted for contamination with chloride salts and for contaminated and uncontaminated samples of hydrogen annealed Manchester Goodfellows chromium.

Hydrogen annealing the chromium from the different batches produced very different results. With the Manchester chromium, a flat adherent scale (Fig. 6) was produced, while those supplied by Harwell still convoluted and spalled, but less severely than the as-received material (Fig. 7). Hydrogen annealing the Ducropur material produced an even lower oxidation rate and the oxide remained flat and adherent (Fig. 8). Examination of the microstructures of the different materials did not reveal any significant differences in the chromium samples. Chemical analysis of these materials is now being undertaken to identify whether a link exists between the oxide morphology and the concentration and mobility of sulphur and chlorine within the sample.

SEM analysis of the scale formed on the as-received Ducropur chromium revealed a flat scale with some evidence of the scratches from the original surface. When this material was hydrogen annealed and then oxidised, the scale was flat but with some nodular growths standing proud from the rest of the oxide. The oxidation rate for the hydrogen annealed material was slower than for the as-received material. This suggests that hydrogen annealing slowed the oxide growth over most of the sample surface rather than promoting the formation of nodules.

TEM analysis of cross-sections of the oxide scale from the hydrogen annealed Ducropur chromium, cyclically oxidised at 950°C, revealed a thin oxide (approx. 2μm) (Fig. 9). The oxide appeared to consist of two regions, the outer having a larger grain size (0.2μm - 1μm) and the inner layer a very small grain size (<0.1μm), as shown in Fig. 10. The oxide consisted exclusively of Cr_2O_3 with no other phases being detected. EDX analysis of the scale detected only Cr, but it is possible that STEM analysis may detect elements which are present at the oxide grain boundaries. The outer edge of electron transparent oxide was flat and was probably from a region away from the growth of nodules. In other regions, where the oxide was not electron transparent, there were regions which contained nodules. With further thinning of the sample it should be possible to examine these regions. The structure of the oxide scale reported for the TEM samples was confirmed by the examination of fractured cross-sections of the oxide in the SEM, where a similar grain structure was noted (Fig. 11).

TEM examination of the scale formed on the Ducropur sample treated with sulphur before oxidation revealed a thicker oxide scale, indicating that the sulphur treatment increases the oxidation rate as well as altering the scale morphology (Fig. 12). Changes were also noted in the underlying chromium metal where recrystallisation occurred to a depth of approximately 20μm below the metal surface. Recrystallisation was never detected on samples which had not been treated with sulphur. Analysis of the electron diffraction patterns from the scale detected only one form of the oxide (Cr_2O_3). The oxide grain size with this scale was variable (0.2μm - 3μm) with the formation of some quite large grains (3μm) (Fig. 13). The scale structure for this material was different from the hydrogen annealed Ducropur chromium, with the inner small grained layer not being detected. However, there is some evidence for the formation of layers within the scale, although the scale is single phase (Cr_2O_3) throughout .

Examination of the as-received Manchester Goodfellows chromium after short oxidation times (10-20 minutes) at 850°C, revealed a very fine grained oxide (50nm-100nm) (Fig. 14). Again, the oxide formed was Cr_2O_3 and no other chromium oxides were detected. Although the oxides formed were convoluted and non-adherent, sulphur and chlorine were not detected within the scale using TEM/EDX techniques, either in solution or as discrete particles.

DISCUSSION

Even with the examination of what appears to be a very simple system, the high temperature oxidation of chromium, significant differences in oxidation behaviour can occur. The materials used were all of high purity, especially compared to commercial alloys, and yet these supposedly identical materials varied by an order of magnitude in oxidation rate under cyclic oxidation and formed significantly different microstructures. It was also noted that the effectiveness of hydrogen annealing in improving oxide adherence and reducing scale convolution varied with sample material. The most likely cause of these variations is the presence of impurities within the metal, as the microstructures of the samples were very similar and they were prepared and oxidised under identical conditions. It is likely that the form in which the impurity is present is as important as the absolute concentration of the impurity. If the impurity was trapped as a stable compound, its chemical activity would be lower and less would diffuse to the surface. If the impurity was present as a compound of intermediate stability, the impurity could reach the surface in significant levels during oxidation but not be significantly affected by hydrogen annealing. While, if the impurity was either in solution or segregated to metal grain boundaries hydrogen annealing would be more effective in reducing its concentration within the alloy. The experiments reported in this paper confirm that some elements have a significant effect upon oxidation, inducing changes in oxidation rate and scale morphology. The effectiveness with which surface contamination with very low levels of sulphur or chloride salts can change the type of scale formed indicates a strong interaction between these elements and the formation of oxide scales. Examination of the hydrogen annealed and sulphur contaminated Ducropur chromium in the TEM revealed that the oxide scale was thicker for the sulphur treated sample, but that the oxide grains were of a similar size. For faster oxidation to occur with the sulphur contaminated samples there must be greater diffusion of reactive species through the scale. As the oxide grain structures and the average grain sizes were not significantly different between the two scales, the most likely explanation of the effect of sulphur upon the growing oxide was that it increased the diffusion rate of either chromium or oxygen ions along the grain boundaries or some other fast diffusion path. However, the levels of sulphur within the scale are below the detection limits of TEM/EDX analysis and therefore direct proof for the presence of sulphur within the scale has yet to be found. It is also possible that sulphur or chlorine act at the metal/oxide interface, reducing the strength of the bond between the metal and oxide. The recrystallisation of the chromium metal for a short distance below the oxide for the samples treated with sulphur shows that low levels of sulphur can induce significant changes within the metal and possibly at the metal/oxide interface. The mechanism by which this addition induces recrystallisation of the chromium is not yet known and its effect upon oxidation and scale adhesion is difficult to predict.

The development of the two different layers in the scale, both with the same crystal structure, on the hydrogen annealed Ducropur chromium is interesting in that it tends to indicate that they have different growth mechanisms. The outer layer probably grows by the outward diffusion of chromium ions, while the inner grows by the inward movement of oxygen.

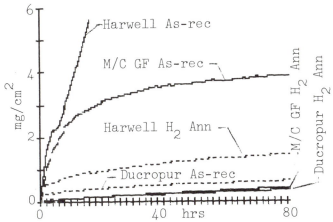

Figure 1: Cyclic oxidation at 950°C (0.1 atm. O$_2$).

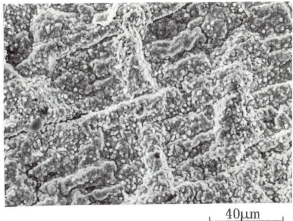

Figure 2: M/C Goodfellows Cr oxidised for 15 min at 950°C.

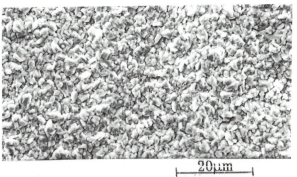

Figure 3: Ducropur Cr oxidised for 100 h at 950°C.

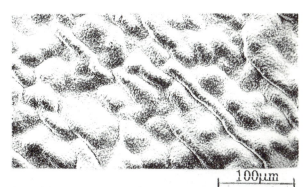

Figure 4: Ducropur treated with sulphur before oxidation at 950°C.

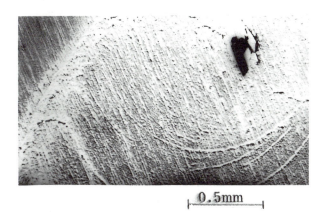

Figure 5: Ducropur Cr treated with sulphur-ethanol before oxidation at 950°C.

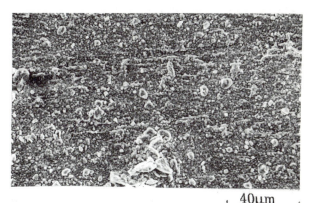

Figure 6: H$_2$ annealed M/C GF Cr oxidised at 950°C.

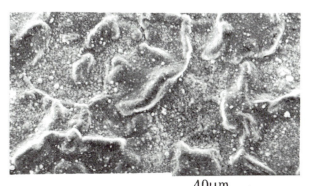

Figure 7: H$_2$ annealed Harwell Cr oxidised at 950°C.

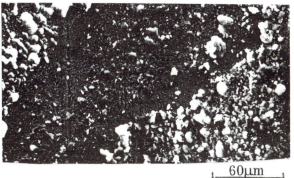

Figure 8: H$_2$ annealed Ducropur Cr oxidised at 950°C.

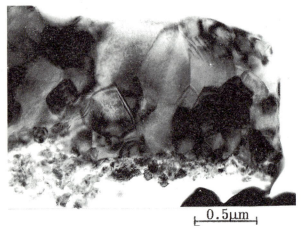

Figure 9: (TEM) H₂ ann. Ducropur cyclic oxidation at 950°C.

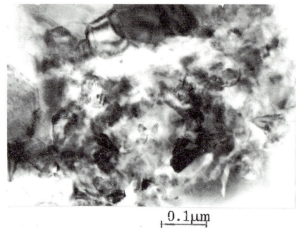

Figure 10: (TEM) H₂ ann. Ducropur cyclic oxidation at 950°C.

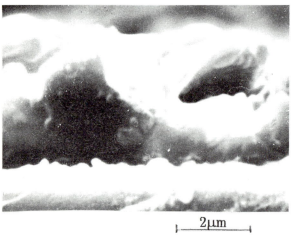

Figure 11: (SEM) H₂ ann. Ducropur cyclic oxidation at 950°C.

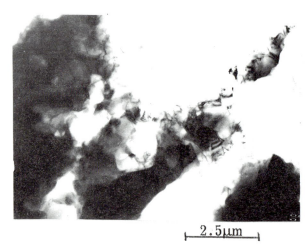

Figure 12: (TEM) sulphur treated Ducropur Cr oxidised at 950°C.

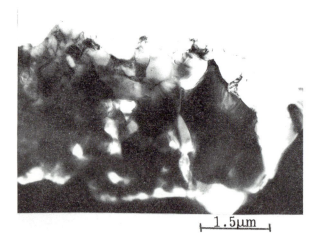

Figure 13: (TEM) sulphur treated Ducropur Cr oxidised at 950°C.

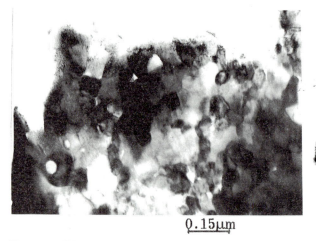

Figure 14: (TEM) M/C GF Cr oxidised for 10 min at 850°C.

If this is correct, then for very slow growing chromia scales approximately four-fifths of the scale growth is due to the outward movement of chromium ions. The other possibility is that it is the remnant of a fine grained transient scale, similar to that reported for samples oxidised for short oxidation times and that the growth of the scale is almost exclusively outward. Further experiments using oxygen-18 are required to identify which possibility is correct.

CONCLUSIONS

1. The oxidation rate and scale morphology of pure chromium vary significantly between different batches of material.

2. Hydrogen annealing of chromium has a significant effect upon oxidation kinetics and oxide morphology but its effectiveness varies with sample batches.

3. The Ducropur chromium and the hydrogen annealed Goodfellows chromium supplied to Manchester produce flat adherent oxides.

4. Contamination of the surface of samples which form flat scales with small amounts of sulphur before oxidation changes the scale to a convoluted poorly adhered oxide.

5. The grain size of convoluted and flat scales are similar.

ACKNOWLEDGEMENTS

The financial assistance of the SERC and the supply of materials by M. Bennett and A. Tuson of the UKAEA and H-P Martinz of Metallwerk Plansee Gmbh are gratefully acknowledged.

REFERENCES

1. D.G. Lees, Oxid. Met. 27, 1987, p.75.

2. I. Melas and D.G. Lees, Mat. Sci. Tech. 4, 1988, p.455.

3. Y.Ikeda, K.Nii and K.Yoshihara, Proc. JIMIS-3(1983), Trans. Jap. Inst. Met.

4. A.W.Funkenbusch, J.G.Smeggil and N.S.Bornstien, Naval Research Contract N00014-82-C-0618, Jan. 1983.

5. D.G. Lees, I.S. Grant and G.W. Lorimer, SERC Proposal Sept. 1982.

6. M. Skeldon, J.M. Calvert and D.G. Lees, Oxid. Met. 28, 1987, p.109.

13 THE OXIDATION OF CHROMIUM

H.M. Flower, P.J. Gould, D.P. Moon* and A.T. Tuson*

Department of Materials, Imperial College, London SW7 2BP, UK.
*AEA Technology, Harwell Laboratory, Didcot, Oxon OX11 0RA, UK.

ABSTRACT

Commercially supplied electrolytic chromium has a number of inherent defects and impurities including a substantial amount of oxygen. Coupons of electrolytically deposited chromium have been heat treated at various temperatures in order to study the development of the metal microstructure and to optimise the removal of oxygen from the metal. The effect of these impurities on the oxidation behaviour of chromium is discussed. *In situ* observations of chromium oxidation have been made in a transmission electron microscope and the results have been presented.

INTRODUCTION

Chromium is a valuable alloying element since the selective oxidation of the chromium to form a chromia scale gives good corrosion protection at high temperatures to many industrially important alloys. Pure chromium has been used by a number of workers as a model for scale formation [1-13], each study attempting to fit a parabolic rate constant to the kinetic data gathered. If these are plotted on one diagram (Fig.1) then a wide variation in the data is seen although the gradients of the lines are roughly similar. It was noted that using different surface preparation techniques had a large effect on the oxidation kinetics [14]. This implies that the state of the metal is important and thus the complete history of the chromium needs to be considered. Examination of previous work on chromium shows that the process/preparation histories of the metal employed by each set of workers are different in at least one respect. This could be the reason for the variation in rate constants and also implies the existence of a microstructure that changes significantly depending on the method of preparation.

Impurities can have a large effect on the behaviour of metals and, recently, there has been some discussion on the effects of non-metallic impurities on the oxidation behaviour of chromium [15,16]. A source of extremely pure material has been needed by previous workers and electrolytic chromium has been widely used because of its high purity. The metal is produced by electrolytic deposition onto an electrode from a bath containing a mixture of chromic acid and sulphuric acid enabling a very low level of metallic impurities to be achieved [17]. Quoted purities of 99.99% are common. Unfortunately the electrolytically deposited metal has inherent defects associated with its method of production [17].

ELECTROLYTIC CHROMIUM

In the present work, examination of a typical coupon of electrolytically deposited chromium obtained from a commercial source exposed these defects. Polishing one face of the coupon revealed a network of fine cracks uniformly spread over the surface, Fig. 2. An X-ray diffraction spectrum taken from the metal, Fig. 3, showed peak broadening of up to 6 times the normal full width at half maximum implying a high degree of internal strain and a small grain size.

A preferred [111] orientation in the direction of deposition was also observed in pole figures determined at Harwell Laboratory, Fig. 4. Transmission electron microscopy showed that the grain size was around 50nm (Fig. 5). A mass spectrometric analysis of the material examined is given in table I. It is seen that oxygen is by far the most major impurity and that, with two notable exceptions, the level of metallic impurities is low. The reason for the high levels of Sb and Sn are not known. Of the non-metallic elements, only sulphur and chlorine are present in large quantities.

As much as 0.1wt% of hydrogen and 0.6wt% oxygen can be absorbed into the metal during deposition [17]. The hydrogen can be quite easily removed by heating the metal to above 400°C [17]. Unfortunately, this causes the oxygen to be redistributed as a fine dispersion of Cr_2O_3 particles [17,18]. These are more difficult to remove. One possible method is to melt the metal and recast. This is never fully effective. The oxide particles reappear at the grain boundaries and triple points in any abundance up to a volume fraction corresponding to the original oxygen content. This is most obvious in Fig. 6. There is also porosity present in the material, but this is not evident from the figure. A more effective method noted

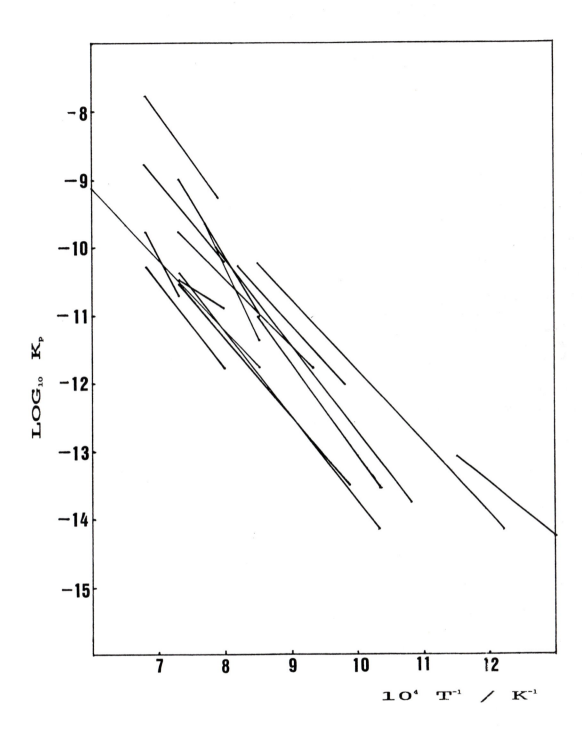

Figure 1: A graph of Arrhenius plots of the parabolic rate constants given in the literature for the oxidation of chromium [1-13].

in previous literature for completely removing oxide particles from the chromium is by annealing in hydrogen at high temperature [17]. In order to investigate this process a series of heat treatments were formulated (see Table II). First, a set of lower temperature anneals were performed in the temperature range 600°C to 1000°C to study the development of the microstructure; these were carried out in hydrogen at 1 atmosphere pressure and also in a vacuum (2×10^{-5} Pa.) to provide a control.

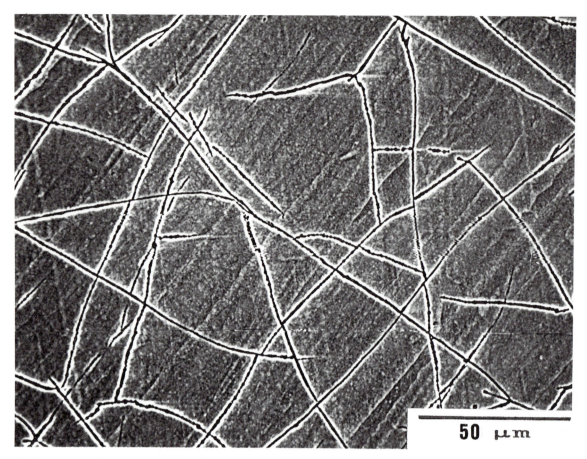

Figure 2: A scanning electron micrograph of electrolytic chromium in the as-deposited state. The electropolished surface is seen to contain a network of cracks. A back scattered electron image.

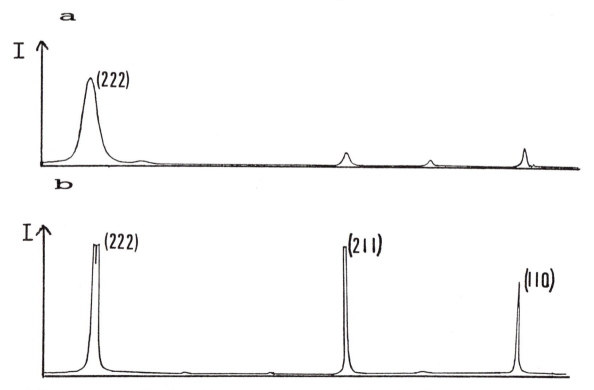

Figure 3: X-ray spectra from: (a) untreated electrolytic chromium and, (b) electrolytic chromium treated in hydrogen for 3 hours at 1500°C showing the high degree of peak broadening in the untreated metal spectrum.

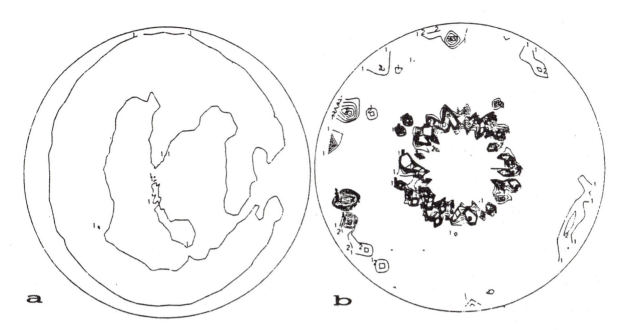

Figure 4: (110) pole figures from: (a) untreated electrolytic chromium and, (b) electrolytic chromium treated in hydrogen for 3 hours at 1500°C. The (110) pole distribution is consistent with a high degree of texture with <111> parallel to the direction of deposition and essentially free rotation around this axis.

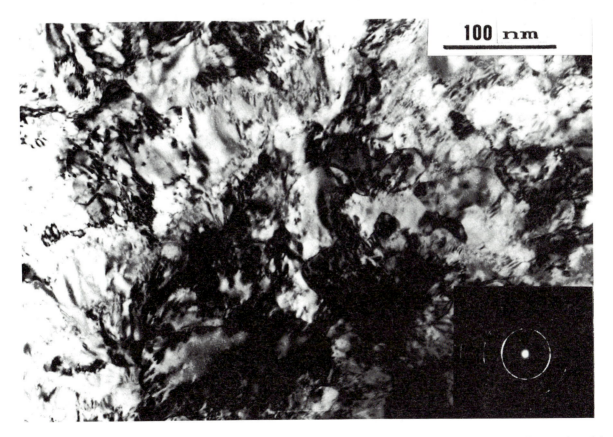

Figure 5: A transmission electron micrograph of as-deposited electrolytic chromium showing the small grain size of the metal. The selected area diffraction pattern is from a larger area than the micrograph.

A second set of heat treatments was also carried out at higher temperatures to study the removal of the particles. These were of limited duration (3 hours) for reasons of safety. The details of these two sets of heat treatments are given in Table II. The results of these heat treatments are shown in the series of transmission electron micrographs in Fig. 7. Up to 1000°C the microstructures of the vacuum annealed and the hydrogen annealed chromium were identical. There is grain growth, but this is retarded by the dispersion of chromia particles spread throughout the material. Fine chromia particles are spread uniformly through the metal with lines of larger chromia particles marking the traces of the cracks. These crack traces are still lines of weakness, as can be seen arrowed in Fig. 7a where the crack has re-opened in the preparation of the foil. Also visible in this figure is a region denuded of the fine oxide dispersion around the lines of larger particles. This implies that a ripening process is occurring with these large particles growing at the expense of the smaller ones. This depletion region increases in width with temperature, until the fine dispersion of chromia particles has disappeared leaving only large blocky particles along the crack traces as shown in Fig. 7b. The grain growth is still inhibited by the pinning effect of these lines of large particles.

It is above 1000°C that the atmosphere becomes important. The vacuum anneals show the particles coarsening until equilibrium is reached. In a hydrogen atmosphere the coarsening process occurs but there is a competing process hydrogen reduction of the oxide particles. Figure 7c is from a coupon treated at 1200°C for 3 hours in hydrogen. The depletion zones around the lines of larger particles can be seen but also visible are depleted regions around the grain boundaries. Presumably hydrogen has diffused into the metal along the grain boundaries and reduced the fine oxide particles near the boundaries. It is obvious that quite a large proportion of the particles have been reduced by this process and, although the lines of large particles are still present, they are not as evident as in previous cases. After 3 hours in hydrogen at 1500°C virtually all of the oxide particles have been removed.

Each hydrogen heat-treated coupon was electropolished and photographed in the SEM using a back-scattered imaging mode. This gave good grain contrast. The grain size of each coupon was measured and the pits in the surface were counted; see Fig. 8 for sample micrographs. It was assumed that the pits were produced by oxide particles being preferentially attacked by the polishing solution and dropping out. If the majority of the pits had been produced by etching or some other process then their numbers would not be expected to decrease with increasing temperature of heat treatment.

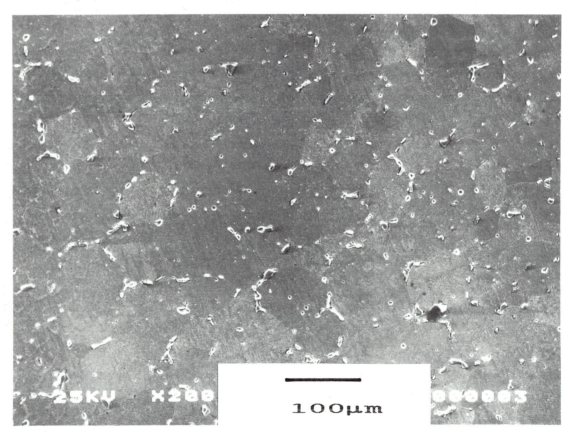

Figure 6: A scanning electron micrograph of electrolytic chromium arc-melted and cast. The oxide particles reform at grain boundaries and triple points.

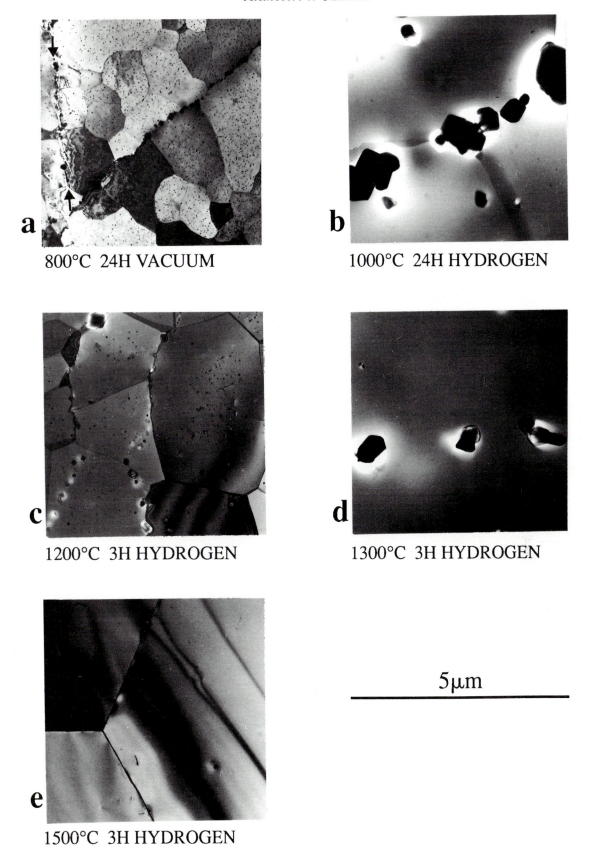

Figure 7: Transmission electron micrographs showing the microstructure of electrolytic chromium after various heat treatments. In (a) the arrows point out a crack that has re-opened in the preparation of the foil.

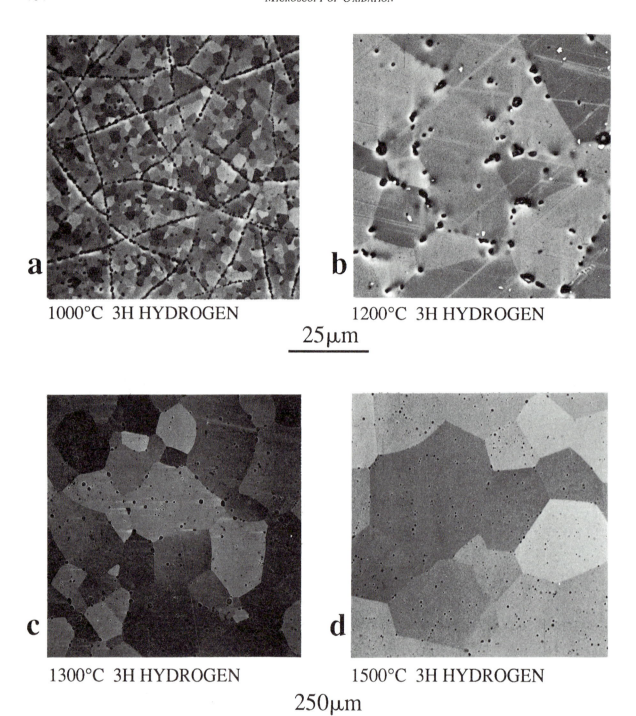

1000°C 3H HYDROGEN

1200°C 3H HYDROGEN

25μm

1300°C 3H HYDROGEN

1500°C 3H HYDROGEN

250μm

Figure 8: Scanning electron micrographs showing a back-scattered image of the electropolished surfaces of the various heat treated coupons. The pits and the grain structure can be clearly seen.

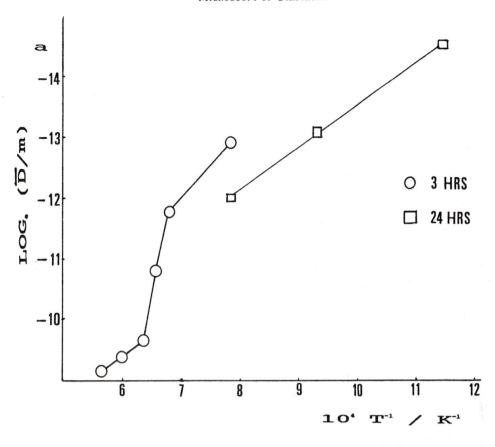

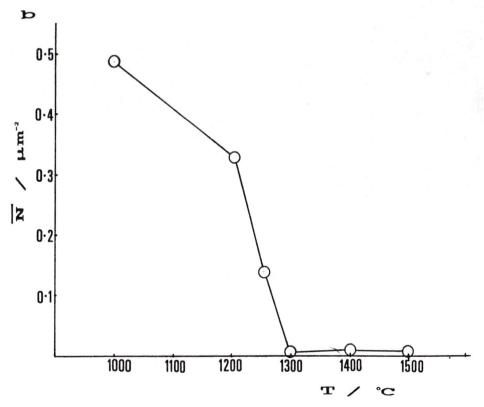

Figure 9: (a) An Arrhenius plot of average grain size against temperature for heat treated electrolytic chromium. (b) A graph of particle number density against temperature for heat treated electrolytic chromium.

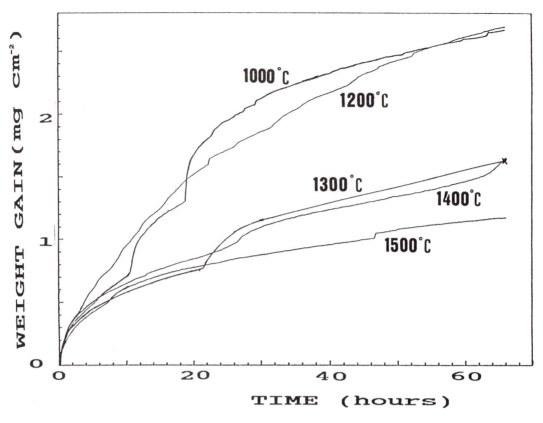

Figure 10: A graph of weight gain versus time for coupons of chromium heat treated in hydrogen for 3 hours at the temperatures indicated and then oxidised for 66 hours at 850°C in 1 atmosphere flowing oxygen.

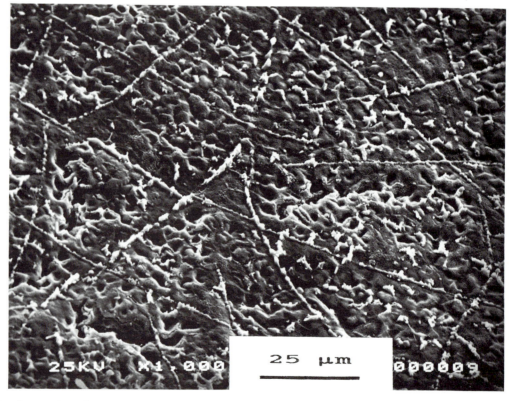

Figure 11: A scanning electron micrograph showing the underside of a spalled oxide fragment from a coupon of electrolytic chromium treated at 800°C in hydrogen for 24 hours before oxidising for 3 hours at 900°C in 1 atmosphere flowing oxygen.

A graph of grain size against temperature is given in Fig. 9a and a graph of surface particle density against temperature is given in Fig. 9b. This has two significant features: the first is an abrupt change in the grain size, occurring at around 1200°C. To either side of this discontinuity the points lie on straight lines of roughly equal gradient. This implies that above 1200°C an inhibitory factor has been removed and the grains are allowed to grow more freely. The graph of particle number density against temperature shows the same abrupt change at around 1200°C. It is most probable that the inhibitory factor preventing free grain growth is the existence of long lines of oxide particles in the material and thus the change in the speed of grain growth would be due to the removal of these oxide particles. From this it can be deduced that the hydrogen atmosphere starts to have a significant effect above 1200°C.

The change in the microstructure is mirrored by the oxidation kinetics for the various heat treatments. Coupons treated in hydrogen at the various temperatures used previously were mechanically polished to a 1 μm diamond finish using only diamond polishing compounds, eliminating possible silicon contamination from SiC grinding papers. The coupons were then oxidised for 66 hours at 850°C in 1 atmosphere of flowing oxygen. The graphs of weight gain versus time are given in Fig. 10. The coupons treated at 1000°C and 1200°C have similar weight gain curves whilst those treated above 1200°C form a second set displaced from the first both in magnitude of weight gain and in shape.

The coincidence between this division and the abrupt change in the oxide particle number density implies that the existence of oxide particles in the material has a deleterious effect on the oxidation rate. It is possible that the lines of particles serve as initiation sites for scale failure, increasing the oxidation rate by introducing a break-away type oxidation regime. This would account for the larger weight gains for the lower temperature annealed samples and for the roughness of their weight gain graphs. For treatments below 1200°C these certainly show only brief periods of parabolic oxidation behaviour whereas at the higher temperatures, these periods are much prolonged. Figure 11 shows the underside of a spalled oxide fragment from an 800°C vacuum annealed coupon and it is evident that the lines of oxide particles, originally in the metal surface have become detached and are now adherent to the underside of the oxide scale. It is clear that the lines of oxide particles do not aid in scale adherence by keying the oxide to the metal as might perhaps have been expected.

If the impurity levels in the treated metal, table I, are compared with those in the untreated material it is noticed that only a few changes have occurred. The largest is the reduction in the oxygen content between 1200°C and 1500°C. This confirms the removal of particles above 1200°C by the hydrogen. By 1200°C the sulphur level has already dropped to a lower level and remains roughly constant. The high chlorine level is unaffected by heat treatment. Of the metallic elements, only the copper level changes with temperature. The source of the copper is thought to be from the furnaces that were used in the heat treatment. Different furnaces were employed for the hydrogen and for the vacuum heat treatments and this is thought to be the source for the variation in the nickel levels also seen.

It is not known what effects the impurities present at a high level will have on the oxidation kinetics; in this study they are masked by the effects of the oxide dispersion. None of the major metallic impurities are among those oxygen-active elements which affect oxidation kinetics via the "reactive element" effect [19]. In addition, although the oxidation is controlled by short circuit diffusion, doping by impurities may slightly alter rates via modification of bulk transport properties. The only concern is the chlorine level and the magnitude of its effect is as yet undetermined.

IN SITU OXIDATION

So far only impurities inherent in the metal have been considered; there has been little mention of impurities introduced onto the metal surface by any surface preparation method used. These could be both chemical and physical in nature. In order to observe the crucial initial stages of oxidation when surface impurities could have their most significant effect it was decided to oxidise chromium in an environmental cell in a Transmission Electron Microscope.

Discs were spark-eroded from coupons of chromium heat treated in hydrogen at 1500°C and prepared for Transmission Electron Microscopy by jet-polishing to electron transparency using a 10% Perchloric acid /90% Methanol electrolyte at 30C, 30V. These thin foils were oxidised in the Gas Reaction Cell of a 1MeV Electron Microscope. *In situ* observations of the growing oxide were made to give insight into the nucleation and growth of the oxide scale. The foils were each oxidised in the temperature range 500 - 900°C in 2 torr flowing oxygen for a few minutes.

A uniform oxide film formed with no strong evidence of preferential nucleation and growth, Fig. 12a, even at existing oxide particles. Figure 12b shows oxidation at an edge of a foil. The metal grain is oriented with (111) parallel to the surface of the foil and normal to the beam direction. As the oxide grows, the metal at the edge is consumed but it is only

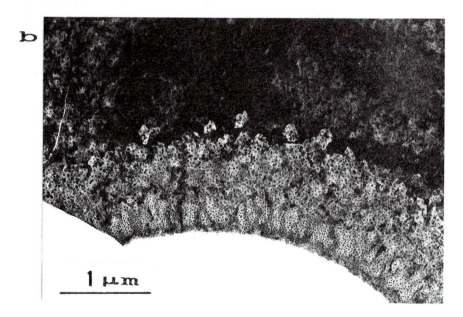

Figure 12: Transmission electron micrographs of chromium thin foils oxidised in the Gas Reaction Cell of a lMeV microscope. (a) Shows no preferential growth at grain boundaries or at the large oxide particle. (b) Shows growth of oxide at the thin edge of a grain oriented with (111) parallel to the foil surface. Note the facetted nature of the oxide/metal interface.

on grains that are close to <111> that the facetting of the metal, seen here, is observed. An oxide texture was also noticed on these grains. It has been mentioned previously that the metal had a strong [111] texture normal to the coupon so that most of the grain surfaces to be oxidised would be on or near (111). The oxide was found to grow with its [0001] axis normal to the foil surface. This is shown clearly on the diffraction pattern in Fig. 13a which has two strong rings made from discrete oxide reflections of the type {hki0} implying an almost free rotation of grain orientation around the [0001] zone axis. As would be expected, the grain size of the oxide is small: around 50nm. Occasionally the oxide pattern is textured and even near epitaxial and only a few reflections make up the rings, Fig. 13b. This suggests that on an ideally

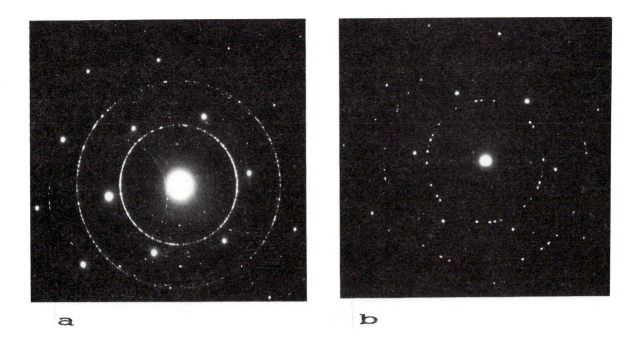

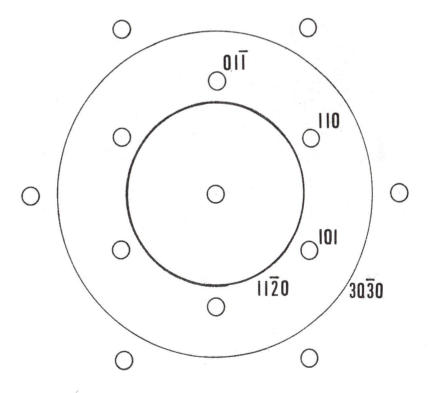

Figure 13: Selected area diffraction patterns of oxide grown on oriented metal surfaces. (a) Is from an area of Fig. 12(b) and shows the ring distribution of the oxide reflections. The oxide is growing with its [0001] axis parallel to the beam direction. (b) Is from another (111) oriented grain and contrasts with (a) to show the range of oxide orientation distributions.

		O	N	Si	S	Cl	Fe	Ni	Cu	Sn	Sb	Other
As Deposited	i	1500	2.7	0.7	86	43	5.6	0.07	9.2	74	710	36.5
	ii	1100	1.8	0.53	83	28	5.2	0.04	8.7	20	280	21.0
1200°C 3 Hours Vacuum		1700	3.8	0.9	5.5	48	5.9	57	52	15	610	33.2
1200°C 3 Hours Hydrogen		1300	4	0.85	12	38	2.9	0.22	280	14	430	28.7
1500°C 3 Hours Hydrogen		48	2.6	3.2	3.8	62	6.7	0.78	340	17	600	33.0

Table I: A table showing the impurity levels in as-deposited electrolytic chromium and in treated chromium. These are ppm levels by weight from mass spectrometry courtesy of N.R.C. Canada.

	24 Hours	3 Hours	1Atmosphere Hydrogen	Vacuum 2 x 10⁻⁵ Pa
600°C	X		X	
	X			X
800°C	X		X	
	X			X
1000°C			X	
	X			X
		X	X	
1200°C		X	X	
				X
1250°C		X	X	
1300°C		X	X	
1400°C		X	X	
1500°C		X	X	

Table II: A table of the various heat treatments carried out on electrolytic chromium detailing the time, temperature and atmosphere used.

clean {111} metal surface true epitaxial oxide growth may be possible although atomistic matching between {111}$_{Cr}$ and {0001}$_{ox}$ is relatively poor and no obvious orientation relationship of "good fit" could be determined. On the real surfaces of electropolished samples, surface "contamination" by electropolishing oxide films and impurity elements (for example Cl from the perchloric acid electrolyte) inhibit the formation of epitaxial oxide nuclei although the strong preference for <111>$_{Cr}$//<0001>$_{ox}$ remains. Small local differences in surface cleanliness or imperfections can lead to the range of oxide orientation distributions observed in the in situ experiments.

The physical state of the surface is also important. Ideally, the oxide should grow in a perfectly smooth plane surface, but no mechanical polishing regime can produce this. There will always be roughness that can affect the oxidation behaviour by acting as crack initiation sites etc. and a deformation layer may also have an effect. In this work, care has been taken to produce as smooth a surface as possible and this has revealed the effects of the oxide particles but any specific results of surface roughness have not been addressed. This also represents a relatively uncontrolled variable in much previous work. The treatments used in this work to rid the chromium of oxide particles result in large grain sizes. This, together with the degree of texture in the metal, can lead to problems with preferential growth on certain grain orientations and these need to be considered as well as the other impurity effects.

CONCLUSIONS

Electrolytically deposited chromium from a commercial source has been examined and found to contain high levels of oxygen and other impurities. The metal has been heat treated to remove these impurities and the development of the metal microstructure has been examined for different temperatures of treatment. Evidence has been obtained indicating that a hydrogen atmosphere will remove oxide particles from the metal only at temperatures above 1200°C. Oxidation kinetics have been measured for each of the treated coupons and it has been found that removing the oxide particles reduces the amount of breakaway oxidation observed thus reducing the oxidation rate constant. The effects of the other impurities in the metal have been discussed. *In situ* oxidation experiments have been carried out in the gas reaction cell of a 1 MeV electron microscope. No preferential nucleation or growth of the oxide has been observed. The oxide that grows on chromium (111) surfaces has a preference for <0001>$_{ox}$//<111>$_{Cr}$ with oxide variants showing a wide range of rotations about the <0001> axis. This range is thought to be due to surface "contamination".

ACKNOWLEDGEMENTS

This work was funded by SERC and AEA Harwell via the CASE award scheme. Laboratory facilities at Imperial College were provided by Prof. D.W. Pashley. The authors would like to thank Dr M.J.Bennett for valuable discussions.

REFERENCES

1. K.P. Lillerud and P. Kofstad, J.Electrochem. Soc. 127, 1980, p.2397.

2. E.A. Gulbransen and K.F. Andrew, J.Electrochem. Soc. 99, 1952, p.402.

3. E.A.Gulbransen and K.F. Andrew, J.Electrochem. Soc. 104, 1957, p.334.

4. C.A. Phalnikar, E.B. Evans and W.M. Baldwin, Jr., J. Electrochem.Soc. 103, 1956, p.429.

5. W.C. Hagel, Trans. A.S.M. 56, 1963, p.583.

6. D. Mortimer and M.L. Post, Corros. Sci. 8, 1968, p.499.

7. L. Caidou and J. Paidassi, Mem. Sci. Rev. Metall. 66, 1969, p.217.

8. D. Caplan, A. Harvey and M. Cohen, Corros. Sci. 3, 1963, p.161.

9. D. Caplan and G.I. Sproule, Oxid. Met. 9, 1975, p.459.

10. D.J. Young and M. Cohen, J. Electrochem. Soc. 124, 1977, p.769.

11. C.S. Giggins and F.S. Pettit, Trans. Met. Soc. AIME 245, 1969, p.2495.

12. C.M. Cotell, K. Przybylski and G.J. Yurek, Proc. Symp. Fundamental Aspects of High Temp. Corrosion II, Ed. D.A.Shores, Electrochem. Soc.,1986, p.103.

13. R.J. Hussey, D.F. Mitchell and M.J. Graham, Werkst. und Korros. 38, 1987, p.575.

14. D. Caplan, A. Harvey and M. Cohen, J.Electrochem.Soc. 108, 1961, p.134.

15. D.G. Lees, Oxid. Met. 27, 1987, p.75.

16. A.W. Funkenbusch, J.G. Smeggil and N.S. Bornstein, Metall.Trans. 16A, 1986, p.1164.

17. A.H. Sully and E.A. Brandes, Chromium: Metallurgy of the Rarer Metals: 1. 2nd Edition, Butterworths (London) 1967.

18. C.P. Brittain and G.C. Smith, J. Inst. Metals 89, 1961, p.407.

19. D.P. Whittle and J. Stringer, Phil. Trans. R. Soc. Lond., A 295, 1980, p.309.

14 EFFECTS OF WATER VAPOUR ON OXIDE GROWTH ON 304 STAINLESS STEEL AT 900°C

A.S. Khanna and P. Kofstad*

Metallurgy Division, IGCARR, Kalpakkam - 60 3102, India.
*Department of Chemistry, University of Oslo, P.B.1033, Blindern, 0315 Oslo 3, Norway.

ABSTRACT

It is well known from the literature that the presence of water vapour may influence and adversely affect the protective oxidation behaviour of some metals and alloy systems. These effects have particularly been demonstrated for iron-based alloys. This phenomenon has been further investigated in this work through studies of oxidation of 304 stainless steel in dry oxygen and in oxygen+2% water vapour at 900°C and particularly through examinations of the morphologies of the oxide scales formed after various lengths of time under different exposure conditions. Possible interpretations of these effects are considered.

INTRODUCTION

The oxidation behaviour of some metals and alloys may be influenced by the presence of water vapour in the oxidizing atmosphere [1-11]. In the case of iron-chromium and iron-silicon alloys and technical steels it is well known that the water vapour adversely affects the protective oxidation behaviour, and higher chromium concentrations are needed in order to obtain protective behaviour in water vapour-containing oxygen than in dry oxygen. The reasons for these effects are not understood.

The aim of this work is to provide further information on the nature of these effects. The work has comprised studies of oxidation rates of coupon specimens of 304 stainless steel in dry oxygen and in oxygen with 2% water vapour and subsequent characterization of reacted specimens. This paper emphasizes results of studies of the morphologies of oxide scales formed under the different exposure conditions for reaction periods up to 200 h. Experimental details have been given elsewhere [10,11].

EXPERIMENTAL RESULTS

The morphologies of the scales formed in the dry and wet gases exhibited marked differences already during the early stages of oxidation. This is illustrated in Figs. 1 and 2 which show SEM micrographs at different magnifications of the scale surfaces for exposure periods up to 50 h in the wet and dry atmospheres, respectively. Thus after exposure for 0.5 h the oxide scale formed in the wet gas consisted of chromia-enriched crystallites which were considerably larger than in the scale formed in the dry gas (cpr. Figs. 1a, b, and c with Fig.2).

On further oxidation (1.5 h) an oxide phase rich in iron began to appear on isolated areas of the surfaces formed in the wet gas. This iron-rich oxide phase also appeared to be porous. This is illustrated in Figs.1d, e and f. The new oxide phase grew with time and eventually completely covered the initially formed chromia-rich layer (Figs. 1j, k, and l from specimens oxidized for 50 h). On the other hand, in the dry gas such overgrowth of an iron-rich phase was not observed; the scale consisted of a uniform layer of rhombohedral chromia crystallites after the same period of exposure (Figs. 2e and f).

Further exposure in the dry gas did not significantly change the morphology of the scale surface. However, after extended exposure (200 h) the scale spalled in isolated areas. Examples of such morphologies are given in Fig.3. For the scales formed in the wet gas for 200 h the outer and porous iron-rich scales spalled heavily and then revealed an inner chromia-rich layer of the scale (Fig. 4). This heavy spallation in the wet gas was also confirmed by weight measurements of the reacted specimens as illustrated in Fig. 5.

These differences in scale growth were also confirmed and revealed through examinations of cross-sections of specimens reacted in the two atmospheres. As shown in Fig. 6 the scales formed in the dry atmospheres were comparatively thin and uniform, while a thicker scale with chromia-rich intrusions into the metal was formed in the wet gas.

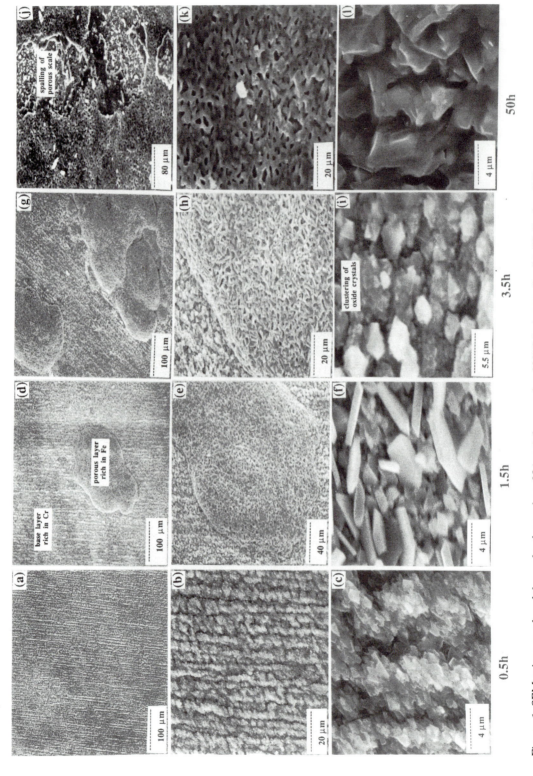

Figure 1: SEM micrographs of the scales formed on 304 stainless steel at 900°C in 1 atm. O_2+2% H_2O after different exposure periods. The scales formed are shown at different magnifications.

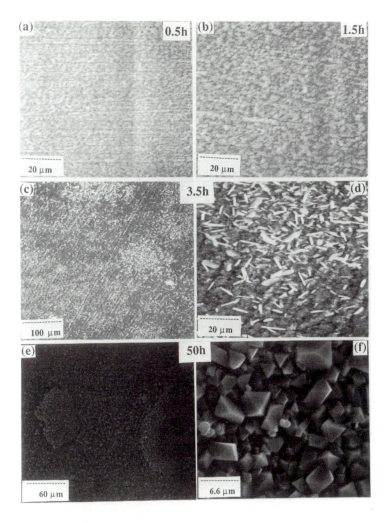

Figure 2: SEM micrographs of the scales formed on 304 stainless steel in 1 atm. dry O_2 at 900°C after different exposure periods.

DISCUSSION

The results have confirmed earlier studies that relatively small amounts of water vapour in the ambient air or oxygen adversely affect the protective oxidation behaviour of iron-based alloys and technical steels. This study shows that initially protective scales - which are enriched in chromium oxide - rapidly lose their protective properties in the presence of water vapour, and the results suggest that the breakdown of the protective scales is accompanied by the formation of an overgrowth of an iron-rich oxide phase on the initially formed scale.

The reason for this behaviour and the change in diffusional properties that apparently causes the loss of protective behaviour can not as yet be satisfactorily interpreted. But as a correlation it may be noted that more recent studies have shown that defect-dependent properties of some oxide systems may be affected by water vapour [12-15]. As regards point defects it is concluded that water vapour serves as a source of proton defects in oxides according to the reaction

$$H_2O(g) + O_0 = 2HO_0 + 2e + 1/2O_2(g)$$

where HO_0 represents a proton associated with an oxygen ion on a normal oxygen site in the oxide. For oxides with low intrinsic defect concentration, e.g. Y_2O_3, Al_2O_3, MgO a.o., it has been shown that such proton defects may significantly alter the defect structure situations and corresponding defect-dependent properties [14,15]. No such studies have been for reported for chromia and no conclusions can as yet be drawn for this oxide.

Figure 3: SEM micrographs of the scales formed on 304 stainless steel at 900°C in 1 atm. dry O_2 after exposure for 200 h.

Figure 4: SEM micrograhs of the scales formed on 304 stainless steel in $O_2+2\%$ H_2O at 900°C after exposure for 200 h.

As regards growth of chromia scales it is now generally concluded that the diffusional transport predominantly takes place by grain boundary diffusion [1]. The fact that the breakdown of the protective properties of the initially formed chromia-enriched scales is accompanied by an overgrowth of an iron-rich oxide phase, suggests that the outward diffusion of iron along grain boundaries in the scale is favoured by the presence of water vapour. The reason for this is a matter of speculation, but it is not unreasonable to ask if the proton defects may affect the defect structure situations in the grain boundary regions and thereby the corresponding diffusional properties.

Although no definite conclusions can be drawn at the present stage, the subject matter is of great importance both as regards basic properties of metal oxides and properties of highly protective scales and should be subject to further studies.

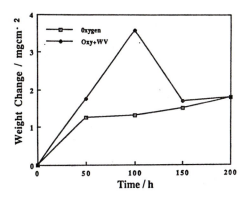

Figure 5: Weight gain as a function of time for the oxidation of 304 stainless steel in dry O_2 and in $O_2+2\%$ H_2O at 900°C. The specimens were weighed before and after the exposures.

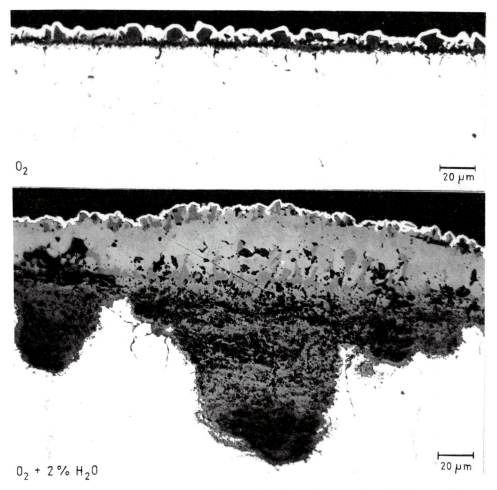

Figure 6: Optical micrographs of cross-sections of scales formed on 304 stainless steel at 900°C after 100 h in dry O_2 and in $O_2+2\%$ H_2O, respectively.

REFERENCES

1. P. Kofstad, High Temperature Corrosion, Elsevier Applied Science, London/New York, 1988.

2. D. Caplan and M. Cohen, J. Met. 4, 1952, p.1057.

3. R.T. Foley, J.Electrochem. Soc. 109, 1962, p.278.

4. A. Rahmel, Werkstoffe u. Korros. 16, 1965, p.837.

5. M. Warzee, J. Hennaut, M. Maurice, C. Sonnen, J. Waty and Ph. Berge, J. Electrochem. Soc. 112, 1965, p.670.

6. Y. Shida, N. Ostuka and H.Fujikawa, Proc. JIMIS 3, 1983, p.631.

7. N. Ostuka, Y. Shida and H. Fujikawa, Oxid. Met. 32, 1989, p.13.

8. I. Kvernes, M. Oliveira and P. Kofstad, Corros. Sci. 17, 1977, p.237.

9. C.W. Tuck, M. Odgers, M. Sachs and K. Sachs, Corros. Sci. 9, 1969, p.271.

10. A.S. Khanna and P. Kofstad, Metals, Materials and Processes, 1, 1989, p.177.

11. A.S. Khanna and P. Kofstad, Proc. 11th Int. Corrosion Congress, Florence, Italy, April 1990.

12. T. Norby and P. Kofstad, High Temp. High Pressures 20, 1986, p.784.

13. T. Norby and P. Kofstad, J. Am. Cer. Soc. 69, 1986, p.784.

15 THE ROLE OF SILICON IN MAINTAINING PROTECTIVE OXIDATION IN A 20%Cr AUSTENITIC STAINLESS STEEL TO HIGH TEMPERATURES

R.C. Lobb and H.E. Evans

Nuclear Electric Plc, Berkeley Nuclear Laboratories,
Berkeley, Gloucestershire GL13 9PB, UK.

ABSTRACT

This paper studies the role of silicon in the prevention of non-protective oxidation in 20%Cr/25%Ni/Nb stabilised stainless steel in CO_2-based gas at 1000°C, using alloy silicon contents in the range 0 to 2.25 wt%. The kinetics of oxidation are studied using gravimetric methods and the morphology of the oxide scale is examined using techniques that include optical and scanning electron microscopy and electron probe microanalysis.

INTRODUCTION

Often economic benefits and the need for improved process efficiency can result in alloys operating in service under more arduous conditions of increased temperature and/or longer times than originally envisaged. It is important that the mode of oxidation does not change under these new circumstances, especially in thin-section components. An example arises with the 20%Cr/25%Ni/Nb stabilised stainless steel used as fuel cladding in the advanced gas-cooled reactor (AGR). Consequently, the high temperature oxidation of this steel has been the subject of detailed investigations over a number of years [1,2], but recent work [3] has established that the normal parabolic kinetics associated with this steel can undergo a change to non-protective oxidation and that the time to reach this transition decreased markedly with increasing temperature above 900°C. This is especially relevant, when under postulated abnormal (fault) conditions within the reactor, it is possible, in a relatively short timescale, for areas of cladding to attain localised temperatures well in excess of those experienced under normal reactor operation (250-900°C).

The role of alloy silicon has been well established within the region of parabolic oxidation. Here, the silicon, which forms a thin, silica interlayer between the chromium-rich outer scale and the metal [4-6], acts as a barrier to the outward diffusion of chromium ions and so controls the reaction rate. The effectiveness of this silica layer in 20%Cr/25%Ni/Nb stainless steel has been shown to depend on the alloy silicon content [7] and for samples oxidised in CO_2-based gas at 900°C, this effectiveness is a maximum at 0.6 wt% Si.

The role of silicon in the prevention of non-protective oxidation is examined here in a series of 20%Cr/25%Ni/Nb stabilised stainless steels exposed to CO_2-based gas at 1000°C. Also, a comparison is made with the oxidation behaviour of the reactor grade material at this temperature.

EXPERIMENTAL

Nine 20%Cr/25%Ni/Nb alloys were cast with alloy silicon contents, obtained by X-ray flourescence analyses, in the range 0 to 2.25 wt%. The alloys were supplied in cold rolled strips 0.38mm thick and 25% cold worked. Specimens of dimension 10 x 20 mm were cut from these strips, cleaned, then given a recrystallisation anneal for 1 h at 930°C in dry H_2. The samples of fuel cladding, of the same dimensions as above, were cut from a production grade, ribbed fuel can with alloy silicon content of 0.58 wt%. The specification range for silicon in this steel is 0.5 to 0.8 wt%.

The oxidation tests at 1000°C were performed in CI Electronics thermobalances enabling continuous monitoring of weight changes and temperature, in a manner described previously [3]. The oxidant was simulated reactor gas of composition CO_2/1-2%CO/300 vpm H_2O/300 vpm CH_4 at 1 atm pressure and 400 mm^{3}s^{-1} flow rate.

After oxidation, scanning electron microscopy (SEM) and energy dispersive X-ray (EDX) analyses were undertaken to determine the surface morphology and, semi-quantitatively, the composition of the surface oxides, respectively. Specimens were then gold sputtered (to mark the outer edge of the oxide and to retain the oxide during preparation), prior to mounting in cross-section and polishing using conventional metallographic techniques. These specimens were examined optically and by SEM and electron probe microanalysis (EPMA).

RESULTS

Gravimetric Data

The early stage of oxidation at 1000°C in each of the alloys was characterised by a decreasing rate of reaction in accord with a parabolic rate dependence, i.e. the relationship, $w^2 = k_w t$, was obeyed, where w is the weight gain, t the exposure time, and k_w the parabolic rate constant. The variation of k_w with alloy silicon content, shown in Fig. 1, indicated a significant reduction with increasing alloy silicon content. A large proportion (almost x 10) of this reduction occurred at <0.5wt% Si, thereafter k_w decreased in the range 0.5 to 2.25 wt% Si, by a factor between 2 - 3 only.

A reproducible observation was that the rate of oxidation eventually increased on continued exposure at 1000°C (e.g. ref.3); the time (t*) and weight gain (w*) at the point of departure from parabolic kinetics were determined in all but the tests on the two highest silicon alloys, which were terminated before this transition was reached. Figures 2a and 2b illustrate a marked variation of t* with alloy silicon content, but the variation of w* was less marked with the suggestion of a minimum value in the range 0.56 to 0.76 wt% Si.

During the cooling cycle to room temperature at the end of each test, virtually zero spallation occurred on the zero Si alloy but the amount of spallation generally increased with increasing alloy silicon level; on the 2.25 wt% alloy severe, and almost complete, oxide spallation was observed.

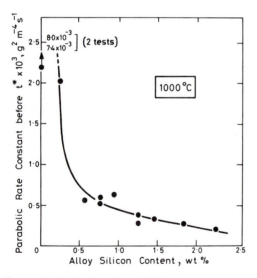

Figure 1: Variation of parabolic rate constant k_w with alloy Si content at 1000°C.

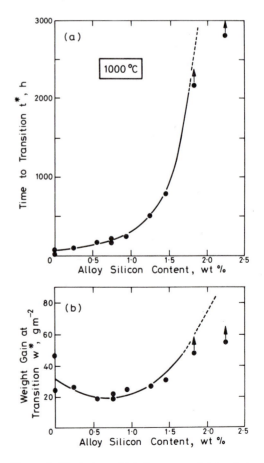

Figure 2: Variation of (a) time to, and (b) weight gain at kinetic transition with alloy Si content at 1000°C.

In the case of a ribbed fuel cladding specimen (0.58 wt% Si) oxidised in the same gas at 1000°C, the transition was observed at $t^* = 195$ h and $w^* = 23.8$ gm^{-2}, which was closely similar to that observed for the 0.56 wt% Si alloy above (see Fig.2).

Scale Examination

Optical micrographs of transverse sections of the surface oxides and underlying metal of selected alloys, which indicate the extent of change in mode of oxidation, are shown in Fig.3. (Note that oxidation times changed from specimen to specimen, increasing from Fig.3a to 3d.) The zero Si alloy (Fig.3a) showed a thick outer oxide with internal oxidation along a broad front, consistent with previous observations of oxide breakdown in 20%Cr/25%Ni/Nb fuel cladding at higher temperatures [3]. In contrast, the 0.56 wt% Si alloy (Fig.3b) for most of its surface exhibited a thinner, uniform surface oxide with underlying intergranular silica intrusions, characteristic of protective oxidation at temperatures between 900 and 1000°C. EPMA showed the outer oxide to be chromium-rich with a thin silica interlayer at the oxide metal interface. With increasing alloy silicon content to 0.94 wt% (Fig.3c), the surface oxide remained uniform but with a suggestion of another, less dense, oxide now forming at the oxide/gas interface; the amount of internal silica attack along grain boundaries also increased. On further increasing the silicon content to between 0.94 to 2.25 wt%, this outermost oxide became more apparent and internal voidage within the bulk metal was now observed. These effects are illustrated in the highest silicon alloy of 2.25 wt% (Fig.3d) which showed a distinct change in oxide morphology although still remaining essentially protective. The surface oxide was distinctly two-phased with an inner uniform layer but with an outer, less dense, layer at the oxide/gas interface.

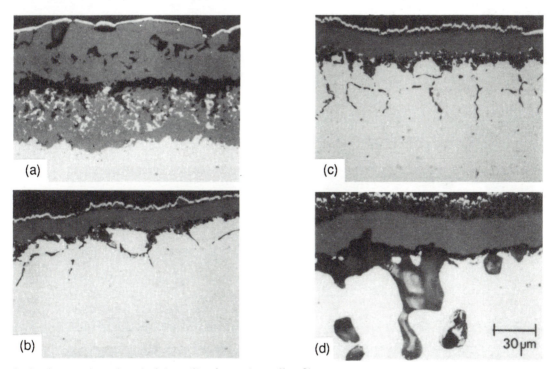

Figure 3: Surface oxide and underlying silica for various alloy Si contents.

Adjacent to the oxide/metal interface, a thick silica layer remained but underlying this, significant voidage was observed. The voids, which were distributed throughout the specimen cross-section in the highest silicon alloy (Fig.4a), were often angular, some were partially filled with silica and many were associated with large, second phase grain boundary particles (Fig.4b).

The surface topography of the outer oxide for two selected specimens are shown in the SEM pictures of Fig.5. A dramatic reduction in the size of crystallite occurred with increasing amounts of alloy silicon content. This change is associated with the replacement of the iron-rich oxide, formed on zero Si alloys (Fig. 5a), with a chromium-manganese spinel at alloy Si levels ≥ 0.5 wt% (Fig. 5b).

For the case of ribbed fuel cladding oxidised at 1000°C, specimens were examined metallographically after exposure periods of 150 and 400 h, i.e. before and after t^*. As expected, the former showed a uniform, protective scale throughout with underlying silica intrusions (Fig.6a) and the latter showed areas of thicker non-protective oxidation (Fig.6b) but, at

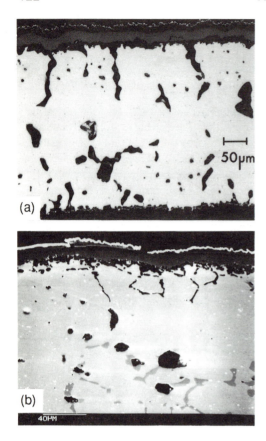

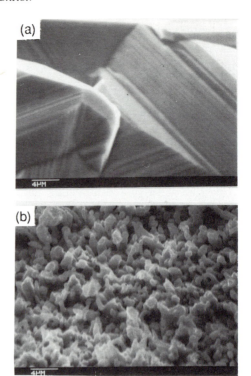

Figure 5: Morphology of surface oxides for selected alloy Si contents.

Figure 4: (a) 2.25wt% Si alloy after 3kh, and (b) 0.94wt% Si alloy after 680h at 1000°C.

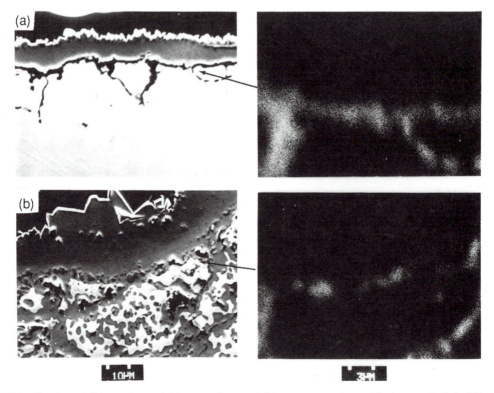

Figure 6: Distribution of Si in regions of (a) protective, and (b) non-protective oxidation on fuel cladding alloy at 1000°C.

this stage, these were confined to the ribs alone. The distribution of silicon in the regions of protective and non-protective oxidation have been examined in more detail using EPMA. It can be seen (Fig.6) that the silica interlayer is less continuous in the non-protective oxide.

DISCUSSION

It is evident that the beneficial role of silicon in protective oxidation is continued to higher temperatures, as exemplified by the delayed incidence of scale breakdown with increasing alloy silicon content (see Fig.2). The improvement is most marked at alloy silicon levels ≥ 0.93 wt% and is probably associated with a thicker silica interlayer. It is recognised [3,6] that this breakdown is coincident with a change in the dominant mechanism of oxidation from outward diffusion of cations through the silica interlayer to inward oxidant movement via pores or microcracks in the scale, although there is still some selective outward diffusion of iron to form an external Fe_3O_4 layer. Clearly, some degeneration of the original protective oxide, and in particular that of the silica interlayer, is involved. Indeed, Fig.6 reveals that the silica interlayer appears to be more fragmented in the regions of breakdown. The reasons for this can only be speculative at this stage but it is possible that the originally amorphous state of the silica layer [5], may, under a combination of high temperature/oxide growth stresses, recrystallise to β-cristobalite/α-quartz allowing easier gas access. Such a transformation has been observed in protective silica coatings on the fuel cladding alloy during oxidation alone at 1300°C[8].

On the other hand, increased alloy silicon levels lead to an increased tendency for oxide spallation during the cooling cycle at the end of each test, e.g. Fig.4a. This increased spallation is likely to be due to an increase in strain energy associated with the thicker chromium-rich scale [9] and, in addition, the thicker silica interlayer may provide an easier fracture path at the oxide/metal interface.

A further detrimental influence of increasing silicon additions is the formation of voids within the alloy. These are not simply a result of the selective oxidation of alloy constituents (Cr, Si) since they do not form at lower alloy silicon levels even though the surface oxide may be of similar thickness. Conventional models of vacancy injection and precipitation are, thus, not applicable. An additional feature of the high-silicon alloys is the presence of large particles at the alloy grain boundaries (Fig. 4b). These particles contain chromium and silicon and are likely to be M_6C or G-phases. A striking feature is that these have a similar distribution and size to the voids observed in the interior of the specimen (cf. Figs.4a & b). It seems likely, therefore, that this effect is another example of void formation by particle dissociation, as first discussed by Evans et al. [10] for similar steels containing $M_{23}C_6$ and M_6C phases. Such dissociation occurs as a result of the diffusion of chromium and silicon to the surface oxide and can be thought of as a Kirkendall effect around the particles but one where vacancies are supplied ultimately by the oxidation process

CONCLUDING REMARKS

The alloy silicon content of 20%Cr/25%Ni/Nb stabilised stainless steel has a significant effect on the maintenance of protective oxidation at 1000°C. Increasing the silicon content from zero to 2.25 wt% progressively delays the onset of non-protective oxidation. However, at large silicon contents, excessive spallation of the surface oxide occurs on cooling and, additionally, extensive void formation develops within the alloy during the oxidation exposure for alloy silicon levels ≥ 0.9 wt%. Clearly, for practical application of these steels, their silicon contents should lie between the extremes of behaviour reported here. The specification range currently used in AGR fuel cladding of 0.5 to 0.8 wt% Si appears to be optimum.

ACKNOWLEDGEMENTS

This paper is published by permission of Nuclear Electric Plc. The authors express their thanks to Mr E.J. Hoare, Mr R.P. Sparry and Mr S.M. Underwood for their experimental assistance.

REFERENCES

1. J.M. Francis and J.A. Jutson, Mater. Sci. Eng., 4,1969, p.84.

2. H.E. Evans, D.A. Hilton, R.A. Holm and S.J. Webster, Oxid. Met., 14, 1980, p.235.

3. R.C. Lobb and H.E. Evans, Proc Conf on Materials for Nuclear Core Applications, British Nuclear Energy Society, London, 1987, 335.

4. J.E. Antill, Werkst. und Korros. 6, 1971, p.513.

5. M.J. Bennett, J.A. Desport and P.A. Labun, Proc Roy Soc, A412, 1987, p.223.

6. M.J. Bennett, Proc 10th Int. Congress on Metallic Corros., Karaikudi (India), CERI, 1987, 3761.

7. R.C. Lobb, J.A. Sasse and H.E. Evans, Mater. Sci. Tech. 5, 1989, p.828.

8. R.C. Lobb and M.J. Bennett, Oxid. Met. 35, 1991, p.35.

9. H.E. Evans and R.C. Lobb, Corros. Sci. 24, 1984, p.209.

10. H.E. Evans, D.A. Hilton and R.A. Holm, Oxid. Met. 11, 1977, p.1.

11. H.E. Evans, Mater. Sci. Tech. 4, 1988, p.1089.

16 THE OXIDATION BEHAVIOUR OF HIGH-Cr CAST IRONS IN A CHLORIDE-CONTAINING IRON ORE SINTERING ENVIRONMENT

H. J. De Klerk and J. Smuts

Research Division, ISCOR Ltd., Box 450, Pretoria 0001, South Africa.

ABSTRACT

The oxidation behaviour of high-Cr iron castings used as so-called fire bars in the manufacture of metallurgical sinter with $CaCl_2$ additions, was studied with scanning electron microscopy techniques. The roles of various constituent elements such as carbon, nickel and aluminium and species from the aggressive environment such as chloride and sulphide in the degradation of the material are illustrated. The oxidation interface region can conveniently be divided into scale, sub-scale and carburised regions. The role of chloride can be simplified as a catalytic effect on the oxidation via the selective attack on carbides in the transition zone between the morphologically distinguishable sub-scale and carburised regions.

It is shown that nickel has a strong influence on the pattern of oxidation by enhancing the formation of sub-scale mostly through its effects on the chromium distribution. The role of carbon is discussed in terms of the increased Cr-depletion of the metallic matrix associated with the formation of Cr-rich carbides both during solidification and during the carburisation process that is associated with chloride-assisted oxidation. The increase in the extent of Cr-depletion with increased exposure is illustrated to explain the transition from predominantly "protective" oxidation during the initial stages of exposure to increased "non-protective" oxidation during the later stages of exposure.

Various microstructural processes associated with oxidation in the chloride-containing sinter environment are illustrated. These include the formation of sigma phase and its decomposition in the carburised region, carbide "growth" and subsequent carbide decomposition at the attack interface and the precipitation of various sulphide species in both the metallic and oxide regions.

INTRODUCTION

The life of fire bars used in a Dwight-Lloyd iron ore sinter plant is determined to a large extent by the oxidation behaviour of the alloy in this type of environment. Material losses caused by oxidation processes can be substantial when non-protective [1] scales are formed and frequently removed in the processing of sinter. The present study was initiated to improve the cost efficiency of the plant by reducing fire bar cost. One of the aspects of the study is the improvement of fire bar life by increased resistance to material loss caused by oxidation processes.

The oxidation resistance of a series of Fe-C-Cr-Ni alloys made up as fire bars was evaluated in an iron ore sintering environment containing $CaCl_2$. Performance was evaluated by measuring certain critical dimensions of the fire bars at regular intervals and determining the regions of highest wear. One of the dimensions found to be the most seriously affected is the head thickness. The oxidation behaviour of this specific site was therefore chosen as an indication of the overall fire bar performance.

During a typical sintering cycle lasting about 2 hours an individual fire bar would be subjected to the following thermal cycle: A 50 mm thick hearth layer ideally consisting of sinter nodules of approximately 25 mm diameter, is loaded onto the sinter strand from a hopper. Next a 450 mm thick layer of moist sinter mix is loaded onto the strand. At this stage the fire bar temperature is close to room temperature. The top sinter layer is ignited in the fire box and a thin layer of reacting sinter is formed. This moving layer approaches the fire bar surface as the sinter process proceeds. During this time the fire bar temperature increases to approximately 1000°C. During the ensuing cooling cycle the fire bar temperature is lowered as it approaches the discharge end of the sinter strand. At the discharge end some regions of the pallet car are still red hot indicating temperatures in the excess of 700°C. During most of this cycle the environment is strongly oxidising and several "aggressive" elements or species are present such as O_2, CO, CO_2, sulphur compounds, KCl "vapour" and water vapour. During the travel back to the "start" of the strand the fire bars cool in air to approximately room temperature. Thus during each sintering "cycle" the fire bar is exposed to a temperature change of approximately 1000°C and a rapidly changing gaseous environment.

Alloy design for high temperature oxidation resistance is based on the control of transport processes in the oxidation product on the component/environment interface [2]. This implies that the transport processes and scale morphology must be known if any improvement in the oxidation performance is to be realised through alloy composition modification.

In this presentation some aspects of the oxidation behaviour and associated microstructural features on the side of the bar where a thick scale layer tends to form are illustrated for four alloys with a base composition of 0,5% C, 0,7% Mn and 26% Cr. Three of the alloys are ferritic matrix alloys containing 0,3% Ni, 2% Ni and 2% Ni + 0,2% Al respectively while the fourth alloy is an austenitic matrix alloy containing approximately 14% Ni and 2% Mn. Examples of different scale morphologies and microstructures of the so-called sub-scale layer and microstructural features in the sub-scale region are shown.

EXPERIMENTAL

Resin mounted specimens from various fire bars were prepared for metallographic examination by light microscopy and SEM. Structural features were recorded using the back-scattered electron (BSE) mode in a Jeol 733 microprobe that was operated at an accelerating voltage of 25 kV or 15 kV. The chemical composition of features were determined by energy dispersive X-ray (EDX) analysis.

RESULTS

It is impossible to deal with all aspects of oxidation for the many relevant alloys. The various samples experienced exposures to the sintering environment ranging from three months to up to two years. Only those examples that are either very interesting or relevant to the rate of component degradation are shown. Most of the micrographs illustrate features that are characteristic of the specific alloy such as the formation and decomposition of sigma phase in the 2 % Ni + 0,2% Al alloy, and the layered oxidation product morphology that characterises the grain boundary oxidation process in the 14% Ni alloy.

In most cases the fire bar/environment interface has the following configuration:
 a surface scale layer of variable morphology and composition
 a sub-scale layer consisting of oxide and matrix "residue"
 a carburised layer that in some cases consists of distinct zones.

Figures 1 to 3 show details of the oxidation product and oxidation front region in the 2 % Ni + 0,2% Al alloy after a long exposure of approximately 2 years. Noteworthy features in these images are: heterogeneity of the scale (Fig. 1); chloride attack along the carbides (Figs. 2 and 3); sigma phase (the light phase in Fig. 2) precipitation and its decomposition to low-Cr ferrite and carbides; precipitation of "secondary" carbides in the ferrite matrix and the specific attack of the ferrite/carbide interface.

Figures 4 to 7 show details of various carbide reactions in transition regions on the 2% Ni alloy after moderate exposures. Figure 4 shows the selective nature of the oxidation process where "penetration" of the oxidation product into the alloy occurs via the carbide networks. The carbide morphology at the attack front in a bar exposed for 7 months is shown in Fig. 5. The carbides at this site is coarsened relative to the carbides in the base microstructure and has a poly-crystalline appearance. This alloy exhibits a definite transformed matrix layer beneath the sub-scale. At the transformation interface illustrated in Fig. 6 there is also a marked change in the carbide morphology compared to the base microstructure. In the transformed zone (to the left) extensive precipitation reactions occurred within the carbide. It appears that the carbide decomposes at this interface by a continuous precipitation reaction. Note the definite interface demarcating two matrix structures. Figure 7a is a BSE image of the sub-scale and carburised region illustrating the coarsening of carbides and the formation of a transformed matrix layer. The transformed matrix layer is associated with the lower local Cr-content of the matrix which in turn is the result of the carburisation process. Figure 7b gives the matrix Cr-content of the carburised layer and sub-scale as function of the the distance from the scale interface.

Figures 8 to 10 show details of the oxidation product morphology in sub-surface regions of the 14% Ni alloy after a rather short exposure of 6 months. Figure 8 illustrates the intergranular nature of the oxidation process and the layered oxidation

(Note that all the micrographs are so oriented that the outer surface is either to the top or to the left of the micrograph)

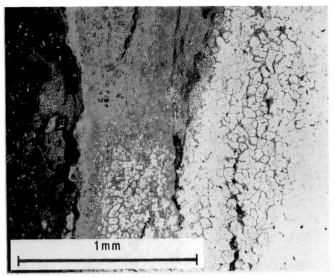

Figure 1: A typical example of the morphology of the fire bar/environment interface region in the 2% Ni + 0,2% Al alloy. It consists of a scale layer and a partially oxidised layer, the so-called sub-scale.

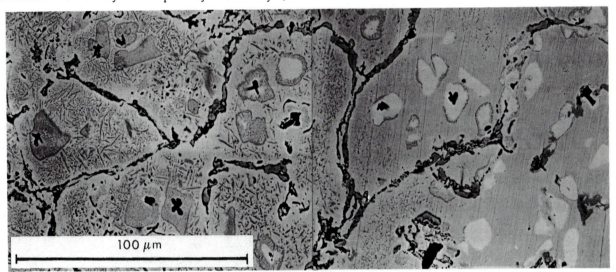

Figure 2: A detail of the interaction of carbide precipitation reactions and chloride attack.

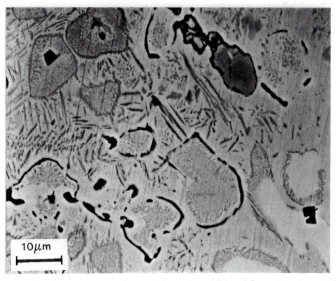

Figure 3: Detail of chloride attack on both primary and precipitated carbides. Note that the oxidation takes place at the ferrite/carbide interface.

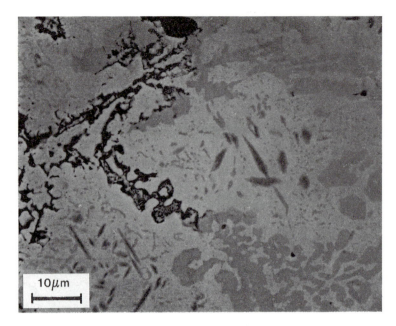

Figure 4: Image of the attack front region in the 2% Ni alloy showing that Cl-attack occurs primarily on the coarse carbide networks followed by oxidation of the finer secondary carbides.

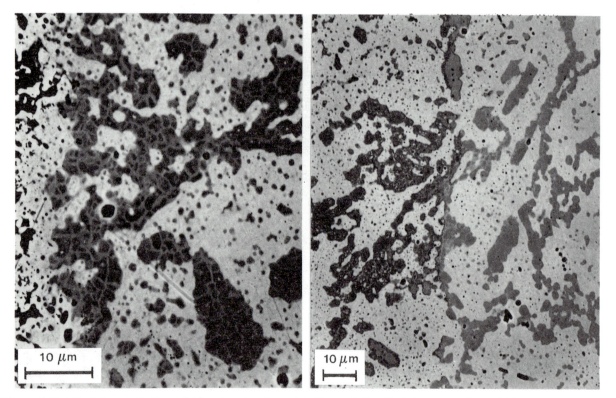

Figure 5: Detail of changes in the carbide microstructure at the attack interface in the 2% Ni alloy.

Figure 6: Detail of changes in the carbide microstructure at the "transformation interface" in the 2 % Ni alloy.

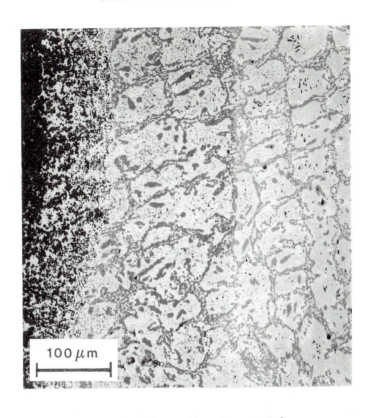

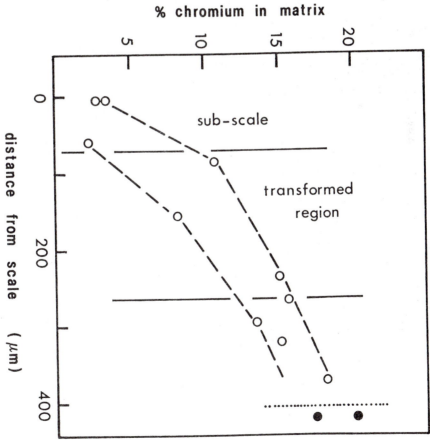

Figure 7: BSE image of the sub-scale and carburised regions in the 2% Ni alloy together with a plot of the matrix Cr-content as function of the distance from the scale interface illustrating the Cr-depletion effect of carburisation and subsequent oxidation processes.

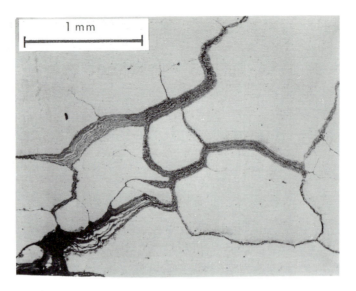

Figure 8: Detail of the extensive grain boundary type attack process in the 14% Ni austenitic matrix alloy.

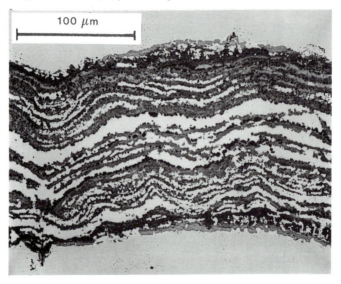

Figure 9: Detail of layered oxidation product formed during the "grain boundary attack" process in the 14% Ni alloy.

Figure 10: Enlarged portion of Figure 9 showing detail of the attack interface at the layered oxidation product.

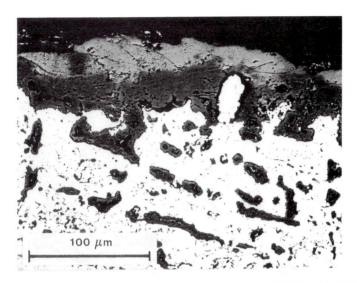

Figure 11: Detail of the double layer scale and sub-scale morphologies in the 0,3% (low) Ni alloy.

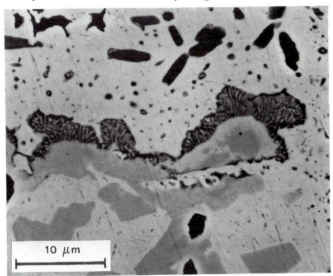

Figure 12: Detail of one of the anomalous interfaces which characterises the carburisation/decarburisation process during oxidation of the 0,3 % Ni alloy.

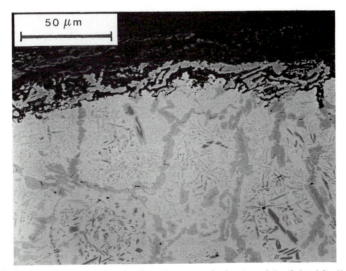

Figure 13: Detail of the scale/sub-scale and carburised region morphologies of the 0,3% Ni alloy at the region where the outer scale layer is continuously removed by abrasion.

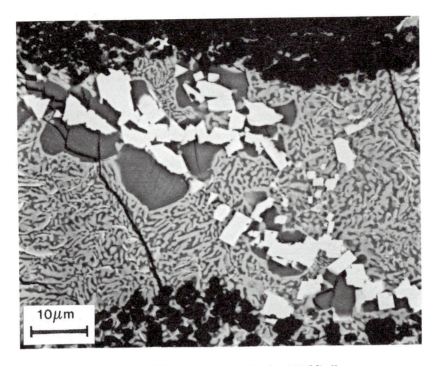

Figure 14: Detail of massive (liquid phase) sulphidation product in the 14% Ni alloy.

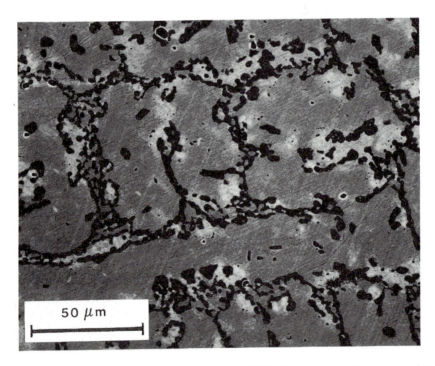

Figure 15: High contrast BSE image of the microstructure of the 2% Ni alloy showing the segregation pattern associated with the carbide distribution. Light coloured areas in the matrix adjacent to the carbides have a Ni-content of up to 6% and a Cr-content of several percent lower than the average matrix composition.

product. Detail of this layered oxidation product is shown in Fig. 9 while the interface between the layered structure and the "base" material is shown in Fig. 10. Note the prominence of the Ni-rich intermetallic phase NiFe (white), and the sulphide $(Ni_{0,5}, Fe_{0,5})S$ (grey, most prominent on the interface) together with the Cr-rich oxide (dark grey to black) layers in the layered structure in Fig. 9. Note also the spherical Cr-sulphides in the matrix adjacent to the oxidation product/matrix interface in Fig. 10 indicating internal sulphidation of the matrix.

Figures 11 to 13 show details of the scale morphology and various carbide reactions in the 0,3% Ni alloy after exposures of one year (Figs. 11 and 12) and 18 months (Fig. 13). Figure 11 shows the typical duplex scale that forms on the side of the bar. Figure 12 illustrates an anomalous interface close to the attack interface that demarcates two regions where carburisation and where decarburisation took place respectively. At this interface the carbide appears to decompose via a discontinuous precipitation reaction [3]. Figure 13 illustrates the morphology of the top or working surface interface region. Note that the sub-scale consists of oxide and metallic matrix in Fig. 11 while in Fig. 13 it consists of carbide and oxide. This difference can mainly be attributed to differences in the transport processes associated with the difference in the scale morphology which in turn is the result of abrasion on the top surface.

Figure 14 shows evidence of massive sulphidation reactions in the 14% Ni alloy. Various Fe-Ni-S phases can be recognised such as NiS (light grey); $(Ni_{0,5}, Fe_{0,5})S$ (dark grey) and Ni_3Fe (white) as well as Fe-rich oxide particles (black). These phases are evidence of the high susceptibility of the alloy to sulphidation.

The BS micrograph in Fig. 15 illustrates the segregation of chromium and nickel that took place both during solidification and subsequently during the oxidation process in the 2% Ni alloy. Ni is mostly concentrated in the light regions while Cr is concentrated in the black carbides. This segregation can at least partly explain the susceptibility of the carbide/matrix interface region to oxidation.

DISCUSSION AND CONCLUSIONS

In general the interpretation of the microstructure of scale layers appears to be rather straightforward: it involves the identification of phases and measurement of dimensions such as layer thickness or particle or void sizes. However the relevance of such measurements in terms of alloy or component behaviour can only be determined if it is related to physical measurements of the degree or rate of component degradation.

It is generally regarded that the formation of the Cr-rich sesqui-oxide is protective [4] and that alloys that form this oxide upon exposure to the specific environment is suitable for application. This approach is probably too simplistic to be applied in the sinter environment because of the possibility of carburisation, sulphidation and chloride-assisted oxidation.

In the case of fire bars the rate of material loss expressed in terms of the change in the width of the fire bar appears to be the most important failure criterium. The rate at which material is lost from this surface is a function of both the rate at which the various "local" oxidation processes take place and the rate at which the surface scale is removed.

The local oxidation processes determine to a large extent the surface scale morphology and is therefore an important aspect of fire bar performance. The microstructures that result from these local oxidation processes vary from one alloy to the other and some features appear to be characteristic of the alloy type. The differences in the oxidation behaviour can largely be attributed to the role that certain constituent elements play in the microstructural evolution of the sub-scale region and their specific effects on the susceptibility to sulphidation and chloride- assisted oxidation.

The role of Ni in the apparently poor performance of Ni-containing materials needs careful consideration. Some of the more obvious explanations of its influence can be found in the increased segregation of alloying elements such as chromium and its influence on the sulphidation resistance.

The role of carbon content can be explained mostly through the effects of carbides on the matrix Cr-content and the effect that carburisation processes which occur concurrently with the oxidation process have on the formation of the sub-scale layer. The carburisation of the base material adjacent to the oxidation front is associated with extensive Cr-depletion of the metallic matrix and leads to non-protective oxidation. It appears that the carbon associated with the carburisation of the base material originates mostly from the oxidation of carbides in the sub-scale region. This means that the degree of carburisation can be controlled to some extent by limiting the carbon content of the alloy to the lowest practical level.

The apparent detrimental effect of relatively small aluminium additions to the cast iron can be explained in terms of the enhanced nucleation of sigma phase on alumina inclusions (see Fig. 2) and the subsequent effect of sigma phase on the carbide morphology.

In summary: The most important criterionfor the development of more suitable alloys for this application is resistance towards chloride-assisted oxidation. The high susceptibility of the Ni-containing alloys to sulphidation processes indicates that the Ni-content of fire bars should be maintained at the lowest practical level.

ACKNOWLEDGEMENT

This paper is published with the permission of the management of Iscor Ltd.

REFERENCES

1. F.S. Pettit, C.S. Giggens, J.A. Goebel and E.J. Felten, Alloy and Microstructural Design, Eds. J.K. Tien and G.S. Ansell, Academic Press (New York), 1976, p.349.

2. S.B. Newcomb, W.M. Stobbs and E. Metcalfe, Phil. Trans. R. Soc. Lond. A319, 1986, p.191.

3. D.B. Williams and E.P. Butler, Int. Met. Rev. 3, 1981, p.153.

4. S.B. Newcomb, W.M. Stobbs and E. Metcalfe, Phil. Trans. R. Soc. Lond. A319, 1986, p.219.

17 CHANGES IN THE SUB-SCALE MICROSTRUCTURE OF A RHENIUM-CONTAINING γ/γ' ALLOY AS A FUNCTION OF ITS OXIDATION AND CREEP

S.B. Newcomb and W.M. Stobbs

Department of Materials Science and Metallurgy,
Cambridge University, Pembroke Street, Cambridge CB2 3QZ, UK.

ABSTRACT

We report here the preliminary results of a TEM study aimed to assess the extent to which the microstructure beneath an oxidation scale formed on a γ/γ' alloy is modified when the material contains rhenium and, in turn, whether or not any such sub-scale modifications are different when the oxidation takes place during creep.

INTRODUCTION

It is now well known that the incorporation of rhenium into high temperature Ni based γ/γ' alloys can improve both their creep and fatigue properties and that these improvements are apparently associated with the segregation of the rhenium to the γ phase. By comparison little is known about the effects of rhenium on the oxidation behaviour of γ/γ' alloys and although there are indications that rhenium can increase the oxidation resistance it is also relevant to examine the sub-scale changes which can occur during oxidation not only because of the inherent interest of the process but also because of the synergistic changes that might occur during oxidation when a load is applied. Any such changes could well clarify the rôle of rhenium in modifying the creep response of these alloys. Our primary goal here has thus been to characterise those changes in the sub-scale microstructure which are associated with oxidation alone and those which are the result of the sub-scale region being crept while the various segregation processes associated with the oxidation process happen in parallel. Of secondary concern has been the examination of the differences in behaviour of the subscale region as a function of whether or not the alloy had been supersolutionised.

EXPERIMENTAL DATA

With the above aims in mind we have examined the scaling behaviours of a directionally solidified Ni based alloy (CM186) containing 2.8wt% Re as well as the γ/γ' alloy (CM247) from which CM186 was developed which contains similar amounts of Al, Co, Cr, Ti, Hf, W, Ta and Mo but no Re. Specimens in the as-cast and aged or supersolutionised and aged condition were oxidised with and without an applied load of 130MPa giving a creep deformation in each state of well under 0.1%. The details of each specimen treatment are given below:

Specimen	Alloy	Condition	Oxidation/Creep
A	DSCM247	1260°C Soln. + Age	50h 980°C Air
B	DSCM186	As Cast +Age	50h 980°C Air. 130MPa
C	DSCM186	As Cast + Age	50h 980°C Air
D	DSCM186	1266°C Soln. + Age	50h 980°C Air. 130MPa
E	DSCM186	1266°C Soln. + Age	50h 980°C Air

The ageing treatment was: 4h at 1080°C, Gas Furnace Quench + 20h at 870°C, Air Cool.

The main emphasis of the work we describe here is centred around data obtained by the examination of the sub-scale regions by the TEM of cross-sectional specimens (prepared by standard methods [1]). However, given the tendency for the oxidation to be generally variable from region to region, sections of the oxidised specimens were also examined optically after a light electrolytic etch in dilute phosphoric acid to bring out the dendritic structure as well as the modifications both to this and to the γ/γ' morphology that had occurred beneath the scale. It should be noted that all the outer oxides tended to spall and we give little data on these here, but it can still be seen from Fig. 1 that both the outward and inward oxidation behaviour of these alloys is indeed grossly heterogeneous. In the region shown for the CM247 alloy (Fig.1a) there is a variable band of near continuous inward oxidation (from 0 to 15μm in depth) as well as dispersed internal oxide (to a further 25μm) in a region of modified matrix structure (apparently devoid of the ordered phase) extending to a depth of about 40μm. Turing to the appearance of the regions beneath the outer scale of the CM186 alloy we find the same degree of heterogeneity, but different trends. Examples of the microstructures found for Specimens B,C,D and E are shown in Figs. 1b-e respectively. The most diverse inward oxidation behaviours were observed for Specimen B for which we find regions showing either localised inward oxidation or generalised coarse internal oxidation, little internal oxidation and here and there very little modification of the matrix. In the absence of load (Fig.1c) the oxidation behaviour of the as cast and aged alloy seemed more similar to that of Specimen A. As solutionised there was little evidence of general inward oxidation and while the region of modified matrix was extensive both with (Fig.1d) and without a load (Fig.1e), the internal oxidation seemed less developed under creep than without a load applied.

ELECTRON MICROSCOPY OF EDGE-ON SAMPLES

Given the gross variations in the oxidation behaviours indicated by the optical data it would be unwise to draw too generalised a set of inferences from our electron microscopical work since not all the scale types have as yet been found in the TEM. Equally, given that we have too little space to show examples of even all the modified sub-scale formats that we have found, summary diagrams of some of our main observations for the five specimens described are shown in Figs. 2a-e.

The sub-scale region of Specimen A did not have one of the localised inward growing scale regions. While the γ' particles were dislocated (Fig.3a) well beneath the modified zone described above, there was a well developed sub-grain structure (Fig.3b) in a ~10μm Al depleted zone where the γ' particles had dissolved. The large (~1μm) Al₂O₃ particles (Fig.3c) found in an approximately 12μm band above this were also enriched in Cr relative to the matrix but neither oriented with respect to the matrix nor to the small Hf (and Ta) enriched core particles that they tended to encompass.

The area of Specimen B examined was, on the basis of the form of the γ' particles well beneath the modified zone, at an intermediate region of the dendritic substructure. At the low creep strain applied the dislocation density observed was generally rather low except in the vicinity of $M_{23}C_6$ colonies (Fig.4a). Considerable modification of the γ' particles was found to a depth of about 12μm beneath the outward growing scale but the interesting feature of the sub-scale alloy was the way in which this γ' modified zone was separated from a region of continuous scaling by a recrystallised band some 3μm in thickness (Fig.4b) where the γ' particles were completely dissolved and the matrix contained little Al. The γ' modified zone was generally depleted in Al while the γ here was also found to be depleted in Re largely because of the dilution effect caused by the reduction in the volume fraction of γ'. Internal Al₂O₃ oxides were again seen in this band (as at A in Fig. 4b) and found to have a high Cr content relative to the bulk as well as 'cores' which were highly enriched in Hf. The larger inward growing oxides at the top of the recrystallised area of the matrix (as at A, at a grain boundary at the top of the recrystallised zone: Fig.4c) were almost pure Al₂O₃ but the other oxides growing above this, while having rather smaller and variable sizes, also had disparate compositions. The main cation present was always Al, but those that were enriched in Cr also contained some Ti and Co, while those that contained high concentrations of Hf and Al tended to contain other elements in only low concentrations. No regions particularly enriched in Re were found though we should emphasise that we have not examined the outer layer scales.

The area of Specimen C which was examined had an inter dendritic appearance and exhibited isolated Al₂O₃ particle formation (Fig.5a) at large depths consistently with the opticals. In such regions there was also a high dislocation density associated with large γ' particles and some tendency to sub-grain formation though this was more clear at the 14-7μm depths at which the γ' had fully dissolved (Fig.5b), as well as above this where there was more dislocation rearrangement (Fig.5c). It was in these uppermost levels of the sub-scale that clusters of small particles (Fig.5d) were found with a cube/cube orientation relation with the matrix (unlike MC) and a lattice parameter consistent with a carbonitride. While PEELS was successfully used to identify the more facetted nitrides formed in Specimen D, as noted below, nitrogen was at too low a level in these particles for it to be quantified. It should be emphasised however that neither the overall chemistry of the particles (they were enriched relative to the local γ in Ti, Hf and Ta but not Re) nor their morphology and orientation relation would be consistent with them being pre-oxidation MC.

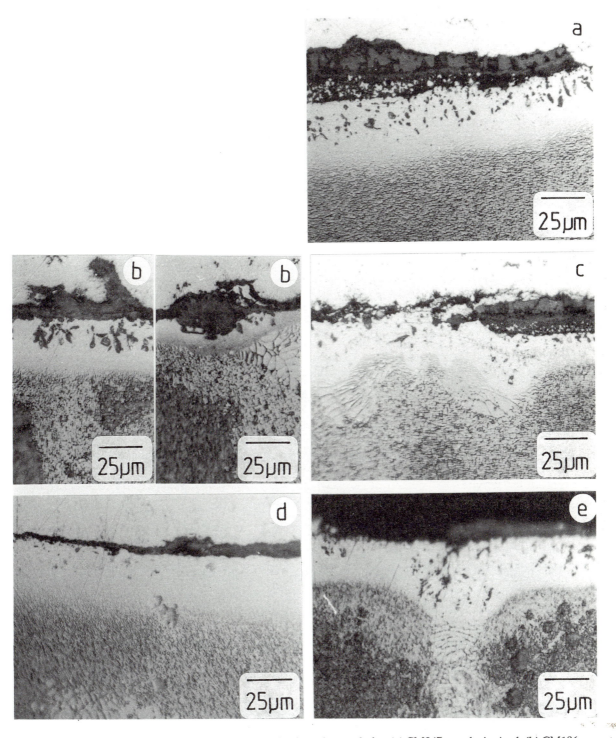

Figure 1: Optical micrographs showing the oxides and sub-scale metals for: (a) CM247 as solutionised, (b) CM186 as cast and crept, (c) CM186 as cast with no load, (d) CM186 as solutionised and crept, (e) CM186 as solutionised and oxidised in the absence of load.

The region of Specimen D examined was found to have a generally fine grained oxide (Fig. 6a) beneath which the alloy was fully recrystallised to a depth of about 4μm with full γ' dissolution to a depth of about 8μm and partly dissolved γ' (Fig.6b) to a further depth of 2-3μm. The oxide compositions in the fully oxidised region showed the same variations that have already been noted for the similarly small oxide grains formed on the as-cast and aged alloy (Specimen B). Larger internal oxides were also found and, in the area examined, these tended to be situated at the first horizontal sub-grain

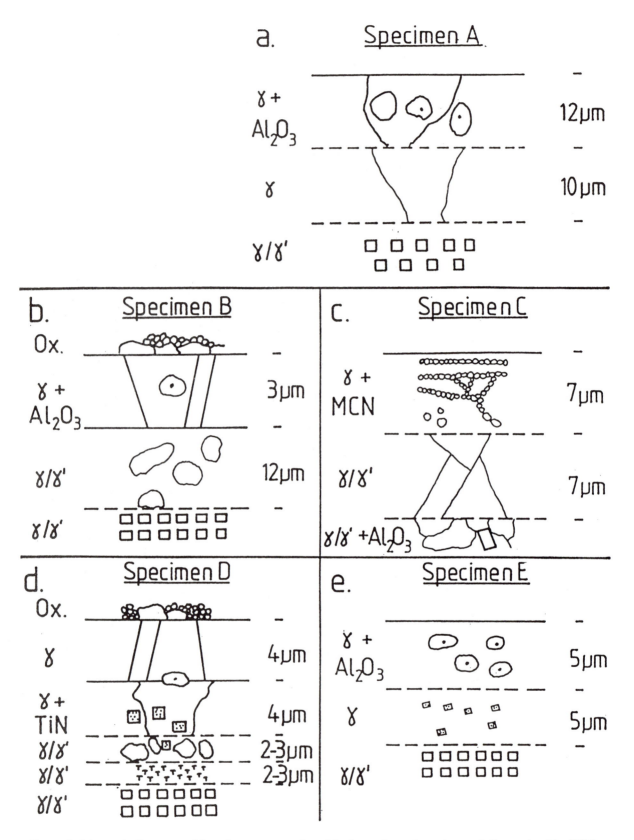

Figure 2: Schematic diagrams of the microstructures found in the regions of specimens A-E examined by TEM.

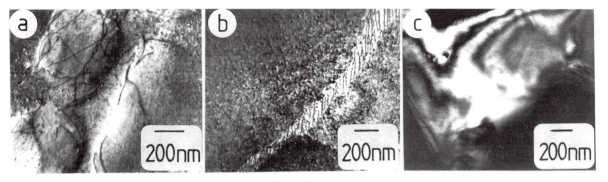

Figure 3: Sub-scale regions of CM247; as cast and aged, A, as described in text.

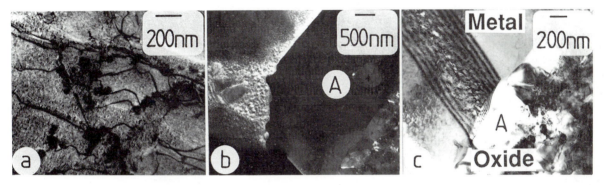

Figure 4: Sub-scale region of CM186 as cast, aged and crept; B.

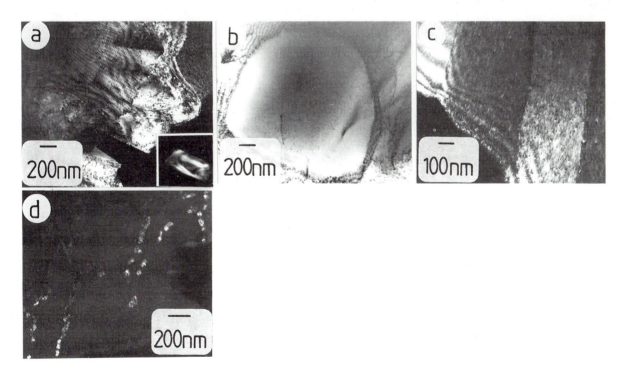

Figure 5: Sub-scale region of CM186 as cast and aged; C.

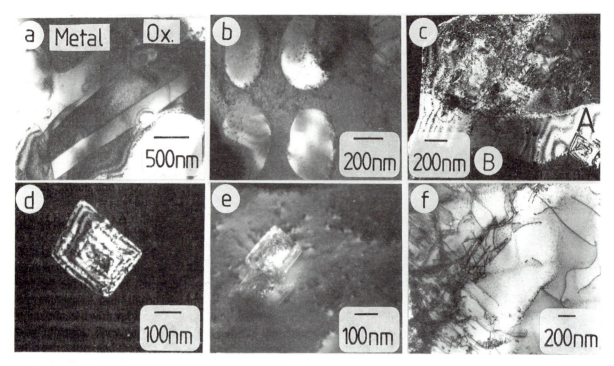

Figure 6: Sub-scale region of CM186 as solutionised, aged and crept; D.

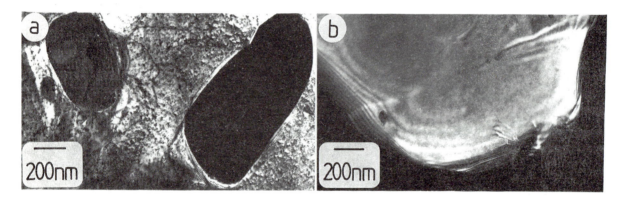

Figure 7: Sub-scale region of CM186 as solutionised and crept; E.

boundary beneath the outer scale. More interesting was the next 4μm region where the γ' was also fully dissolved but the grain boundaries were less facetted. Here there were a number of large (200nm) nitrides (as characterised using PEELS) with a very high Ti content which exhibited a cube/cube orientation relationship with the matrix. Examples of such particles are marked A in the matrix dark field image in Fig.6c from which their regular cuboidal shape can be visualised. This is further demonstrated when they are imaged in their own reflections, as in Fig.6d. Their size tended to be smaller beneath the boundary marked at B below which also the dislocation content of the alloy was high and the sub grain boundaries were less well developed. When this area was imaged using a TiN reflection it became apparent that the dislocations tended to be decorated by rather small nitrides that necessarily thus too exhibited the same orientation relation as the larger cuboidal particles (see Fig.6e). A further feature of this region of Specimen D which apparently distinguished it from the crept as-cast and aged alloy was that an isolated band of high super dislocation activity was found (as shown in Fig.6f) well beneath the oxide where the γ' was not obviously affected by the oxidation.

The area examined of the supersolutionised specimen (E) which was oxidised without an applied load differed in a number of ways from its loaded counterpart (Specimen D). As for the as-cast specimens, the uppermost regions beneath the outer scale were not recrystallised but the dislocation content in the 10μm zone in which the γ' was fully dissolved was fairly high despite not being in morphological groupings reflecting an origin in their having been needed by larger γ' particles. The internally oxidised particles seen in the upper part of this zone were also particularly large and often contained several Hf enriched core particles (see Figs.7a and 7b).

DISCUSSION

We have discussed elsewhere [2] the way a material that is initially of a uniform composition can develop localised instabilities beneath a scale which then lead to the promulgation of different oxidation mechanisms adjacently to one another with the local stabilisation of different sub-scale composition gradients. While the rhenium in CM186 appears to be uniformly distributed in the γ phase of the supersolutionised specimens some of the other elements exhibited rather variable compositions from point to point (as a function of a retained dendritic substructure) on the initially oxidised surface for the supersolution temperature used. In consequence it was clear both from the optical and from the TEM data described here that there are a number of ways the details of the oxidation process can differ in different regions. Further work will be required in the correlation of the microstructures found on the TEM scale with those that can be seen optically before we can be confident to generalise on the full range of behaviours which can occur. It is for example surprising that the optical evidence would generally suggest rather deeper modifications to the γ/γ' microstructure than were found using electron microscopy.

Nevertheless we may note that while Hf and Cr do contribute to the outer scaling in ways that are well recognised in the literature, Hf can also become oxidised internally (thus reducing any passivating effect it might have) to form particles acting as nuclei for Al_2O_3 formation internally. The effect of even very light creep deformation is also clear in encouraging recrystallisation in the upper regions of the alloy thus allowing improved diffusion pathways both into and and out of the alloy. The rôle of Ti in forming nitrides, as has been observed by other authors [e.g.3], seems more confusing in that there is no obvious correlation of the process with the initial state of the alloy. On the other hand perhaps our most interesting result is that it would not appear that Re plays an active part in changing the form of the sub-scale microstructure so that the indications which have been obtained of improved oxidation resistance for Re containing alloys must relate to its segregation to the outer scales. Remembering that there is a necessary dilution of the sub-scale in Re due to the absence there of γ' our results would still be consistent with this conclusion.

ACKNOWLEDGEMENTS

We thank Prof. D. Hull for the provision of laboratory facilities and the SERC for financial support. We also thank Rolls Royce plc (Leavesden) for providing specimens as well as R. Wing and M. Newnham for useful discussions.

REFERENCES

1. S.B. Newcomb, C.B. Boothroyd and W.M. Stobbs, J.Microsc. 140, 1985, p.195.

2. S.B. Newcomb and W.M. Stobbs, Mat. Sci. and Technol. 4, 1988, p.384.

3. J. Litz, A. Rahmel and M. Schorr, Oxid. Met. 30, 1988, p.95.

SECTION FOUR
Growth and failure of alumina scales

18 THE NUCLEATION SITES OF γ-Al$_2$O$_3$ CRYSTALS IN THERMAL OXIDE FILMS ON ALUMINIUM

K. Shimizu, A. Gotoh, K. Kobayashi,
G.E. Thompson* and G.C. Wood*

*Department of Chemistry, Faculty of Science and Technology,
Keio University, 3-14-1 Hiyoshi, Yokohama 223, Japan.
*Corrosion and Protection Centre,
University of Manchester Institute of Science and Technology,
Manchester M60 1QD, UK.*

ABSTRACT

Electropolished aluminium specimens were thermally oxidized in air at 515°C for various periods of time. Thermal oxide films were then stripped from the aluminium substrate and examined thoroughly in the transmission electron microscope with particular attention directed at the sites at which γ-Al$_2$O$_3$ crystals nucleate and the relationship of such sites to surface features of the aluminium substrate. It was found, for the first time, that 'easy paths' for the diffusion of oxygen, or the nucleation sites of γ-Al$_2$O$_3$ crystals are not distributed randomly over the electropolished aluminium surface, but form preferentially in the amorphous oxide layer grown over the pre-existing metal ridges. Thus, the diffusion of molecular oxygen through cracks in the amorphous oxide layer represents the most expected and realistic basis for explaining the local growth of γ-Al$_2$O$_3$ crystals in thermal oxide films on aluminium.

INTRODUCTION

Except for the cases where bare aluminium surfaces, prepared carefully under ultra-high vacuum conditions, are exposed to oxygen at high temperatures, the thermal oxidation of aluminium always starts from the surfaces covered with an amorphous oxide layer, 2-3 nm thick, which is generally called an 'air-formed ' film. If aluminium supporting an air-formed film is heated in air or oxygen at temperatures above 450°C, thickening of the amorphous oxide layer is observed initially but, after some induction periods, γ-Al$_2$O$_3$ crystals start to grow at the amorphous oxide-metal interface. Previous studies [1, 2] have established that the growth of γ-Al$_2$O$_3$ crystals do not proceed by crystallization of the initially formed amorphous oxide layer, but by the inward diffusion of oxygen through highly localized 'easy paths' in the amorphous oxide layer. However, the nature of the 'easy paths' and the transport mechanism of oxygen through these 'easy paths' has remained the subject of discussion .

In the present paper, thermal oxide films were stripped from the aluminium substrate and examined thoroughly in the transmission electron microscope with particular attention directed at the sites at which γ-Al$_2$O$_3$ crystals nucleate in order to gain further insight into these problems.

EXPERIMENTAL

Annealed aluminium foil of 99.99% purity and 0.1 mm thickness was used for all experiments. This foil is polycrystalline with grain sizes around a few hundred microns, but exhibits a preferred (100) orientation. About 90% of the surface is covered with grains of (100) orientation, with the rest having orientations close to (110) or (311), as determined by the etch pit method.

The foil was cut into small pieces of dimensions 10 x 50 mm. They were then electropolished individually in a perchloric acid-ethanol bath at temperatures below 10°C with a constant current density of 100 mA/cm^2. After electropolishing, the specimens were rinsed thoroughly in absolute ethanol and, finally, dried in a warm air stream. Prior to thermal oxidation in air at 515°C, the electropolished foils were given a brief etch in a solution consisting of 1.5 ml of HF, 10 ml of H$_2$SO$_4$ and 90 ml of H$_2$O for 30 s, rinsed thoroughly in distilled water and then dried in a warm air stream.

Thermal oxide films were stripped from the aluminium substrate using a mercuric chloride/methanol solution and examined in a JEM 2000 FX II transmission electron microscope operated at 100 kV.

RESULTS AND DISCUSSION

Figure 1 shows transmission electron micrographs of the stripped thermal oxide films grown on various faces of aluminium in air at 515°C for 30 min. A characteristic darker furrowed pattern is observed in the amorphous region of the thermal oxide film grown on the (110) face; relatively long, straight furrows, spaced about 70 nm, are observed clearly. Gamma-alumina crystals with sizes up to about 0.1 μm are also observed in the thermal oxide grown on the (110) face. The crystals are rod-shaped, with their long axes aligned perpendicular to the straight furrows. The selected area electron diffraction pattern (inset, Fig. 1a) indicates an epitaxial relationship between the aluminium substrate and the γ-Al_2O_3 crystals in such a way that $(110)_{ox}//(110)_{Al}$. Direct counting of the number of crystals gives an average nucleation density of about 2.2×10^{10}/cm^2. Darker corrugation patterns are observed in the amorphous region of the thermal oxide films grown on other unspecified aluminium faces (Fig. 1b,c). Gamma-alumina crystals are also observed and are of pentagonal (Fig. 1b) or rod-shape (Fig. 1c). The average nucleation density of the γ-Al_2O_3 crystals is estimated to be about 3.2×10^9 and 1.0×10^{10}/cm^2 for the films shown in Fig. 1b and c respectively. Electron diffraction analysis has shown an epitaxial relationship between the aluminium substrate and the γ-Al_2O_3 crystals in these films. However, no further attempts were made to determine the orientation of these aluminium substrates.

An examination of these micrographs reveals immediately an important correlation between the nucleation sites of the γ-Al_2O_3 crystals and the background contrast observed in the amorphous region of the thermal oxide film. It is apparent that the γ-Al_2O_3 crystals are not distributed randomly over the surface, as suggested by Beck et al. [2], but are sited along the fine bands of darker contrast in the furrowed or corrugation patterns. Figure 1b also indicates clearly that the triple points in the darker corrugation pattern, where three furrows meet, are the most favourable sites for the nucleation of the γ-Al_2O_3 crystals.

Figure 2 shows a transmission electron micrograph of the stripped thermal oxide film grown on the (100) face in air at 515°C for 1 h . After 1 h oxidation, the growth of discrete γ-Al_2O_3 crystals is observed in the thermal oxide films grown on the (100) face. The crystals are of sizes up to about 0.1 μm and exhibit very irregular shapes. No specific shape can be assigned for the crystals grown on the (100) face of aluminium. Furthermore, a network of fine cracks is observed in each γ-Al_2O_3 island. The electron diffraction pattern (inset, Fig. 2) indicates a random orientation of the γ-Al_2O_3 crystals. The average nucleation density of the γ-Al_2O_3 crystals is estimated to be about 7.7×10^9/cm^2. A close examination of the micrograph indicates that the γ-Al_2O_3 crystals are again not distributed randomly over the surface, but are present along the darker hexagonal cell pattern observed only faintly in the amorphous oxide film.

The darker furrowed or corrugation patterns observed in the amorphous region of the thermal oxide films have previously been explained as being due to the local and periodic variation in film thickness by Doherty et al. [1]. However, cross-sectional transmission electron microscopy of the aluminium substrate and its air-formed film or thermal oxide film, allied with a novel specimen preparation technique of ultramicrotomy, has shown recently that the explanation given by Doherty et al. is inappropriate [3]. It was found that the electropolished aluminium surfaces are not microscopically smooth, but there appears to be a network of ridges whose pattern is dependent on the grain orientation of the aluminium surface. Furthermore, it was found that the amorphous thermal oxide layer is of uniform thickness perpendicular to the local metal surface. No local regions of thicker film formation or periodic variation in film thickness, as suggested by Doherty et al. [1], were observed in the amorphous thermal oxide layer.

It is evident that the background contrast observed in the amorphous region of the stripped thermal oxide film is not due to the local variation in film thickness, but is due to the presence, in the electropolished aluminium surfaces, of a network of ridges, as first suggested by Randall and Bernard [4]. It is readily understood that when the amorphous thermal oxide films grown on the ridged aluminium surfaces are stripped from the substrate and examined in plan in the transmission electron microscope, i.e. normal to the plane of the entire film, the film grown over the ridges is viewed obliquely, thus giving rise to the relatively dark local contrast in the image.

This interpretation for the origin of the background contrast observed in the amorphous region of the stripped thermal oxide films, together with the present observations given above, leads to an important conclusion concerning the sites for the 'easy paths' for the diffusion of oxygen or nucleation of the γ-Al_2O_3 crystals. It appears that such sites form preferentially in the amorphous oxide layer grown over the pre-existing metal ridges. This suggests strongly that the 'easy paths' for the diffusion of oxygen are cracks which develop in the initially formed amorphous oxide layer. During thermal

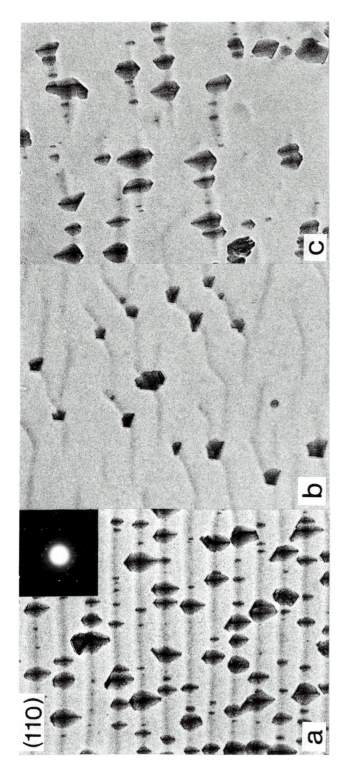

Figure 1: Transmission electron micrographs of the stripped thermal oxide films grown on various faces of aluminium in air at 515°C for 30 min.

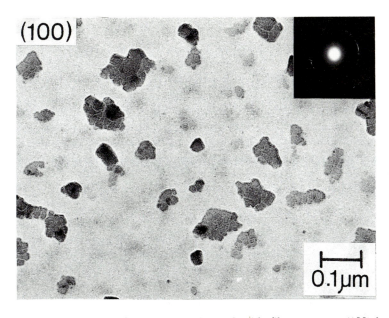

Figure 2: Transmission electron micrograph of the stripped thermal oxide film grown on (100) face of aluminium in air at 515°C for 1 h.

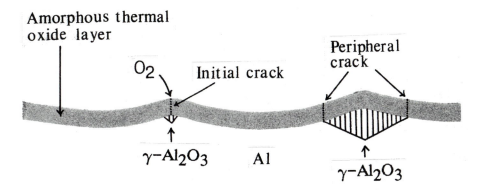

Figure 3: A schematic picture representing the nucleation and growth of γ-Al$_2$O$_3$ crystals in thermal oxide films on aluminium.

oxidation, the amorphous thermal oxide layer is subjected to tensile stresses due to the difference in the thermal expansion coefficient between the oxide and aluminium. Cracking of the amorphous oxide layer represents a route through which the stresses are relieved. There is little doubt that the ridges or triple points of the ridges, where high local stresses are expected, are the most preferable sites for the film cracking. Once cracks develop in the amorphous oxide layer, molecular oxygen has ready access to the metal-oxide interface. The direct reaction between the molecular oxygen diffusing through the cracks and the exposed bare aluminium surface leads to the immediate formation of γ-Al_2O_3 crystals at the amorphous oxide-metal interface as indicated by Henry et al. [5] and by Eldridge et al. [6]. Thereafter, the volume expansion associated with crystalline alumina formation leads to the development of high local strains in the amorphous oxide layer, particularly at the peripheries of the crystals, which would, in turn, contribute to enhanced cracking of the amorphous oxide layer. The peripheral feeding of the molecular oxygen through such cracks allows the lateral and continuous growth of the γ-Al_2O_3 crystals, since new cracks are always being formed at the peripheries of the expanding crystal boundaries as shown schematically in Fig. 3.

Recently, Eldridge et al. [6] have employed an O^{18}/SIMS technique to specify the transport mechanism of oxygen through the 'easy paths' in the amorphous oxide layer. It was suggested that the anionic diffusion of oxygen through the 'easy paths' in the amorphous oxide layer is responsible for the growth of γ-Al_2O_3 crystals in the thermal oxide films on aluminium. If anionic diffusion of oxygen is assumed, it is difficult, however, to understand why the 'easy paths' for the diffusion of oxygen should form preferentially in the amorphous oxide layer covering the metal ridges. Thus, the diffusion of molecular oxygen through the cracks which develop during thermal oxidation in the amorphous oxide layer covering the metal ridges seems the most realistic and acceptable basis for explaining the local growth of the γ-Al_2O_3 crystals in the thermal oxide films on aluminium.

ACKNOWLEDGEMENT

Thanks are due to the Royal Society for the provision of a Guest Research Fellowship to Dr. K. Shimizu.

REFERENCES

1. P.F. Doherty and R.S. Davis, J. Appl. Phys. 34, 1963, p.619.

2. A.F. Beck, M.A. Heine, E.J. Caule and M.J. Pryor, Corros. Sci. 7, 1967, p.1.

3. K. Shimizu, G.E. Thompson, G.C. Wood, A. Gotoh and K. Kobayashi, submitted to Oxid. Met.

4. J.J. Randall and W.J. Bernard, J. Appl. Phys. 35, 1964, p.1317.

5. R.H. Henry, B.W. Alker and P.C. Stair, Solid State Commun. 42, 1982, p.23.

6. J.I. Eldridge, R.J. Hussey, D.F. Mitchell and M.J. Graham, Oxid. Met. 30, 1988, p.301.

19 SIMS INVESTIGATIONS OF THE TRANSPORT PHENOMENA IN CHROMIA AND ALUMINA SCALES ON ODS ALLOYS

W.J. Quadakkers, W. Speier*, H. Holzbrecher* and H. Nickel

Institute for Reactor Materials,
*Department for Chemical Analysis,
Research Centre Jülich,
P.O. Box 19 13, D - 5170 Jülich, FRG.

ABSTRACT

The oxidation behaviour of the ODS alloys MA 754, a chromia former and MA 956, an alumina former, was investigated at 1000 and 1100°C. Main emphasis was placed on the studies of scale growth mechanisms using SIMS analysis of oxide scales grown during a two-stage oxidation where an ^{18}O-enriched gas was used in the second stage. It was found that the oxide scales on both materials grow mainly by oxygen transport, but a significant amount of outward scale growth occurs, especially for the chromia former. Indications have been found that not only differences in the oxidation behaviour of various heats of one material but even differences between the chromia and alumina forming alloys can be attributed to differences in the amount of outward scale growth.

INTRODUCTION

Although nickel- and iron-based Oxide Dispersion Strengthened (ODS) alloys were first developed around twenty years ago, the present demand for materials which can operate at higher temperatures than can normally be achieved with conventional alloys has led to a renewed interest in ODS type alloys. In spite of the clear disadvantages of this group of materials imparted e.g. by the complex powder-metallurgical production process and the anisotropy of mechanical properties, the combination of excellent creep strength and superior oxidation/corrosion resistance of ODS alloys offer the possibility for design of highly stressed components in aggressive environments for higher temperatures than today's standards [1]. This allows an increase of process efficiency in e.g. chemical engineering and energy conversion.

For their high temperature oxidation/corrosion resistance nickel- and iron-based ODS alloys rely on the formation of slowly growing and strongly adherent chromia or alumina scales. The excellent properties of oxide scales on ODS alloys are imparted by the oxide dispersion, usually yttria, which is present in small amounts of around 0,5 - 1 wt%. The mechanism for the significant improvement in oxide scale properties, especially growth rate and adherence, has been subject of a large number of publications. The various mechanisms proposed have been summarized in several seminar papers (see e.g. [2,3]) and will therefore not be extensively discussed here.

Regarding the practical application of ODS alloys at high temperatures, prediction of the oxidation behaviour during long service life based on relatively short time laboratory experiments is of great importance. A reliable extrapolation of scale growth kinetics and spalling behaviour requires a precise knowledge of the oxide growth mechanisms especially the transport phenomena in the scales. The most widely used method for studies of growth processes of oxide scales is that in which inert metal markers are placed on the metal surface prior to oxidation. The location of the markers after oxidation is then determined from scale cross-sections. However, because of the relatively large size of the marker compared to the thickness of chromia and especially alumina scales, markers can give only rough information of the transport phenomena in these scales. Young and de Wit [4] showed that the location of the inert marker strongly depends on the method by which it is applied and therefore marker experiments can lead to erroneous conclusions on the scale growth mechanisms. Brückmann and Gil [5] observed that a vapour deposited gold film on cobalt tended to be changed into liquid-like droplets during oxidation. The location of the gold marker within the scale strongly depended on the droplet sizes and therefore did not allow any reliable information on the oxide growth processes.

Far more accurate information can be obtained from two-stage oxidation experiments in which the atmosphere in the first or second oxidation period is enriched by ^{18}O [4,6]. The location of the ^{18}O in the scale can be determined by nuclear-reaction analysis or secondary-ion mass spectroscopy (SIMS). The two-stage oxidation method in combination with other experimental methods e.g. for studying scale composition and morphology reveals much more detailed and reliable information on transport phenomena in oxide scales than the marker method. Recently Quadakkers et al. [7] used this method for studying the growth mechanisms of chromia and alumina on the ODS alloys MA 754 (Ni-20Cr-0.5Y$_2$O$_3$) and MA 956 (Fe-20Cr-4.5Al-0.5Y$_2$O$_3$) at 1000°C. The behaviour was compared with that of model alloys with the same base composition, but without Y$_2$O$_3$. It was found that in none of the investigated cases was the growth exclusively determined by either cation or anion transport. The scales always grew by a combination of both diffusion processes, but for the ODS-alloys the amount of cation transport was significantly smaller than for the corresponding yttria-free alloy. The oxygen transport in the oxide scales on the ODS alloys occurred via oxide grain boundaries; the transport path for the small amount of cation movement could not be determined with certainty.

Ramanarayanan et al. [8] also studied the growth processes of the oxide scales on the two mentioned ODS alloys, using Pt markers. In both cases they found the marker at the scale-gas interface.

The aim of the present study was a detailed investigation of the growth processes of oxide scales of the ODS alloys MA956 and MA754 in the temperature range 1000 to 1100°C using the two-stage oxidation technique. By combination with detailed analysis of scale structure and composition using SIMS and SEM/EDX, the transport properties are correlated with scale structure and morphology.

EXPERIMENTAL

The ODS alloys MA 956 and MA 754 (supplied by Inco Alloys Limited, Hereford, U.K) were used in the solution-annealed condition. The alloy microstructures are shown in Fig. 1. One heat of each material was used throughout the investigations.

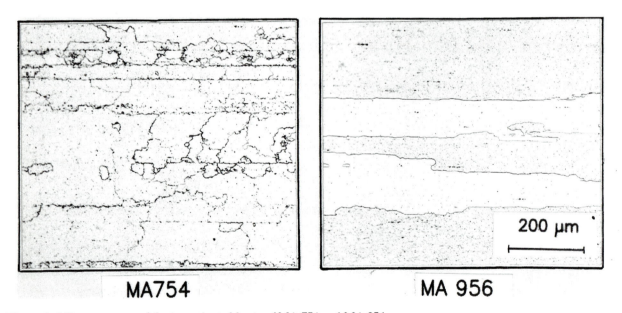

Figure 1: Microstructure of the investigated heats of MA 754 and MA 956.

The detailed analysed chemical composition is given in Table 1. For studies on heat-to-heat variation, four further heats of MA 956 were used in cyclic oxidation experiments and morphology studies.

All alloys were cut into rectangular specimens 20 x 10 x 2 mm, and the surfaces were polished to a 600 grit finish and cleaned in acetone before use. The specimens for two-stage oxidation were polished down to a 1 μm finish using diamond paste.

Three types of oxidation experiment were carried out: 1. Cyclic oxidation up to around 1000 h at 1000 and 1100 °C (56 minutes heating, 4 minutes cooling). 2. Long term oxidation up to 5000 h at 1000 °C in which the specimens were furnace

Table 1. Chemical compositions of the investigated materials

Alloy Composition (wt%)

Alloy/ Heat	Fe	Ni	Cr	Al	Ti	C	S	N	Y
MA 754 (AKA)	0.57	bal	20.6	0.19	0.43	0.063	0.001	0.128	0.42
MA 956 (AKB)	bal	-	19.1	4.25	0.31	0.028	0.008	0.051	0.35

cooled to room temperature at regular time intervals for weight measurements. 3. Short time experiments (<10 h) for studying the time dependence of scale morphology. In all cases the test atmosphere was synthetic air with a moisture content of 9 mbar.

In the two-stage oxidation experiments the specimens were placed in a gas-tight quartz tube and moved into the furnace after the operating temperature had been reached. In the first stage, the specimens were oxidized in synthetic air. Then the quartz tube was evacuated (required time around one minute) and ^{18}O-enriched air was introduced (^{18}O-enrichment was 10 % of total oxygen content). The total oxidation times were chosen such that the eventually formed scales had a thickness of around 0.5 μm. The isotope distribution in the scales formed during the two-stage oxidation was determined by depth profiling using secondary ion mass spectroscopy (SIMS). The equipment used was a Cameca IMSMF with a Cs ion source.

Structural and chemical analyses were carried out using optical microscopy, scanning electron microscopy (SEM), energy dispersive X-ray analysis (EDX) and X-ray diffraction. A few selected specimens were analysed by Rutherford Back Scattering (RBS).

OXIDATION KINETICS

Figure 2 shows linear plots of weight gain vs. time during cyclic oxidation of MA 956 and MA 754 at 1000 and 1100°C. Whereas the chromia forming alloy clearly shows scale spallation especially at the higher temperature, no evidence of scale exfoliation was observed at both temperatures for the alumina forming material.

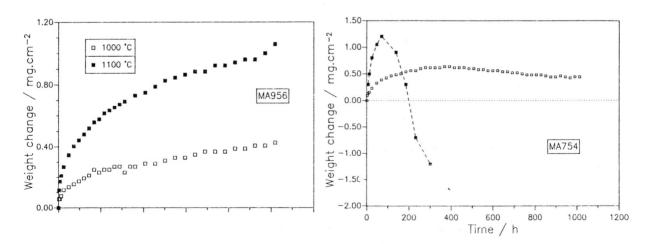

Figure 2: Weight changes vs. time during cyclic oxidation of MA 754 and MA 956 at 1000 and 1100°C.

Figure 3 shows the weight changes of the two materials as a function of time during long term oxidation at 1000°C. The quantitative differences between the materials are similar to those observed during cyclic oxidation. Figure 4 shows typical differences in scale thickness in metallographic cross-sections.

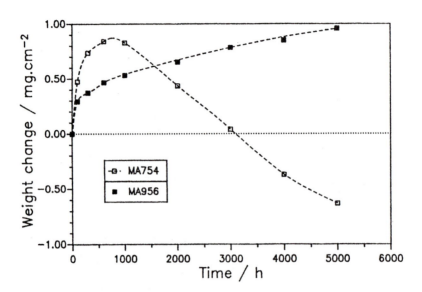

Figure 3: Weight change vs. time during long time oxidation at 1000°C.

TWO-STAGE OXIDATION AT 1000 AND 1100°C

Figures 5 and 6 show the oxygen isotope distributions in the chromia and alumina scales formed during two-stage oxidation at 1000 and 1100°C. As the ^{18}O-enrichment was only 10 %, the measured curves were recalculated to profiles which would have been obtained if pure ^{18}O had been used.

In the scales on MA956 at both temperatures, the ^{18}O is mainly concentrated near the scale-alloy interface. The shape of the curves clearly shows the transport phenomena occurring [7,8]: the scale grows nearly exclusively by oxygen grain boundary transport. Whether the small amount of outward growth occurs via lattice or grain boundary diffusion cannot be determined with certainty from the two-stage oxidation. A further discussion of this question will be given after the morphological studies. Also a possible explanation for the small, but reproducible kink in the outer part of the ^{18}O profile at 1100°C will be discussed later.

For the chromia scales on MA 754 the growth mechanism is not fundamentally different from that observed for the alumina on MA 965: the scale growth also occurs to a large extent by grain boundary oxygen diffusion. However the relative amount of outward scale growth is larger than observed for MA 956. The relative contribution of outward scale growth tends to increase with decreasing temperature.

SCALE COMPOSITON AND STRUCTURE

MA 754

In all cases, X-ray diffraction analysis revealed the formation of single-phased chromia layers. Only after repeated scale spallation could small amounts of a NiCr-base spinel phase be detected. A typical scale morphology after short time oxidation is given in Fig. 7. The scale consists of a flat oxide base on which small nodules are growing. The nodules coarsened with extended exposure time [7]. EDX analysis always showed the presence of titanium in the nodules. Due to the limits of depth resolution, it could not be detected with certainty whether they also contain other elements, such as nickel and yttrium. The coarsening of the nodules can be seen very clearly in the metallographic cross-sections of long term oxidation specimens in Fig. 4.

SIMS depth profiling indicated that yttrium and nickel are enriched in the outher part of the scale (Fig. 8). For a correct understanding of the SIMS-profiles it should be mentioned, that the measured ion intensities are only directly correlated with element concentrations if the sputtering occurs in an unchanged "matrix". The high intensity of nickel near the scale-alloy interface is due to the fact that the sputtering medium changes from oxide to metal.

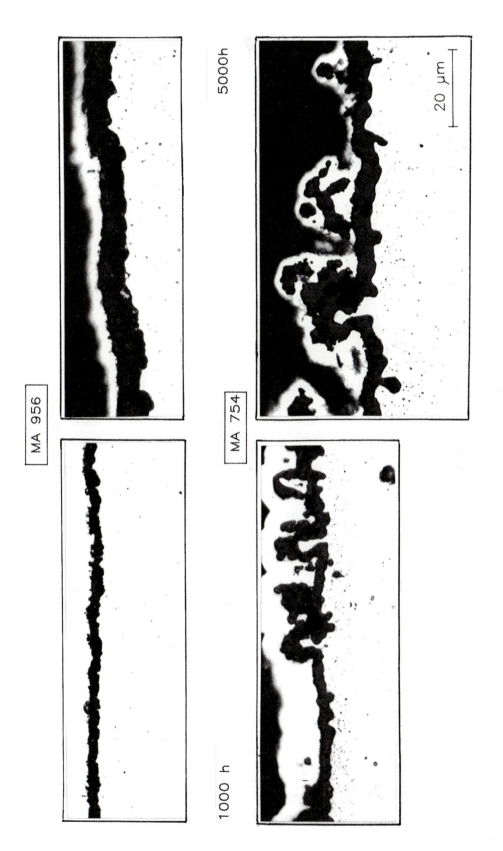

Figure 4: Cross-sections of surface scales on MA 754 and MA 956 after 1000 h (left) and 5000 h (right) oxidation at 1000°C.

MA 956

The alloy formed single-phase alumia layers under all experimental conditions as far as could be detected by X-ray diffraction. At first sight, the scale morphology after short-time oxidation seems to the quite different from that of the chromia layers on MA 754. However, especially at higher magnifications, it can be seen that also in this case the scale consists of a flat base on which very small nodules are growing [7] as can be seen in Fig. 9. Within the detection limits of EDAX, the base scale was pure Al_2O_3. In the nodules, again quite high amounts of titanium were detected, sometimes in combination with yttrium and small amounts of iron. The enrichment of the two last mentioned elements is confirmed by the SIMS-depth profiles (Fig. 10) and could also be detected by RBS.

The nodule formation which leads to a "roughening" of the surface is probably responsible for the small "kink" in the oxygen profile at 1100°C (Figs. 5 and 10). Initially, the surface sputtering incorporates a combination of inward growing

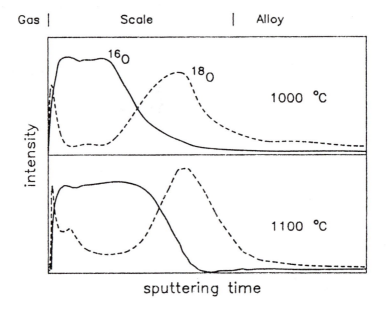

Figure 5: Oxygen isotope distributions in the oxide scales on MA 956 after two stage oxidation.

    1100°C: 2 hours first stage, 4 hours second stage

    1000°C: 3 hours first stage, 9 hours second stage

Measured profiles were corrected to an ^{18}O-enrichment in the second stage of 100% oxygen.

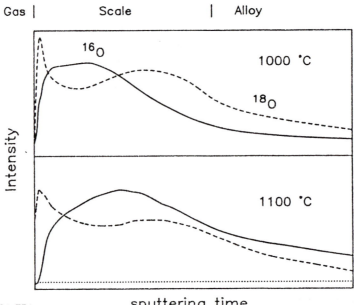

Figure 6: Like Fig. 5 for MA 754.

scale and outward growing nodules. Only after a certain sputtering time is the base scale measured. The correlation of yttrium and iron with the outer part of the ^{18}O-profile before the "kink" occurs, indicates that these elements are enriched in the nodules.

A typical example of the scale composition after long time oxidation is shown in the scale cross-section in Fig. 11. Light coloured, stringer-like phases which sometimes reach from one interface to the other are clearly visible. EDX analysis revealed the presence of titanium and sometimes yttrium. In some cases the presence of small amounts of iron was indicated. By relief polishing of the metallographic specimens, it could be shown that the light phases were present on the oxide grain boundaries (Fig. 11).

DISCUSSION AND CONCLUSIONS

The results have shown that the chromia and alumina scales on the commercial ODS alloys MA 754 and MA 956 at 1000 and 1100°C grow mainly by oxygen grain boundary transport, but especially in the case of the chromia scales a significant amount of outward growth was detected. For both oxides the outward growth does not occur via cation lattice diffusion as has been proposed for FeCrAlY alloys [10]. A comparison of the SIMS results with the morphology studies clearly shows that the outward growth does not occur over the whole scale but only locally by the formation of small nodules on top of the "base scale" which grows by oxygen diffusion. A comparison of the nodule composition as indicated by EDX analysis and SIMS depth profiling with the cross-sectional analysis of the alumina scales strongly suggests that like the oxygen transport the outward scale growth also occurs via oxide grain boundaries.

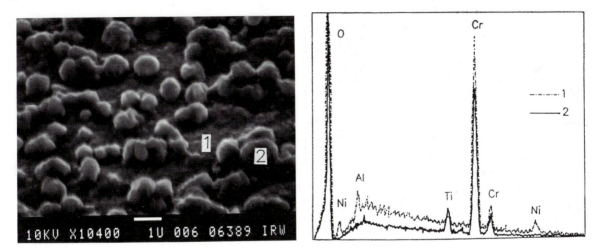

Figure 7: Morphology and EDX-analysis of oxide scale on MA 754 after 5 minutes oxidation at 1100°C.

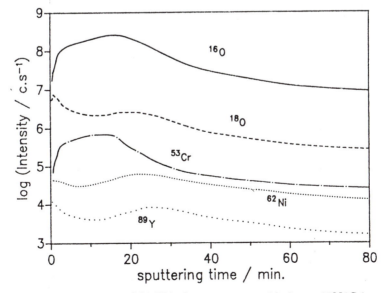

Figure 8: SIMS-depth profiles of oxide scale on MA 754 after two-stage oxidation at 1100°C (compare Fig. 6). Oxygen profiles were not corrected to 100% ^{18}O-enrichment.

Bennett et al. [11] considered the reduction of outward scale growth in chromia scales on yttrium-implanted stainless steel as the main reason for reduced scale growth and consequently improved scale adherence compared to yttrium-free alloys. Ramanarayanan et al. [8] draw similar conclusions for the difference in growth and adherence of chromia on MA 754 and Ni-30Cr based on their Pt marker studies. For the alumina scales on MA 956 and Fe-25Cr-5Al they did not find a correlation between scale properties and growth mechanisms.

Quadakkers et al. [7], showed on similar materials, using the more accurate oxygen tracer method, that both for alumina and chromia the improved scale properties of ODS alloys compared to those of yttria-free alloys are correlated with the transport phenomena in the scales. Not only the decreased growth and improved adherence but also the more selective oxidation of ODS alloys compared to yttria-free material can be explained by assuming that reduced outward scale growth is the dominating effect imparted by the oxide dispersions.

Bennett et al. [11] and Moon [12] showed that implanted yttrium tends to segregate to oxide grain boundaries in chromia scales. It was claimed that due to this segregation the rapid diffusion paths for cations are blocked. Similar mechanisms have been proposed for yttrium-containing alumina formers by Cotell et al.[13]. Segregation studies were not carried out

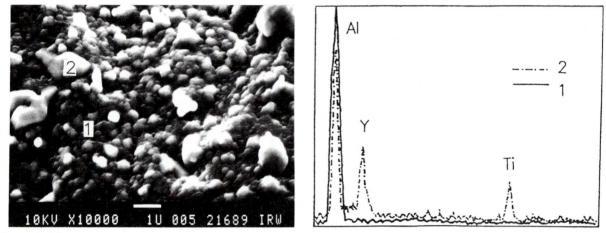

Figure 9: Morphology and EDX-analysis of oxide scale on MA 956 after 1 hr oxidation at 1100°C.

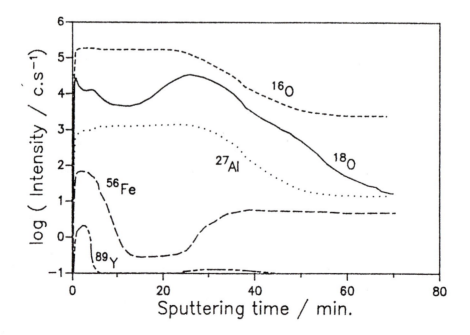

Figure 10: SIMS-depth profiles of oxide scale on MA 956 after two-stage oxidation at 1100°C (compare Fig. 5). Oxygen profiles were not corrected to 100% ^{18}O-enrichment.

in the present work, but the observed local occurrence of yttrium-containing phases in the alumina scale cross sections (Fig. 11) makes an yttrium grain boundary enrichment quite likely.

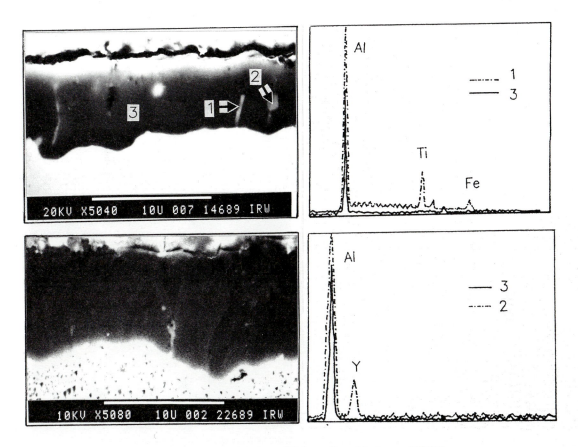

Figure 11: Cross-section of oxide scale on MA 956 after 1000 h cyclic oxidation at 1100°C.

Whether the blocking of cation diffusion paths by yttrium occurs directly or indirectly, e.g. by gettering impurities such as sulphur, cannot be concluded with certainty. Smeggil et al. [14] claimed that poor oxide scale adherence is caused by segregation of sulphur impurity to the scale alloy interface; yttrium and yttria improve scale adherence by gettering sulphur. This mechanism does not, however, explain the changes in growth rate and the selectivity of oxidation imparted by yttria dispersions. Quadakkers et al. [15] and Khanna et al. [16] found indications that sulphur impurites in NiCrAl-alloys enhanced the outward growth of the alumina scales. This would imply that the above mentioned effect of yttria on the scale transport properties might be partly due to an indirect effect such as the gettering of sulphur.

Irrespective of the exact mechanism of cation path blocking, the effective reduction of outward scale growth seems to be a necessary requirement for optimum scale properties with respect to growth and adherence. If this is the case, then the often observed, reproducible differences in long time oxidation behaviour between various heats of one type of ODS alloy might also be correlated with differences in outward scale growth.

Figure 12 shows scale morphologies of five different heats of alloy MA 956 after one hour oxidation at 1100°C. A comparison with the long term cyclic oxidation resistance (Fig. 13) shows a clear correlation between the extent of nodule formation and the oxidation kinetics. Heat BLJ exhibits practically no outward growth and consequently the slowest oxidation kinetics. Heat BJA possesses by far the fastest oxidation rate and at the same time the most pronounced outward scale growth after one hour. Heat AKB, which was used for the experiments described in the previous sections, shows an intermediate behaviour with respect to oxidation kinetics as well as nodule formation.

A detailed discussion of the reasons for the differences in growth kinetics of the various heats will be given in a separate publication. Here it should only be mentioned that the main factors are alloy microstructure, yttria distribution and the amount of titanium which is not tied up in compounds such as Ti(CN).

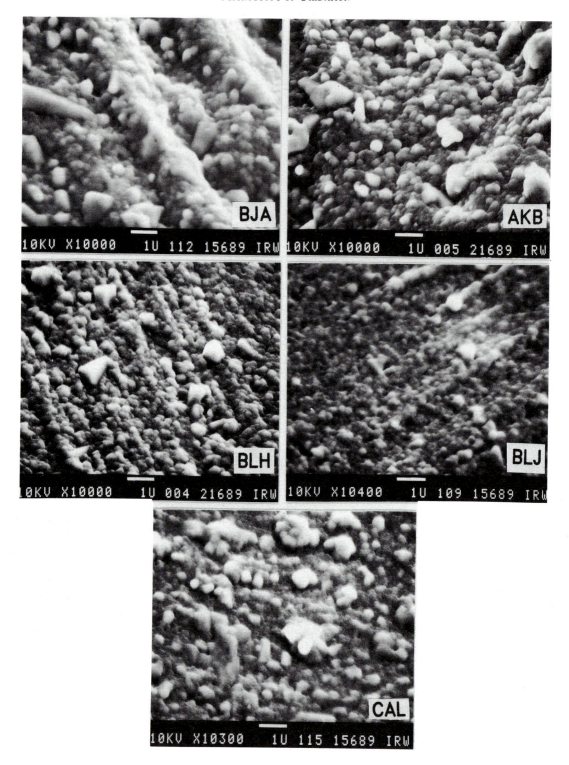

Figure 12: Nodule formation on various heats of MA 956 after 1 h oxidation at 1100°C.

If the differences in scale properties (growth, adherence) of ODS alloys compared to yttria-free alloys with the same base compositions are related to reduced outward scale growth [7] as well as the differences in oxide growth rates between various heats of one alumina forming ODS alloys, one might speculate that the differences between MA 956 and MA 754 could be largely caused by differences of the growth phenomena with respect to outward growth, although the two materials form scales of completely different compositions. If one compares the cross sections of the oxide scales of the two materials in Fig. 4 taking the previous morphological studies into account, the scale on MA 754 seems at first sight to be much thicker than that on MA 956, in agreement with the gravimetric studies (Fig. 3). One should consider, however,

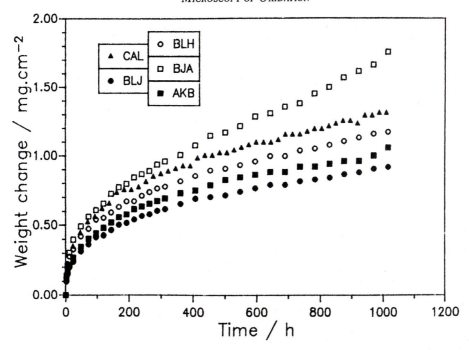

Figure 13: Weight changes vs. time during cyclic oxidation of the various heats of MA 956 shown in Fig. 12.

that the scale on MA 754 consists of an inward growing "base oxide" on which nodules, originating from outward growth, are growing. If one compares this "base oxide" with the nearly exclusively inwardly growing alumina scale on MA 956, the difference in thickness of the scales on the two materials is small. This indicates that it should be possible to design ODS alloys which form chromia scales in which the outward scale growth is as effectively blocked as in the oxide on MA 956. In that case the chromia growth rate and adherence should not be significantly different from that of the alumina on MA 956.

Nickel-chromium based alloys generally possess better high temperature strength than iron-based materials. Alumina-forming nickel based alloys reqiure quite high aluminium contents for obtaining reliably protective alumina scales. As high aluminium contents often have a negative influence on mechanical properties and workability, chromia forming ODS alloys which are optimized in the way mentioned above could be of great benefit in many technical applications. The main temperature limitation might be due to the formation of volatile CrO_3 at temperatures above 1100°C.

ACKNOWLEDGEMENTS

The authors are grateful to their colleagues in KFA/IRW for their assistance in carrying out the experimental work. Mr Baumanns for performing the oxidation experiments, Mr Hoven and Mr Gutzeit for metallographic investigations, Dr Wallura and Mr Els for SEM/EDX analysis and Dr Schulze and Mr Lersch for X-ray diffraction. Mr K. Schmidt from KFA/ISI is gratefully acknowledged for the RBS measurements.

REFERENCES

1. G. Korb, New Materials by Mechanical Alloying Techniques, Ed. E. Arzt and L. Schulz, DGM, Oberursel 1990, p.175.

2. J. Jedlinski, High Temperature Corrosion of Metals, Report Research Centre Julich, FRG, Jül-Conf-76, Eds. W.J. Quadakkers, H. Schuster and P.J. Ennis, August 1988.

3. The Role of Active Elements in the Oxidation Behaviour of High Temperature Metals and Alloys, Ed. E. Lang, Elsevier Applied Science, London, 1989.

4. E. Young and J.H. de Wit, Solid State Ionics, 14, 1985, p.6.

5. A. Brückmann and A. Gil, High Temperature Corrosion of Metals, Report Research Centre Jülich, FRG, Jül-Conf-76, Ed. W.J. Quadakkers, H. Schuster and P.J. Ennis, 1988, p.44.

6. K.P.R. Reddy, J.L. Smialek and A.R. Cooper, Oxid. Met., 17, 1982, p.429.

7. W.J. Quadakkers, H. Holzbrecher, K.G. Briefs and H. Beske, Oxid. Met. 32, 1989, p.67.

8. T.A. Ramanarayanan, R. Ayer, R. Petkovic-Luton and D.P. Leta, Oxid. Met. 29, 1988, p.445.

9. S.N. Basu and J.W. Hallovan, Oxid. Met. 27, 1987, p.143.

10. A.M. Huntz, G. Ben Abdelrazik and G. Moulin, Applied Surface Science, 28, 1987, p.364.

11. M.J. Bennett and A.T. Tuson, Harwell Laboratory Report AERE R 13309.

12. D.P. Moon, Harwell Laboratory Report AERE R 12870, 1987.

13. C.M. Cotell, K. Przybylski and G. Yurek, J. Electrochem. Soc. (extended abstract), 86 - 1, 1986, p.530.

14. J. Smeggil, Mater. Sci. and Eng. 87, 1987, p.261.

15. W.J. Quadakkers, C. Wasserfuhr and A.S. Khanna, H. Nickel, Mat. Sci. and Technol. 4, 1988, p.1119.

20 ALUMINA SCALE GROWTH AT ZIRCONIA/MCrAlY INTERFACE: A MICROSTRUCTURAL STUDY

L. Lelait, S. Alpérine* and R. Mévrel*

LEM CNRS/ONERA, B.P. 72, 92322 Chatillon Cedex, France.
*ONERA Materials Science Department, B.P. 72, 92322 Chatillon Cedex, France.

ABSTRACT

One of the first requirements for the application of thermal barrier coatings on hot stage turbine components is a high bond strength between the metallic bond coat and the ceramic undercoat. In the case of yttria partially stabilized zirconia top coat plasma sprayed on a NiCrAlY bond coat, oxide scale growth at the ceramic/metal interface is a major contribution to the enhancement of interfacial thermomechanical resistance. In order to better understand this phenomenon, microstructural observations of the alumina scales formed at 1100° C and 1200° C under air, between low pressure plasma sprayed NiCrAlY and air plasma sprayed ZrO_2 - 8. 5 wt. % Y_2O_3, have been performed by classical and analytical transmission electron microscopy on transverse thin foil specimens. The evolution of the oxide grain morphology from the metal/oxide to the oxide/oxide interface has been studied. The oxide grains are small and equiaxial near the oxide/zirconia interface, coarser and elongated near the MCrAlY/oxide interface, suggesting that the scale growth principally takes place at the latter interface. Segregation of yttrium at oxide grain boundaries has been detected either as a solute or as $Y_2Al_5O_{12}$ garnet precipitates. Significant quantities of zirconium are also present inside the oxide grains. The oxide growth seems to be dominated by a classical grain boundary oxygen diffusion mechanism. The presence of zirconium inside the alumina grains also suggests that a secondary growth mechanism has taken place: Al_2O_3 partially forms at the oxide/zirconia interface by chemical reduction of ZrO_2 by Al. The comparison between the microstructures observed and that of alumina scales grown under similar conditions on bare MCrAlY alloys gives some insight on how the ceramic top-coat modifies NiCrAlY high temperature oxidation mechanisms.

INTRODUCTION

One of the critical requirements for the application of thermal barrier coatings on hot stage turbine components is a high bond strength between the metallic bond coat and the ceramic overlayer [1]. In the case of yttria partially stabilized zirconia top coat plasma sprayed on a metallic bond coat, oxide scale growth at the ceramic/metal interface is a major contribution to the interfacial thermomechanical resistance of the coating. Plasma sprayed MCrAlY layers (M = Ni and/or Co) have been extensively used as high temperature resistant coatings to lengthen the service life of gas turbine parts operating under oxidizing atmospheres, and in some cases, in the presence of corrosive liquid phases such as sulfates [2]. They are widely used as thermal barrier bond coats, advantage being taken not only of their oxidation resistance, but also of their high temperature plasticity [3]. Although the intrinsic oxidation resistance of MCrAlY alloys has been the subject of numerous papers [4,5], it is to be noted that only limited efforts [6,7] have been dedicated to the study of the interfacial oxide scale growth at the ceramic/MCrAlY interface of thermal barrier coatings.

In fact, one can imagine more than one reason for which oxidation mechanisms could differ in the former and the latter case: the ceramic topcoat could lower the value of the oxygen pressure (PO_2) at the oxide scale external surface; zirconia top coats could also be a secondary source of Y diffusion (and primary source of Zr diffusion) in the growing oxide scale, thereby modifying the boundary conditions for the well-known "active element effect". On this ground, it is interesting to examine if and how the presence of a zirconia top-coat modifies the oxide scale growth mechanisms on MCrAlY alloys. For this purpose, a study was initiated to investigate the microstructure of alumina scales formed by oxidation of YPSZ thermal barrier coatings and to compare it with published data concerning alumina scales grown under similar conditions on a bare MCrAlY alloy [4]. The microstructure of the oxide scales was principally studied by transmission electron microscopy on transverse thin foils. The evolution of the size and morphology of the oxide grains was particularly looked at, since it constitutes an indication of the scale growth mechanisms (diffusing species). Analytical studies were also performed with the hope of observing eventual diffusion of zirconium and/or yttrium in the oxide grains or grain boundaries, whether coming from the metallic bond coat or the zirconia top coat.

MATERIALS AND METHODS

The substrates used are 25 x 40 x 2 mm³ coupons of Hastelloy X superalloy (Ni-22Cr-18.5Fe-6Ta-9Mo-1.5Co-0.6W-0.1C-0.5Si wt.%), which were sand-blasted (Al_2O_3 220 mesh, 2 bars) and degreased (vapor phase trichloroethylene). The substrates were first coated with a 100μm thick Ni-23Cr-6Al-0.5Y(wt.%,AMDRY 963) alloy deposited by low pressure plasma spraying. These samples were then heat treated during four hours at 1080° C under secondary vacuum for bond coat diffusion and covered with a 300 μm thick layer of air plasma sprayed ZrO_2 - 8.5 wt.% Y_2O_3 (MEL SCZ8). Plasma spraying operations were performed by Heurchrome. Oxidation of the specimens was obtained during 100h annealing at 1100 and 1200°C in air followed by air quenching.

Transverse thin foils for transmission electron microscopy (TEM) were prepared following a procedure derived and adapted from the one described by King et al. [8]. This procedure is explained in full detail elsewhere [9]. Jeol 200CX and 4000FX microscopes were used to observe three regions across the oxide layer: near the oxide/zirconia interface, in the central zone of the scale, and near the oxide/NiCrAlY interface. In each case, the distributions of the alumina grain sizes and anisotropy were determined by image analysis routines from a population of at least 100-150 grains. Grain size determination refers to the diameter of equivalent circular grains (computed from an area/perimeter ratio); anisotropy of the grains refers to the ratio of their Feret radii.

Chemical microanalysis was performed on the thin foils by EDS technique (KEVEX device equipped with a"Quantum"diode) using a probe diameter of about 10nm.Finally, elementary microanalysis was performed using a genuine high sensitivity and high spatial resolution ion probe instrument (SIMS, probe diameter ≈80nm) designed at O.N.E.R.A.The general description of this instrument and its performance can be found in [10].

EXPERIMENTAL RESULTS

During annealing an α-alumina scale grows between the NiCrAlY bondcoat and the zirconia topcoat. After 100h at 1100°C (resp. 1200° C) the thickness of the scale is around 4 μm (resp. 6.5 μm).

Oxide Scale Morphology and Microstructure

Typical bright field micrographs of the alumina scale grown at 1200° C, in its inner region (close to the NiCrAlY/Al_2O_3 interface) can be seen in Fig. 1a. Micrographs taken in the median and external regions (close to the zirconia/Al_2O_3 interface) can be seen in Figs. 1b and c respectively. The morphology of the oxide scale significantly changes throughout its thickness. Close to the NiCrAlY, the alumina grains are elongated (typical anisotropy between two and four), columnar and display numerous intergranular pores, whereas near the Al_2O_3/ZrO_2 interface, they are more isotropic (typical anisotropy between one and two) and pore free.

In the inner region, besides the microvoids mentioned above, defects such as dislocation pile-ups can be seen, as well as low angle grain boundaries (Fig. 1d). Very fine spinel oxides (Ni(Al,Cr)$_2$O$_4$ according to SAD measurements and EDS analysis) have also been detected in this region. At the zirconia/alumina interface there is a fine dense microstructure of small equiaxial grains (Fig. 1c): (Al,Cr)$_2$O$_3$, Ni(Al,Cr)$_2$O$_4$ spinels and NiO. They are associated with small zirconia cubic grains already present in the as sprayed specimen. The coexistence of zirconia grains with aluminum based oxides is a consequence of the "roughness" of the oxide/oxide interface.

Alumina Grain Size Distribution

The distributions of grain sizes, determined from TEM observations are reported in Fig. 2; they show that the average grain size regularly increases from the scale external surface to the oxide-metal interface. After 100h at 1100°C (resp. 1200° C), the distribution of alumina grains near the alumina/zirconia interface is relatively sharp, the grain sizes varying from 0.01 to 0.08μm (resp.0.02 to 0.25μm). In the inner region, the size distribution is more scattered, the size limits being 0.02 and 0.35 μm (resp. 0.04 and 1.00 μm). A qualitative comparison between these values and published data concerning the grain size distribution of an alumina scale grown by oxidation at 1100° C during 100h on a bareNi-24Co-23Cr-7.8Al-0.2Ywt.% alloy [4] is possible (although the MCrAlY composition differs somewhat from ours). The average grain size in the latter case is significantly higher than in the former one. In fact, as can be seen in Fig. 2, the grain size distribution of the oxide grown on bare NiCoCrAlY at 1100° C is very similar to that of the oxide scale grown at 1200°C at the NiCrAlY/zirconia interface; so are the respective thicknesses of the oxide scales after 100 h.

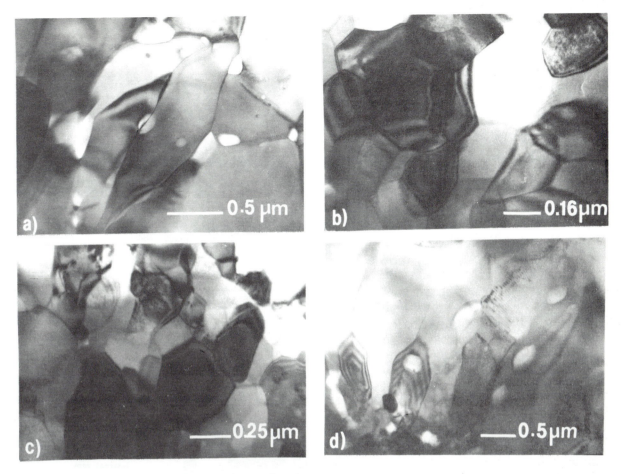

Figure 1 : Microstructure of the alumina scale grown during 100h at 1200°C ; (a) near NiCrAlY/Al$_2$O$_3$ interface, (b) median zone, (c) near Al$_2$O$_3$/ZrO$_2$ interface, (d) typical defects near NiCrAlY/Al$_2$O$_3$ interface

Chemical microanalysis

Chemical microanalysis experiments (EDS and SIMS) show that the oxide scale grown at 1200°C at the zirconia/ NiCrAlY interface is composed of pure α-alumina. The alumina scale is bordered, on the NiCrAlY side, by a fine layer of pure chromia (≈ 0.24 μm thick), visible by EDS (Fig. 3) as well as by SIMS (Fig. 6b). Moreover, most Al$_2$O$_3$ grains contain in their lattice small amounts of chromium.

As shown in Fig. 4 (EDS), the presence of zirconium is detected almost everywhere in the alumina scale, both in the bulk of the oxide grains and in the oxide grain boundaries (within the limitation of a 10 nm EDS probe). A rough estimate of the zirconium content in the Al$_2$O$_3$ grains (using the thin foil approximation) is a few weight percent. It is worth pointing out that the zirconium content decreases further and further away from the alumina/zirconia interface.

Figure 5a and b present detailed energy dispersive X-ray spectra surrounding the high energy Y and Zr peaks, collected from regions near the NiCrAlY/Al$_2$O$_3$ and Al$_2$O$_3$/zirconia interfaces respectively. In each figure, one can see the superimposition of two spectra: one recorded with the 10 nm probe centered in the bulk of a grain, the other one with the probe across a grain boundary. On the NiCrAlY side of the alumina scale (Fig. 5a), an yttrium signal is clearly visible, strongly reinforced in the grain boundary. On the other side of the alumina scale (Fig. 5b), no such reinforcement is visible. Even though the size of the probe is large compared to the thickness of the analyzed grain boundary, this experiment unambiguously reveals yttrium segregation at the oxide grain boundary near the NiCrAlY/Al$_2$O$_3$ interface: the yttrium concentration as indicated by this analysis is of course lower than the real yttrium concentration inside the grain boundary, because of the probe size "convolution" effect. The presence of yttrium in the oxide scale is also detected in the form of small Y$_3$Al$_5$O$_{12}$ garnets lying in some grain boundaries of the inner region. Some coarser particles can be seen on SIMS yttrium map (Fig. 6c). The range of yttrium segregation is however too narrow to be properly imaged by SIMS, even with a high spatial resolution probe (size ≈80 nm).

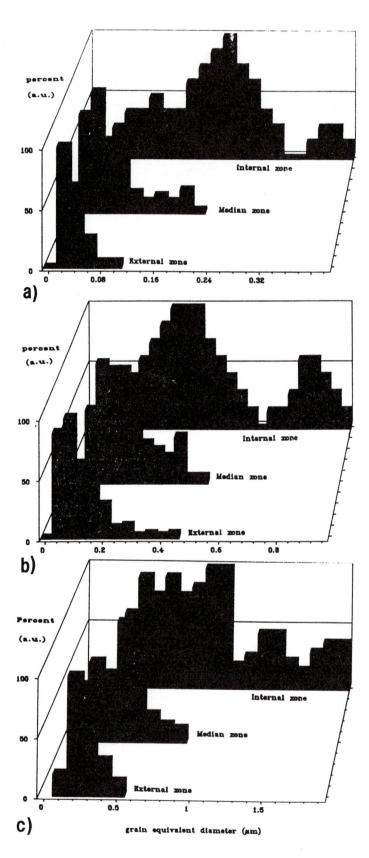

a)

b)

c)

Figure 2 : Histograms of grain size distribution ;
(a) 100h at 1100°C,
(b) 100h at 1200°C,
(c) 100h at 1100°C on bare NiCoCrAlY (from [4]).

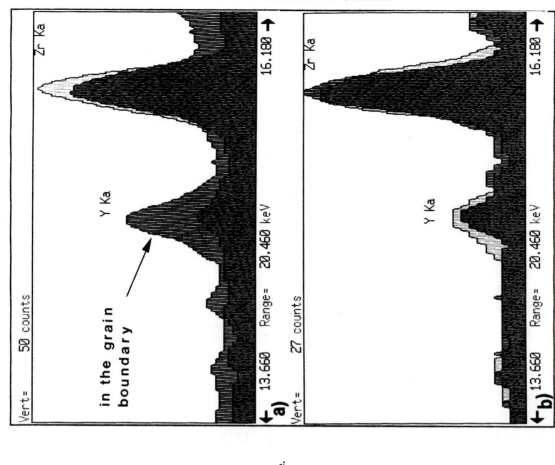

Figure 5 : EDS spectra in the Y and Zr high energy peaks region ; superimposition of spectra recorded in the grain bulk and at a grain boundary ; (a) near NiCrAlY/Al₂O₃ interface, (b) near ZrO₂/Al₂O₃ interface.

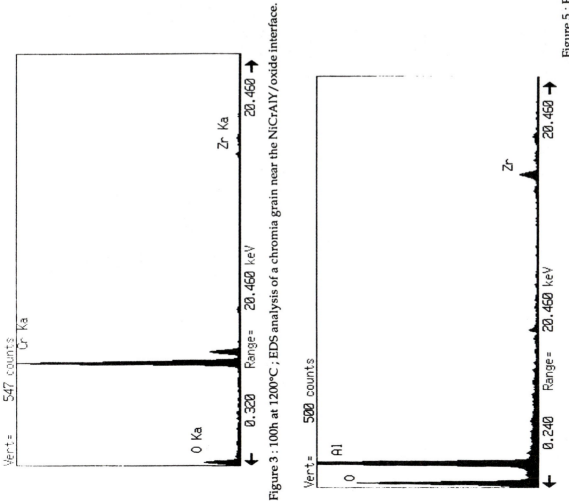

Figure 3 : 100h at 1200°C ; EDS analysis of a chromia grain near the NiCrAlY/oxide interface.

Figure 4 : EDS analysis in an alumina grain showing the presence of zirconium.

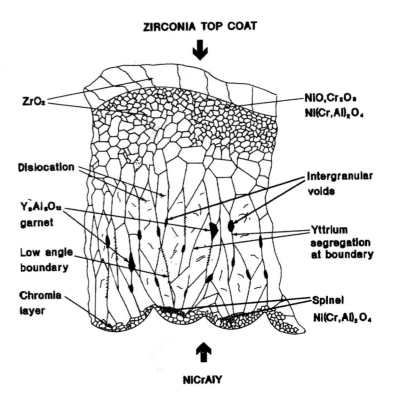

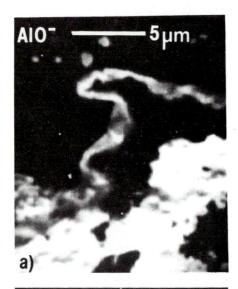

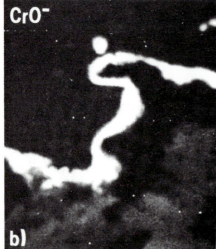

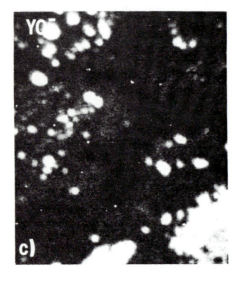

Figure 7 : Schematic representation of the oxide scale grown during 100h at 1200°C at the ZrO_2/NiCrAlY interface.

Figure 6 : SIMS imaging of the oxide scale grown during 100h at 1200°C ; (a) AlO⁻ secondary ions image, (b) CrO⁻ image, (c) YO⁻ image.

DISCUSSION

The principal microstructural features of the alumina scales grown at the ZrO_2/NiCrAlY interface are schematically represented on Fig. 7. The mechanisms of formation of the oxide scale in our NiCrAlY-YPSZ system can be discussed on the basis of these microstructural observations and also from a comparison between the microstructures of the oxide scale grown on the duplex system and on bare MCrAlY [4]. The structural modifications which could be attributed to the presence of a zirconia top coat essentially consist of a modification of the grain size distribution inside the oxide scale (smaller grains) and inward diffusion of zirconium (and to a lesser extent yttrium) inside the alumina grains. The microstructure of the oxide scales grown at the zirconia/NiCrAlY interface needs to be discussed in terms of oxide growth mechanisms, oxide composition and scale plasticity.

Alumina Growth Mechanisms

The presence of large elongated grains near the metal/oxide interface indicates that the oxide growth occurs predominantly at this interface (at 1100°C as well as at 1200° C). Besides, the presence of yttrium segregation in the oxide grain boundaries — either as $Y_3Al_5O_{12}$ precipitates or on a finer scale — strongly suggests that the oxide growth mechanism is one of oxygen diffusion at the oxide grain boundaries (Al diffusion in the grains volume being slow). Indeed, the presence of an "active element" at the oxide grain boundaries —although not detected by Choquet by EDS on oxide thin foils, due to instrumental limitations — must enhance grain boundaries oxygen diffusion, as suggested by this author [11].

In the case of oxide growth at a MCrAlY/Zirconia interface, however, this oxide growth mechanism could well not be the only one involved. The presence of significant amounts of zirconium (and lesser amounts of yttrium) inside the oxide layer, with a positive concentration gradient from the inner to the outer part of the oxide scale, shows that, at 1100 and 1200° C, part of the alumina layer has formed by reduction of the partially stabilized zirconia layer, which is thermodynamically allowed at these temperatures (at 1200°C, $\Delta GZrO_2 = -18.0$ kJ.mole^{-1} $\geq \Delta GAlO_3 = -19.4$ kJ.mole^{-1}). Zirconium and yttrium atoms thus "trapped" in the alumina layer can further diffuse towards the inner zone of the scale. The secondary growth mechanism, which takes place at the oxide/zirconia interface is probably of limited extent (as indicated by the grain morphology typical of a scale growth at the metal/oxide interface), since it is limited by the slow volume diffusion of Al atoms through the oxide scale.

Other Oxides

The presence of a thin chromia layer at the NiCrAlY/oxide interface, after 100h at 1200°C, is also somewhat unexpected (Cr_2O_3 being less stable than Al_2O_3 at this temperature). It probably forms towards the end of the oxidizing treatment after important aluminum depletion of the NiCrAlY has taken place preventing the further formation of a continuous alumina layer. This argument is supported by the relatively low initial aluminum content of the bond coat alloy (6.5 wt. %) and by TEM. EDS measurements in the NiCrAlY near the metal/oxide interface (confirming important aluminum depletion during Al_2O_3 scale formation). Some chromium has also diffused and substitutes for aluminum in the Al_2O_3 grains.

Other oxides than alumina and chromia have been identified in the oxide scale. The presence of nickel oxides and very small spinels in the outermost region of the scale is not surprising: they were formed during the early stages of oxidation, when a stationary oxidation regime is not yet established, and pushed back as the oxide growth proceeds. On the other hand, the presence of $NiAl_2O_4$ spinels near the MCrAlY/oxide interface is not so easily explained. One notices that the small spinel grains are localized in convoluted regions of the alumina scale. The growth of such oxides could thus be due to a stress induced modification of the oxides respective stability. Another possible cause of this spinel formation could be a local Al depletion in an alumina encapsulated metallic "pouch".

Plasticity

The presence, in relatively large quantities, of dislocation pile-ups inside the oxide scale grown at 1100°C and 1200° C is not easy to interpret. Choquet et al. [4] demonstrated that the high value of residual stresses ($\approx -5.7 \pm 0.4$ GPa) measured on an alumina scale grown at 1100° C on a bare MCrAlY corresponded to the calculated thermal stresses, this means that the level of oxide growth stress, at the oxidation temperature, is likely to be lower than the flow stress of alumina at this temperature. It is thought therefore that plastic deformation inside the scale could originate during cooling, at intermediate temperature, as a result of thermal stresses. According to [12] the flow stress associated with prismatic slip varies from about 500 MPa to 1 GPa when the temperature varies between 1200 and 900° C. Another possible cause for the generation of dislocations inside the alumina layer could be a stress relaxation phenomena occurring during thinning of the sample.

CONCLUSION

The microstructure of oxide scales grown at 1100 and 1200° C at the NiCrAlY/zirconia interface of thermal barrier coatings has been studied by classical and analytical transmission electron microscopy, on transverse cross sections, as well as by high resolution SIMS. The oxide grain morphology, comparable to that of alumina scales grown on bare MCrAlY at 1100° C, is typical of oxide growth at the metal/oxide interface. Segregation of yttrium at oxide grain boundaries has been observed, suggesting that the scale mainly forms by oxygen grain boundary diffusion. Significant amounts of zirconium inside the alumina grains have also been observed sustaining the hypothesis that the oxide scale has also partially grown by reduction of the zirconia top coat. The large quantities of defects observed in the alumina scale formed at 1200° C prove that at these temperatures, a significant plasticity of yttrium, zirconium and chromium doped alumina can be obtained. These microstructural studies show that the scales developed by oxidation at 1100 and 1200°C on thermal barrier coatings have most of the required characteristics for an efficient fracture toughness enhancement of the bond coat/top coat interface.

ACKNOWLEDGEMENTS

The authors gratefully acknowledge the help of F. Hillion for experimental SIMS analysis and J.L. Pouchou for E.D.S. spectra interpretation .

REFERENCES

1. R. A. Miller, J. L. Smialek and R. G. Garlick, in Advances in Ceramics, Vol. 3, Eds. A. H. Heuer and L. W. Hoobs, TheAmerican Ceramic Society, Colombus, OH, 1981, p. 241.

2. S. Stecura, NASA Technical Memorandum 86905.

3. J-M. Veys, A. Rivière and R. Mévrel, First Plasma-Technik Symp., Lucerne, Ed. H. Eschnauer, Vol. 2, 1988, p. 115.

4. P. Choquet and R. Mévrel, Mat. Sci. Engng, A1 20/A121, 1989, p. 153.

5. T. E. Ramanarayanan, M. Raghavan and R. Petkovic-Luton, J. Electrochem. Soc., 131 (4), 1984, p. 923.

6. D. S. Shur, Ph. D. Thesis, Case Western University, Cleveland USA, 1984.

7. G. Tremouille, J-L. Derep, R . Portier, Proc. of the 11th Int. Therm. Spray. Conf., Montreal, Canada, September 1986, p.445.

8. W. E. King, N. L. Peterson and J. F. Reddy, Proc. Int. Cong. on Metallic Corrosion, Toronto, Canada, June 1984, p. 4.

9. L. Lelait, S. Alpérine, C. Diot and R. Mévrel, Mat. Sci. Engng, A1 20/A121, 1989, p. 475.

10. G. Slodzian, B. Daigne, F. Girard and F. Boust, Proc. of the 6th Int. Conf. Secondary Ion Mass Spectrometry, Eds. A. Benninghoven et al., John Wiley, Versailles, France, Sept. 1987, p. 189.

11. P. Choquet, Thèse d'état, Orsay, France (1988).

12. P. Castaing, J. Cadoz and S. H. Kirby, J. Phys., 42, C3-6, 1981, p. 43.

21 MORPHOLOGY OF OXIDE SCALES ON Ni₃Al EXPOSED TO HIGH TEMPERATURE AIR

J.H. DeVan*, J.A. Desport and H.E. Bishop

AEA Technology, Harwell Laboratory, Didcot, Oxon OX11 0RA, UK.
*Oak Ridge National Laboratory, Oak Ridge TN 37831, USA.

ABSTRACT

Microscopic observations of oxide scale growth on Ni₃Al-based alloys are described to explain the effect of temperature on scale morphologies and the beneficial effect of 8% chromium on oxidation properties at 600 to 760°C. These observations are also used to examine the role of internal oxidation processes and remedial effects of chromium on the susceptibility of Ni₃Al to grain boundary embrittlement in air at 600 to 760°C.

INTRODUCTION

The long-range ordered alloy Ni₃Al affords attractive mechanical and corrosion properties for elevated-temperature applications. Although brittle intergranular fracture at ambient temperatures impeded its initial development, this problem was overcome by microalloying with boron, a grain boundary segregant [1]. A subsequent ductility problem manifested at 500 to 800°C, was traced to stress-induced embrittlement of grain boundaries by oxygen. The loss of ductility in air was accompanied by a change in the fracture mode from transgranular to intergranular. This oxygen effect was ultimately eliminated by the addition of 8% chromium to the base Ni₃Al composition [1]. Adding zirconium or hafnium increases the creep and tensile strength at high temperatures.

Given the critical role of chromium in resisting embrittlement by oxygen, it is important to examine in detail the effects of chromium on the oxidation properties of Ni₃Al. This paper compares the oxidation behaviour in air of the chromium-modified composition with the base Ni₃Al alloy in the temperature range 600 to 900°C. Experimental data are drawn from thermogravimetric and associated metallographic studies by the authors as well as from microscopic studies conducted by other workers at Oak Ridge National Laboratory and Harwell Laboratory.

EXPERIMENTAL

Compositions of the alloys investigated are listed below (in at. %) and their fabrication and processing are described elsewhere [1].

Heat Number	Al	Cr	Hf	Zr	B	Ni
IC-136	22.9	-	0.5	-	0.24	bal
IC-219	16.8	8.0	0.4	-	0.10	bal
IC-326	17.1	8.0	-	0.5	0.10	bal

Oxidation studies of IC-136 and IC-219 were conducted on rectangular coupons 12 x 6.2 x 0.55 mm thick in mullite tube furnaces. All major surfaces of the coupons were mechanically polished before testing, ending with a 1-μm diamond abrasive. Coupons were heated to the designated oxidation temperature in pure argon, and the argon was then replaced with an air atmosphere constantly refreshed at a rate of 300 cm³/min. Specimen exposures ranged from <10 min to 2 weeks, with weight changes continuously recorded during the longer duration tests. After oxidation, specimen surfaces and cross sections were examined by some or all of the following techniques: optical microscopy, scanning electron microscopy (SEM), secondary ion mass spectroscopy (SIMS), Auger electron spectrography (AES), and X-ray diffraction. Additional information was obtained from microscopy on fractured surfaces of crack-growth specimens of heats IC-136 and IC-326 performed under the direction of Charles Hippsley at Harwell Laboratory.

METALLOGRAPHY OF "STEADY-STATE" OXIDE SCALES

Two oxidation regimes were identified in thermogravimetric studies, one relating to exposure temperatures of 900°C and above, the other to temperatures from 600 (the lowest studied) to 825°C. Figure 1 shows the appearance of the higher-temperature oxide scale formed on Ni_3Al, which consisted of a very thin, continuous Al_2O_3 inner layer, an even thinner continuous $NiAl_2O_4$ outer layer, and occasional mounds of NiO external to the $NiAl_2O_4$. The Al_2O_3 inner layer extended into the alloy along grain boundaries connecting to the surface and at the points of greatest penetration, encapsulated islands of ZrO_2 or HfO_2. The oxidation characteristics of the chromium-containing alloy (IC-219) at 900°C and above were quite similar to those of Ni_3Al in terms of both weight change and scale morphology. Chromium appeared to be associated with the outer spinel layer, and the outer mounds of NiO were reduced in number.

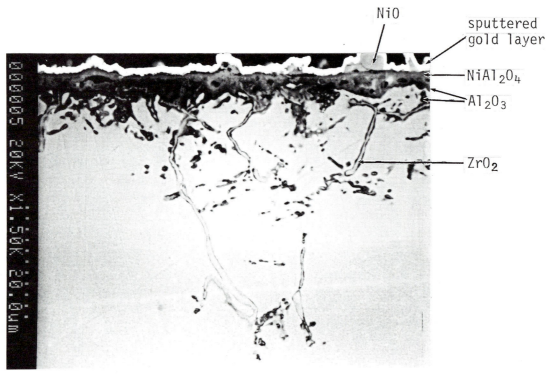

Figure 1: Backscattered electron micrograph of typical scale morphology on Ni_3Al containing 0.5% Zr after exposure to air for 168 h at 1050°C. (1500 x)

At the lower temperatures (≤825°C) the oxide scales were considerably thicker than the scales described above, and the weight changes of the specimens were correspondingly greater. A typical scale formed on Ni_3Al exposed for 196 h at 760°C is shown in Fig. 2. In this case the scale consists of a relatively thick NiO outer scale and an inner layer comparable in thickness to the outer except at prior metal grain boundaries, where the inner oxide depth noticeably recedes (Fig. 2). Depending on the exposure time and temperature, an extremely thin $NiAl_2O_4$ layer and, in some cases, a thicker metallic nickel layer formed along the boundary between the inner and outer layer. The outer layer was readily identified as NiO, but the inner layer was more difficult to resolve. Except when the thin $NiAl_2O_4$ boundary layer was present, X-ray diffraction gave no patterns for an aluminum-containing oxide, and we were unsuccessful in isolating sections of the inner scale for electron diffraction studies. However, from electron microprobe examination of the scales and underlying metal, and observing that the inner scale was ferromagnetic, we eliminated all thermodynamically possible combinations of Ni, Al, and O except one - disordered Ni-Al with submicroscopic Al_2O_3 grains. Failure of X-ray diffraction to show any patterns for Al_2O_3 suggests that the crystallite size of the Al_2O_3 particles must be <6 nm.

The oxides formed on the chromium-containing alloys at lower temperatures were notably thinner than those on Ni_3Al and weight changes were correspondingly smaller. However, as shown in Fig. 3, the morphology of the oxide scale was similar - a nickel-rich outer layer and an internal oxidation zone consisting of an aluminum-depleted disordered alloy with submicroscopic Al_2O_3 precipitates. As in the case of Ni_3Al, the weight changes accompanying air oxidation of the chromium-containing alloy were greater at 760°C than at 900°C; however, the magnitude of increase at the lower temperature was considerably smaller when chromium was present.

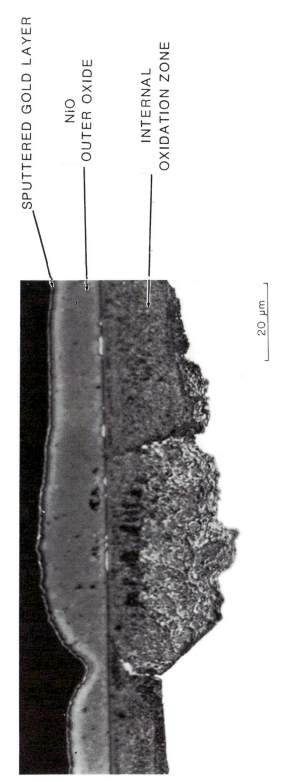

Figure 3: Photomicrograph of Cr-modified Ni₃Al (IC-219) exposed 236 h to air at 760°C. (1000X)

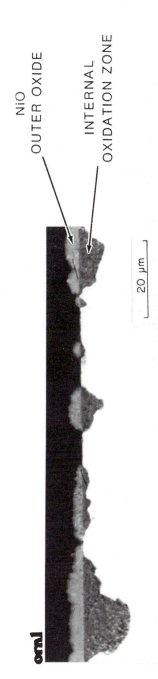

Figure 2: Photomicrograph of Ni₃Al (IC-136) exposed 195 h to air at 760°C. (1000X)

To examine the interface between the inner scale (internal oxidation zone) and the underlying alloy, an Ni₃Al specimen (oxidized at 760°C) was deeply acid-etched using an iodine-methanol mixture. This treatment selectively attacked the alloy along the desired interface and very quickly parted the oxidized region and underlying metal. SEM examination of the two previously joined interfaces revealed patches of filamentary Al_2O_3 layers on the underside of the inner oxide and resolvable Al_2O_3 freestanding particles on the etched metal surface. The thin Al_2O_3 layers were most well formed along prior metal grain boundaries that had intersected the original metal surface. These findings indicate that oxidation conditions along the front of the advancing internal oxidation zone are beginning to support the growth of a continuous oxide layer in place of dispersed Al_2O_3 particles, a condition that is achieved much earlier above prior metal grain boundaries. Thus, although the rates of oxidation at 600 to 825°C became parabolic and showed an Arrhenius dependence on temperature, they may not be truly steady-state rates. Still lower parabolic rate constants may be achieved at longer times.

In addition to exposure time, residual cold work is another factor that can affect the transition between internal oxidation and layered growth. Two of the edges across the specimen had been sheared after the final anneal, and the oxide morphology on these edges was very different from that on the as-annealed faces. The internal oxidation zone seen on the annealed edge (Fig. 1) was completely missing on the sheared edge, which showed only an extremely thin Al_2O_3 film. This suggests that local cold working and associated dislocations provided short-circuit diffusion paths to accelerate the movement of aluminum to the exposed surface.

METALLOGRAPHY OF SHORTER-TERM OXIDE SCALES

Studies of oxidation processes at shorter times (≤10 min) were conducted to gain an understanding of the mechanisms of the embrittlement of Ni₃Al in air at 500 to 850°C. Initial tests were conducted in air at 760°C for 5 to 30 min, and parallel examinations were made of the shoulder regions of fractured tensile specimens that had been exposed to air for comparable periods at 760°C. X-ray diffraction of Ni₃Al specimens even after 5 min showed the formation of NiO and a disordered Ni-Al alloy, but again there were no peaks that could be associated with an aluminum-containing oxide. SIMS profiles were obtained by sputtering the near-surface region with argon and oxygen ions, respectively. The outermost region of the surface (≤0.15 μm) was enriched in nickel and oxygen and below this was a thicker region enriched in oxygen and, to a lesser extent, aluminum. A tapered cross section of an Ni₃Al specimen exposed for 5 min at 760°C was prepared for imaging analysis of the near-surface region using a liquid metal (gallium) ion source combined with SIMS [3]. The latter examination showed that the exposed surface was covered by a submicron NiO outer layer and a thicker internal oxidation zone that reached a uniform depth of ~1.7 μm with deeper penetration along grain boundaries, as shown in Fig. 4.

The problem of oxygen-induced embrittlement at a growing crack tip concerns even an earlier stage of oxidation than that addressed above. Using electropolished TEM disks, Horton et al. [5] compared the surface characteristics of Ni₃Al (similar to IC-136) with Ni₃Al containing 8% chromium (similar to IC-326) after 45 s of oxidation in air at 600°C. Evidence of internal oxidation was found in both the Cr-modified and unmodified alloys. Although the addition of 8% chromium to Ni₃Al reduced the depth of oxygen penetration in the disks by a factor of two, there was no indication that a uniform oxide layer had developed. Nickel-enriched mounds with NiO on the surface were formed on both the standard and Cr-modified Ni₃Al. Al_2O_3 or $(Cr,Al)_2O_3$ formed preferentially between the mounds and along grain boundaries, rejecting nickel from both areas. Related studies of oxide scales formed on Cr-modified Ni₃Al (IC-326) were conducted by Hippsley et al.[4]. In this case, specimens were exposed to air for 5 h at 600 to 760°C. X-ray diffraction analysis of a specimen exposed at 760°C showed strong peaks corresponding to NiO. Much weaker peaks could be attributed to Cr_2O_3, but there were no diffraction patterns corresponding to aluminum-containing oxides. SIMS depth profiles of this specimen indicated at least two oxide types: one close to the surface enriched in nickel and the other, lying below, enriched in chromium, aluminum, zirconium, and possibly boron. Another depth profile of the outer oxide using XPS showed that it contained an excess of nickel over the stoichiometric level in NiO and a level of ~15 at. % aluminum. Chromium was detected by XPS within the first 1 to 2 μm of this layer but not below this level. The oxide scale formed on the chromium-modified alloy after 5 h at 650°C was very similar to that at 760°C, as determined by SIMS and XPS results, but at 600°C oxidation was very superficial and discontinuous.

Analyses of oxides formed on intergranular fracture surfaces after cracking are generally similar to the above results for free surfaces. SIMS depth profiles on fracture surfaces from 700°C air tests [4-6] showed slight nickel enrichment at the oxide air interface for both the Cr-modified and unmodified alloy. However, nearer the crack tip in the Ni₃Al alloy a continuous nickel ribbon containing no detectable oxygen or aluminum was observed along the original grain boundary.

OXYGEN ION IMAGE

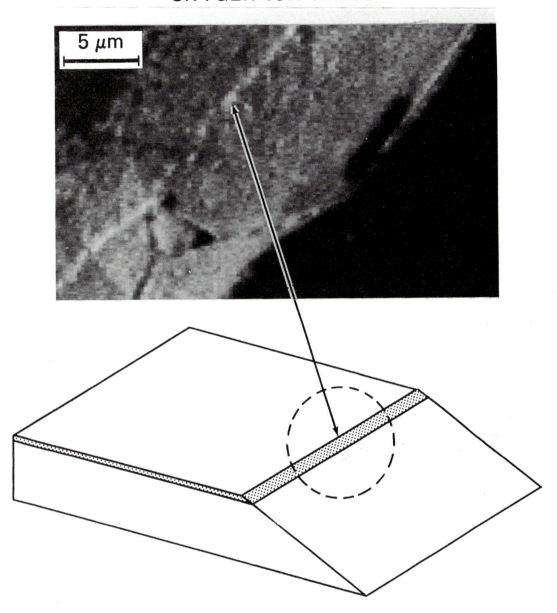

Figure 4: SIMS image of oxygen distribution near Ni$_3$Al (IC-136) surface after 5 min exposure to air at 760°C.

On either side of this nickel ribbon and continuing into the grain boundary ahead of the crack tip was an oxygen diffusion zone enriched in aluminum [6]. In contrast, no layers or diffusion zone could be detected near the crack tip of the Cr-modified alloy fractured at 760°C [4].

If these oxide morphologies are compared with oxide scales developed after longer times at 600 to 760°C, there are important similarities in terms of internal oxidation and chromium effects. In all cases, the dominant alloying elements, Ni and Al, appear to react similarly in the unmodified and Cr-containing alloy, forming kinetically favoured NiO through which oxygen diffuses to react with aluminum to form particulate Al$_2$O$_3$. The presence of chromium is manifested by the mixing of Cr$_2$O$_3$ in the Al$_2$O$_3$ region below or, in earlier stages, surrounding the NiO.

EFFECT OF CHROMIUM ON OXIDATION

In previous reviews [7,8] covering the beneficial effects of chromium on the oxidation of aluminum-containing alloys, two possible factors have been cited:

1. Through the rapid formation of a protective Cr_2O_3 layer, chromium acts to reduce the flux of oxygen to the underlying alloy and accelerates the growth of a coherent Al_2O_3 layer below the Cr_2O_3 layer.

2. When multiple reactive species are present (i.e. Cr and Al), the faster reaction of one of the species (Cr) with oxygen sacrificially prevents depletion of the remaining species (Al) and thereby promotes layered growth of the oxides of the remaining species.

Both factors accord with experimental observations in alloy systems containing much lower aluminum and higher chromium contents than the alloys studied here. However, they do not appear applicable to the effect of chromium on the oxidation of Ni_3Al alloys. At temperatures of 900°C and above, there is essentially no effect of 8% chromium on the oxidation of Ni_3Al in air, and oxide morphologies differ only in the replacement of small amounts of aluminum by chromium in the Al_2O_3 and $NiAl_2O_4$ scale components. At higher temperatures, the aluminum content in these alloys is obviously sufficient to support the growth of a highly resistant alumina scale with or without chromium. As the temperature decreases, conditions change from layered growth of alumina to internal oxidation with correspondingly increased oxidation rates for both the unmodified and Cr-modified alloys. In this case the 8% chromium addition markedly reduces the oxidation kinetics compared to the unmodified alloy. However, the oxidation kinetics of the modified alloy are still faster at 760°C than at 900°C. Furthermore, the oxide scale is similar in morphology to the unmodified alloy, i.e. alumina appears to be in particulate form with no evidence of a coherent chromium-containing layer, and a NiO outer layer occurs with or without chromium, although the growth rate is significantly reduced when chromium is present.

The metallographic findings provide no direct evidence of how chromium limits the extent of internal oxidation. Microprobe analyses of the unmodified alloy showed no indication of an aluminum concentration gradient either across the internally oxidized region or on approaching it from the metal interface below it. This discounts any possible role for chromium as a "sacrificial" reactant (factor 2 above), not a surprising finding in view of the relative concentrations of aluminum and chromium. One possible effect of chromium could be the formation of a microscopically unresolvable barrier layer that initially impedes the adsorption of oxygen and thereby accelerates the growth of an alumina layer ahead of the internal oxidation front. Such a chromium-enriched layer might also account for the slower growth of the NiO outer layer. Auger analyses have in fact shown traces of chromium on the gas exposed surface of the outer NiO oxide layer. Once the NiO layer is established, however, growth of any Cr-containing external layer must immediately stop, since chromium from the underlying metal would be oxidized before it could move through the NiO. Another role of chromium may relate to slowing the movement of oxygen through the internally oxidized zone or nickel through the NiO layer. For example, the presence of chromium could reduce the extent of supersaturation of oxygen and thus the concentration driving force for oxygen diffusion or, alternatively, it could simply reduce the mobility of oxygen in this region.

Similar speculations can be advanced to explain the beneficial effects of chromium on susceptibility to grain boundary embrittlement in air. In this case, the mechanisms must limit the ingress of oxygen into grain boundaries intersecting the exposed free surfaces. Again, metallographic findings on fracture surfaces are inconclusive as to the role of chromium; however, they do give evidence that diffusion of oxygen ahead of the crack tip occurs in the case of Ni_3Al [6] but not in the case of the Cr-containing alloy [4].

CONCLUSIONS

1. Above 825°C oxide scales on Ni_3Al alloys consist of a relatively thin $NiAl_2O_4/Al_2O_3$ continuous layer joined to a deeper Al_2O_3 grain boundary network in the alloy near the exposed surface. ZrO_2 and HfO_2 are also observed as grain boundary reaction products contained within the Al_2O_3 network. The addition of 8% chromium to the alloy does not affect the growth rate or morphology of the scale.

2. Below 825°C oxidation of Ni_3Al proceeds by the outward growth of a uniform NiO layer and by penetration of oxygen into the alloy below the original alloy surface. The addition of 8% chromium markedly slows the growth rate of the NiO layer and the depth of internal oxidation but does not produce a coherent Cr-containing oxide scale.

3. Internal oxidation below 825°C initially proceeds faster in the grain boundaries than in the adjoining metal matrix. However, the higher rate leads ultimately to the growth of an Al_2O_3 film on the boundary, and at longer times the penetration of oxygen in the near-boundary region is actually reduced compared to the matrix.

4. Microscopic examinations to date have been inclusive in pinpointing the mechanism by which chromium attenuates oxidation and air-accelerated crack growth in Ni$_3$Al at 500 to 760°C. The effect of chromium is manifested without an observable change in the oxide morphology and without the development of a robust Cr-containing barrier layer seen in alloys with higher chromium and lower aluminum contents. It is uncertain whether the chromium effect is manifested through a transient barrier film that shortens the transition from internal oxidation to layered Al$_2$O$_3$ growth or by directly affecting the transport processes normally controlling internal oxidation.

5. Short-term oxidation exposures. as depicted by examinations of TEM disks, fracture surfaces, and crack tip regions, produce many of the same internal oxidation features seen in longer-term tests. Again the effect of chromium is manifested by a reduction in the extent of internal oxidation rather than by a fundamental change in oxide morphology.

ACKNOWLEDGEMENT

This research was sponsored by the Fossil Energy AR&TD Materials Program, U.S. Department of Energy, under contract DE-AC05-840R21400 with Martin Marietta Energy Systems Inc. and by the Underlying Research Programme of the UKAEA.

REFERENCES

1. C.T. Liu, V.K. Sikka, J.A. Horton and E.H. Lee, ORNL-6483, Oak Ridge National Laboratory report, August 1988.

2. J.H. DeVan and C.A. Hippsley, Oxidation of High-Temperature Intermetallics, Eds. T. Grobstein and J. Doychak, The Minerals, Metals, and Materials Society Warrendale, 1988, p.31.

3. D.P. Moon, A.W. Harris, P.R. Chalker and S. Mountfort, Harwell Laboratory report AERE R 13097, April 1988.

4. C.A. Hippsley, M. Strangwood and J.H. DeVan , Harwell Laboratory report AERE R 13655, in press.

5. J.A. Horton, J.V. Cathcart and C.T. Liu, Oxid. Met. 29, 1988, p.347.

6. C.A. Hippsley and J.H. DeVan, Acta Metall. 37, 1989, p.1485.

7. G.C. Wood and F.H. Stott, High-Temperature Corrosion, NACE-6, Ed. R. A. Rape, National Association of Corrosion Engineers, Houston, 1983, p.227.

8. P. Kofstad, High-Temperature Corrosion, Elsevier Applied Science, New York, 1988, p.372.

22 STRUCTURAL STUDIES OF INTERNAL OXIDES IN NICKEL ALUMINIUM ALLOYS

G.J. Tatlock, R.W. Devenish and J.S. Punni*

Department of Materials Science and Engineering,
University of Liverpool, P.O. Box 147, Liverpool L69 3BX, UK.
*AEA Technology Dounreay, Thurso, Caithness KW14 7TZ, UK.

ABSTRACT

Internal oxides produced during the oxidation of Ni 2%Al in air at 1000°C have been studied with a combination of high resolution transmission electron microscopy and microanalysis of ion beam thinned samples. Acicular particles in the shape of rods and platelets of both Al_2O_3 and NiAl spinels were examined from different regions of the internal oxidation zone. A variety of faults and defects were characterised in the spinel particles; and their possible role during the oxide development is discussed.

INTRODUCTION

Most nickel-aluminium alloys depend for their protection at high temperatures on the development of a continuous layer of alumina. However at low aluminium concentrations, below ~6% by weight in a binary alloy, aluminium rich internal oxides are formed in the subscale region. The exact figure depends on oxidation temperature, oxygen pressure etc. [1]. The general features of internal oxidation have been reviewed extensively [2-4] and analytical expressions have been derived to describe the kinetics of internal oxidation [e.g. 5, 6]. However, it is only more recently that the types and distribution of internal oxides have been studied in detail [e.g. 7, 8]. For dilute alloys, the overall picture is one of nickel-aluminium spinel particles and rods formed under an outer nickel oxide layer with Al_2O_3 present near the advancing internal oxidation front. The alumina particles transform to spinel oxides as the oxidation front advances. Unfortunately as techniques for the examination of individual particles, often submicron in diameter, become readily available, this simple picture becomes more complex. For example, as shown by Tatlock et al. [5], both spinel and alumina particles are often found in close proximity over a greater depth through the internal oxide region than previously thought. Also the internal structure of the particles is very complex with twins and other linear and non-linear faults present in the spinel oxides. The further investigation of these defects forms the basis of this present paper.

EXPERIMENTAL

The alloy was prepared by melting together nickel (99.9% purity) and aluminium (99.95% purity) in an induction furnace. The charge was reduced to a 1 mm thick sheet by alternate annealing and rolling. Electron probe microanalysis was used to determine the overall composition as Ni 2.2%Al by weight. Oxidation was carried in airout at 1 atm. pressure for times up to 200 h and thin TEM samples were cut and mechanically polished prior to ion beam thinning with 4kV Ar ions. Most of the lattice imaging and microanalysis was carried out on a Philips 400T TEM equipped with a Link EDX detector and a Gatan EELS spectrometer.

RESULTS AND DISCUSSION

Figure 1 shows an SEM micrograph of a cross-section through a Ni 2%Al sample after oxidation in air for 4 h. The internal oxides appeared to form continuous rods or platelets lying approximately perpendicular to the sample surface. TEM examination of the longitudinal and transverse cross-section samples confirmed the SEM observations, although many other small, less acicular particles were also observed between the rods (Fig. 2). These could be isolated particles or sections through rods growing parallel to the growth front, although this latter explanation seems unlikely. Unfortunately the presence of a large number of small disconnected particles between the rods would cast doubts of some of the previous analyses using deep etched samples [7] where all the material between the continuous rods was removed by etching. Any attempt to quantify the amounts of oxides present would then be subject to considerable error.

The complex microstructure of the spinel particles is illustrated in the micrograph in Fig. 3. A wide variety of twin boundaries and other linear and non-linear defects is usually observed. Some of these may be formed during growth of

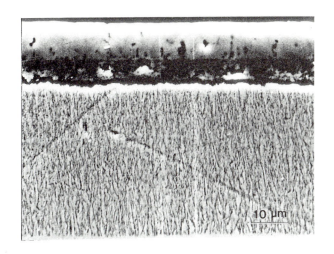

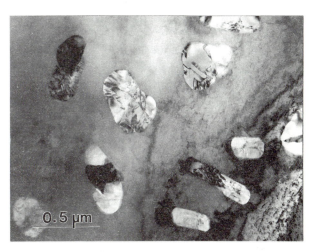

Figure 1: SEM image of a cross-section through the outer scale and internal oxidation zone in Ni 2% Al oxidised in air for 4 h at 1000°C.

Figure 2: TEM image of isolated spinel oxide particles in the internal oxidation zone.

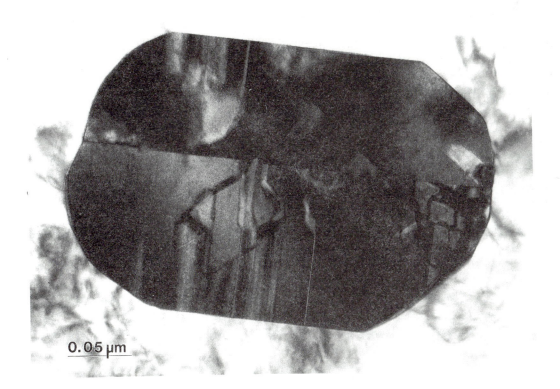

Figure 3: Complex faults and defects in a nickel-aluminium spinel oxide particle.

the particles since large compressive stress may build up due to the volume change on formation of the oxides. Other defects may accommodate composition variations within the spinels, which may be represented as $NiO.n\,Al_2O_3$ where n is thought to vary from 1 to 1.4 for Ni-Al depending on temperature and other growth conditions[10].

The spinels often adopt a parallel sided blade like morphology with sides parallel to {111} and twin boundaries running along their length. However no dominant orientation relation was observed between particles and matrix, unlike the Cu-Al system, for example, where spinel particles adopted a truncated tetrahedron morphology with a central twin boundary and a specific particle matrix orientation[11]. The nucleation and growth of a series of twinned regions along one side of a spinel blade, giving rise to a set of steps along the edge, can also explain the apparent changes of direction along the blade when viewed at low magnification.

A doubly twinned spinel particle is shown in Fig. 4a together with an appropriate diffraction pattern in Fig. 4b. The twin plane is the usual close packed {111} plane giving a characteristic diffraction pattern when viewed along <110>. Although the twin boundaries appeared to be planar for most of their length, a few isolated areas appeared to be stepped as shown in Fig. 4c. Images in the overlap region containing both matrix and twin then show a distinctive periodicity of three times the 111 spacing when viewed along [110].

Other non planar faults are also observed in the spinel particles. One example of such a fault which threads through two twin boundaries is shown in Fig. 5a. The fact that it is continuous across twin boundaries would suggest that the fault leaves the oxygen sub lattice unaltered. Hence the most likely explanation for the fault is an antiphase boundary which separates regions of different stacking of the Ni and Al sub-lattices. Lattice images of two regions across boundaries viewed along [011] are shown in Fig. 5b and c. A shift in the 200 fringes of 1/4[100] is evident across the boundary in Fig. 5c. This would be consistent with an antiphase boundary if associated, for example, with a further shift of e.g. 1/4[010] given a total translation of 1/4[110] and a detailed interpretation of all these features will be given elsewhere[12].

One other type of planar defect commonly present in the spinel particles is shown in Fig. 6a. This micrograph was taken with the beam tilted away from the <110> direction so that the domain-like nature of the defects could be observed. The most striking features of these images are the lines of bright dots often associated with the faults; and two higher magnification examples from different areas of the sample shown in Fig. 6a are shown in Fig. 6b and c. The distribution of the dots is reminiscent of the contrast associated with crystallographic shear structures as noted, for example, in the work on tungsten oxides by Iijima [13]. A change in stacking of the (111) planes in the spinel structure would formally lead to microtwinning or anti-phase structures, if the oxygen sub-lattice was preserved. However, defects on the oxygen sub-lattice may lead to crystallographic shear planes of the type required and are often found growing in from the edge of a crystal as shown in Fig. 6c.

The appearance of the white dots is relatively insensitive to defocus under typical lattice-fringe imaging conditions and in crystallographic shear structures is due to a region with relatively open channels through the crystal. A more detailed analysis of these defects is being undertaken and will be reported elsewhere [12]. However, if these defects correspond to a disordered region of the oxygen sublattice in the spinel particles and they also run along the rods of spinel oxides then they may act as fast diffusion paths for oxygen. It has already been observed that the depth of internal oxidation of Ni-Al alloys tends to be independent of alloy composition over the temperature range 800-1100°C[14]. This was explained by enhanced diffusion of oxygen along the interface between internal oxide rods and the alloy matrix. However, the effect would be further enhanced if oxygen were able to diffuse more rapidly through faulted regions of the spinel oxides themselves.

ACKNOWLEDGEMENTS

Useful discussions with Drs C.J. Kiely and R.C. Pond and the financial support of SERC are gratefully acknowledged.

Figure 4a: A doubly twinned spinel particle.

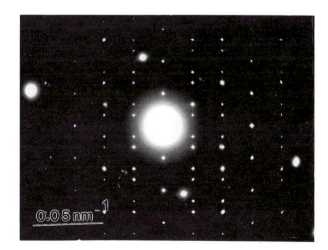

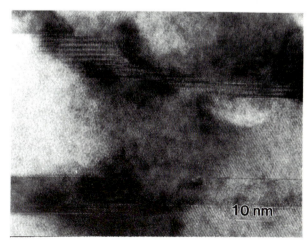

Figure 4b: A <110> diffraction pattern from the matrix and twin.

Figure 4c: Stepped regions along the twin boundary.

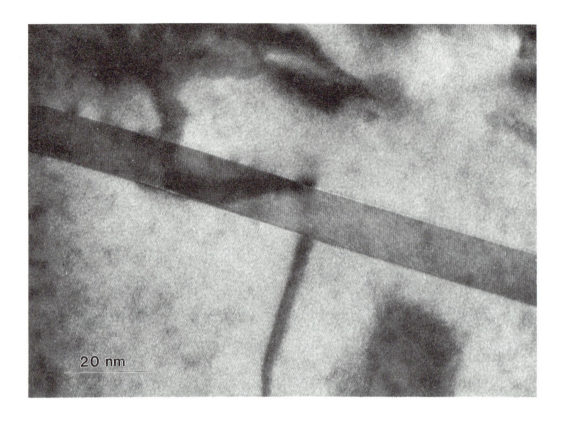

Figure 5a: A high resolution image of twin boundaries in a spinel oxide.

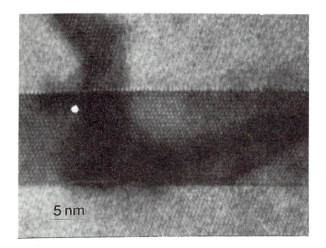

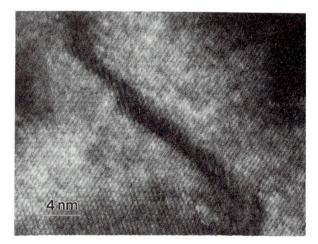

Figure 5b: Greater detail of an area of Fig. 5a showing a fault threading through the twin interface.

Figure 5c: A 200 lattice fringe image across a faulted region showing a shift of the fringes across the boundary.

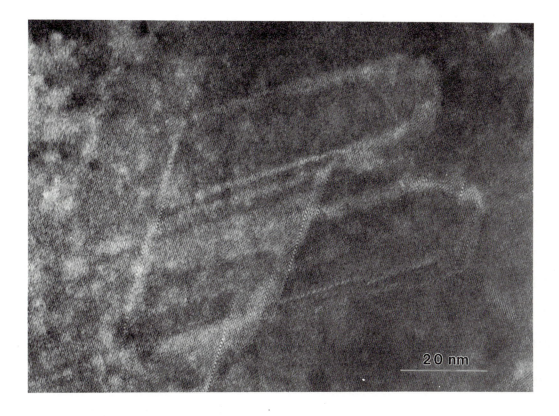

Figure 6a: Possible crystallographic shear domains in a NiAl spinel oxide.

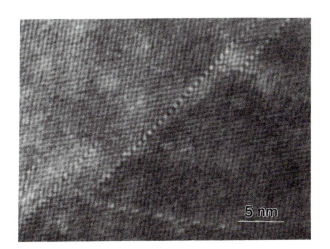

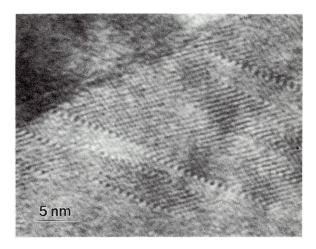

Figure 6b: Greater detail of an area of Fig. 6a showing the characteristic rows of bright dots along specific directions.

Figure 6c: Faulted domains growing in from a grain boundary.

REFERENCES

1. F.S. Pettit, Trans AIME 239, 1967, p.1296.

2. R.A. Rapp, Corrosion 21, 1965, p.382.

3. J.L. Meijering, Adv. Mater. Res. 5, 1971, p.1.
4. J.H. Swisher, Oxidation of Metals and Alloys, Ed. D.L. Douglass, ASM 1971, p.235.

5. C. Wagner, Z. Elektrochem. 63, 1959, p.772.

6. F. Maak, Z. Metallkunde 52, 1961, p.545.

7. H.M. Hindam and W.W. Smeltzer, J. Electrochem. Soc.127,1980, p.1622.

8. F.H. Stott, Y. Shida, D.P. Whittle, G.C. Wood and B.D. Bastow, Oxid. Met. 18, 1982, p.127.

9. G.J. Tatlock, A.G. Baxter, R.W. Devenish and J.S. Punni, Inst. Phys. Conf. Ser. No.78, 1985, p.455.

10. C.B. Carter and H. Schmalzried, Phil. Mag. A52, 1985, p.207.

11. P. Pirouz and F. Ernst, Proc. Conf. on Metal Ceramic Interfaces, Santa Barbara, Ed. E. Ruhle and A.G. Evans, 1989.

12. G.J. Tatlock, R.W. Devenish and C.J. Kiely (1990) - in preparation.

13. S. Iijima, J. Solid State Chem. 14, 1975, p.52.

14. D.P. Whittle, Y. Shida, G.C. Wood, F.H. Stott and B.D. Bastow, Phil. Mag. A. A46, 1982, p.931.

23 METALLOGRAPHIC CHARACTERIZATION OF THE TRANSITION FROM INTERNAL TO EXTERNAL OXIDATION OF Nb-Ti-Al ALLOYS

R. A. Perkins and G. H. Meier*

Lockheed Missiles & Space Co., Palo Alto, CA, USA.
*University of Pittsburgh, Pittsburgh, PA, USA.

ABSTRACT

Protective alumina scales are formed on Nb-Al alloys by the addition of third elements that reduce the solubility and diffusivity of oxygen in the alloy and produce microstructures in which the diffusivity of Al is high. Titanium is the most effective third element addition and stabilizes a high temperature B2 phase on which alumina scales can be formed at an atomic fraction of aluminium (N_{Al}) as low as 0.32. At temperatures below 1300°C, the alloy transforms to a two phase microstructure of γ-TiAl and σ-NbAl$_2$ on which the formation of alumina scales is more difficult. A transition from external to internal oxidation of Al occurs and the alloy must be modified in composition and structure to promote the formation of alumina scales at lower temperatures. The effects of Cr, V, and Si additions, Al content, Nb:Ti ratio, and temperature on the transition from internal to external oxidation of Al have been studied by optical metallography and scanning electron microscopy. Reaction kinetics are correlated with alloy and oxide scale microstructures and the conditions under which protective alumina scales can be formed by the selective oxidation of Al are defined.

INTRODUCTION

The oxidation behavior of Nb-base alloys was the subject of extensive research in the period of 1955-1970. The results as summarized in a review by Stringer [1] reveal that most of the effort was devoted to the additions of elements that would alter the major oxide of Nb (Nb$_2$O$_5$) to reduce the rate of oxidation without having a deleterious effect on mechanical behavior. The efforts were not successful and it became apparent that alloying of Nb to produce a surface oxide that was not based on Nb$_2$O$_5$ would be required. A detailed study conducted by Svedberg [2] in 1975-76 revealed that useful resistance to oxidation could be provided by the formation of alumina or oxide scales having a rutile structure. Unfortunately, such scales were formed only on hard, brittle intermetallic compounds such as NbAl$_3$.

Wukusick [3] studied the effect of Ti additions to Nb-Al alloys as an approach to reducing low temperature "pest" attack. He found that alloys in the Nb-Ti-Al system would form protective alumina scales at temperatures above 1300°C but formed non-protective titania-based scales at lower temperatures. Best resistance to oxidation was exhibited by alloys in the γ-TiAl phase field saturated in Nb (Fig. 1, Region A). Behavior at lower temperatures was improved by adding Cr to the alloy but this lowered the melting point and decreased toughness of the alloys.

A fundamental study of the factors controlling the selective oxidation of Al from Nb-base alloys was conducted by Perkins et al. [4-6] from 1987-89. They found that the addition of a third element to Nb-Al alloys was required to form protective alumina scales. The most effective addition was found to be Ti which produced a high temperature B2 phase in which the diffusivity of Al was sufficiently rapid to permit formation of alumina by selective oxidation. Best performance was found for Nb-Ti-Al alloys in the cross hatched area "B" of Fig. 1. These alloys have a single phase B2 structure above 1300°C and a 2-phase γ+σ structure below this temperature. The change in microstructure results in a transition from external to internal oxidation of Al with decreasing temperature as shown in Fig. 2. At 1300-1600°C, the alloys form alumina in air and oxidize with the same parabolic kinetics and rate constants as NiAl. Below 1300°C, they oxidize with parabolic kinetics but at a rate 2-3 orders of magnitude greater than that of an alumina former. The effect of variations in alloy composition and microstructure on this transition and the effect of oxide scale composition and structure on kinetics is summarized in this paper. All compositions are given in at.% .

EFFECT OF Al

The transition from internal to external oxidation of Al in Nb-Ti-Al alloys occurs with increasing Al content for a given

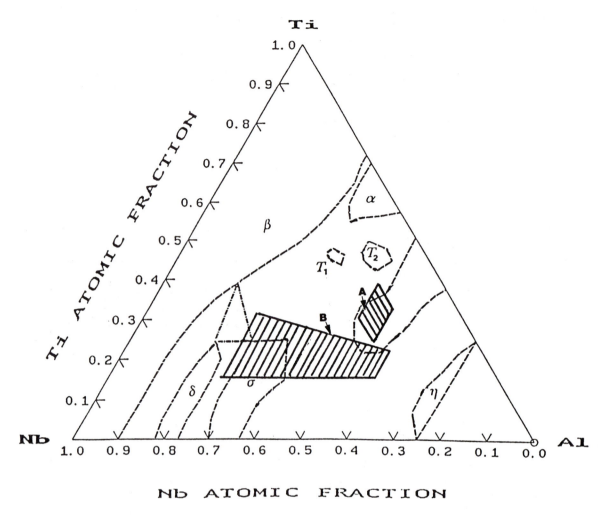

Figure 1: Ternary phase equilibria in the Nb-Ti-Al system at 1100°C (Ref. 7).

temperature. At 1400°C where the alloy has a single phase B2 structure, the transition occurs at the lowest value of N_{Al} (0.44) for the ternary system as shown by the curve of weight gain vs Al content in Fig. 3. At 1100°C where the alloy has a 2-phase $\gamma+\sigma$ structure, a much higher concentration of Al(>0.55) is required to form an external alumina scale. The ternary alloys oxidize with linear kinetics at a rate independent of Al content when the Al is oxidized internally.

The ternary alloy must be modified by the addition of other elements to form alumina scales below 1300°C at lower Al concentrations. The Wagner analysis of the transition from internal to external oxidation of an active element indicates that the critical concentration of the solute element for external scale formation is inversely proportional to the square root of the diffusivity of the solute and directly proportional to the square roots of the solubility and diffusivity of oxygen in the alloy [8]. Diffusivity of Al could be increased by stabilizing the B2 phase to lower temperatures. Unfortunately, no additions have been found that will stabilize the B2 structure below 1300°C at Al concentrations above 30%.

Alumina has not been formed on β-type alloys at low Al concentrations. This appears to be the result of the very high solubility and diffusivity of oxygen, particularly in the β-phase. The inability to form alumina on the 2-phase $\gamma+\sigma$ alloys also appears to be the result of a high solubility and diffusivity for oxygen. The γ-phase with 46.5% Al in the ternary base is internally oxidized selectively as shown in Fig. 3 (dark phase in alloy with 0-5.2% Cr). The light colored σ-phase is not oxidized internally. Instead, it is taken up in the growing scale and becomes oxidized in the scale to form a voluminous transient oxide. Rapid internal oxidation results in loss of Al needed to cut off the growth of the transient oxide. It is concluded that the compositions must be adjusted by adding elements to reduce the solubility and diffusivity of oxygen in the alloy, regardless of microstructure, in order to form alumina as temperature is reduced.

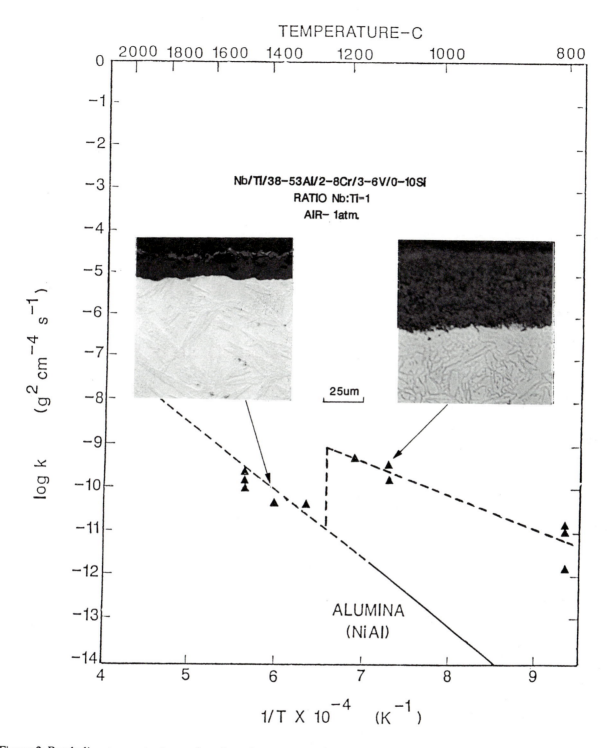

Figure 2: Parabolic rate constants as a function of temperature for Nb-Ti-Al-Cr-V-Si alloys.

EFFECT OF Cr, V, and Si

The critical N_{Al} to form an external alumina scale in air at 1400°C was reduced significantly by the addition of Cr and V. As shown in Fig. 3, oxide scale thickness and depth of internal oxidation decreased with increasing additions of Cr and V. A continuous alumina scale was formed on alloys with 34% Al by the addition of 4.3% V while discontinuous alumina was formed with additions of Cr up to 10.2%. Continuous alumina scales were formed on alloys with 38-40% Al by the addition of either 5% V or Cr. Vanadium was more effective than Cr in reducing the N_{Al} required to form alumina. Metallographic observations show that the amount of transient oxidation before a continuous alumina film is formed also is reduced by the addition of Cr. Vanadium, on the other hand tended to increase the amount of transient oxidation. Additions of Cr + V appeared to be more effective than either alone. Alumina formation was enhanced by V while transient oxidation was reduced by Cr (Fig. 3.)

Additions of Cr and V were not effective at temperatures below 1200°C and alloys with 34-44% Al modified with up to 8% Cr + 6% V formed thick oxide scales that spalled on cooling (Fig. 4). The oxides were a mixture of rutile and $AlNbO_4$ (Fig. 5). Decreasing the Nb:Ti ratio from 1.33 to 0.88 resulted in formation of Al_2TiO_5 as an external scale with $AlNbO_4$ and $TiO_2 \cdot Nb_2O_5$ as underscales. The amount of internal oxidation increased as the Ti content of the alloy increased (Fig. 5). The Nb:Ti ratio controls the alloy microstructure and new phases with reduced oxidation resistance appear with either very low or very high Ti in the alloy. The ratio has little effect on the transition to form alumina scales and primarily controls rates of transient oxidation at temperatures below 1300°C. A ratio of 1-1.3 provides a stable $\gamma+\sigma$ microstructure at low temperatures and is preferred for best oxidation resistance.

The rate of oxidation below 1300°C has linear kinetics and increases rapidly with increasing temperature, reaching a maximum at 1100-1200°C dependent on the amount of Cr + V added (Figs. 4, 6). Increasing Cr + V decreased both the maximum rate and the temperature for the maximum rate. Oxidation rate decreased rapidly above 1200°C, shifting to parabolic kinetics as a continuous alumina scale formed to cut off growth of the transient oxide (Fig. 6). The lowest temperature at which continuous alumina could be formed on Nb-Ti-Al-Cr-V alloys was 1300°C.

The minimum temperature for alumina formation was reduced significantly by adding Si to the alloys. As shown in Fig. 7, continuous alumina-base scales with little transient oxidation were formed in air at 1100°C by the addition of 5-10% Si to Nb-Ti-Al alloys with 40-45% Al. The alloys oxidized with parabolic kinetics but at a rate 2 orders of magnitude greater than that of a pure alumina former at 1100°C (Fig. 2). Precise compositions of the scales has not been determined, but it is likely that a mullite-type scale has been formed instead of pure alumina. Breakaway oxidation with a shift to linear kinetics was observed after 10-20h at 1100°C. The addition of Cr and V to the Si modified alloys increased the time to breakaway to 80h but did not eliminate breakaway. Breakaway occurred at the edges and corners of samples and appeared to be related to the selective internal oxidation of a Si-rich phase (Fig. 7). The precise cause of breakaway to linear kinetics has not been established.

SUMMARY

The transition from internal to external oxidation of Al in Nb-Ti-Al alloys is controlled by alloy microstructure and composition. Titanium stabilizes a B2 phase in alloys with >44% Al above 1300°C on which alumina can form readily. Elements such as Cr, V, and Si are added to reduce the solubility and diffusivity of oxygen in the alloy, thereby enhancing alumina formation at lower concentrations of Al. Alumina scales cannot be formed at temperatures below 1300°C where a two phase $\gamma+\sigma$ microstructure exists unless Si is added. Silicon modified alloys, however, break away to linear kinetics in <100h at 1100°C as a result of selective internal oxidation of a Si-rich phase.

ACKNOWLEDGEMENT

This work was supported by the Air Force Office of Scientific Research under contract F49620-86-C-0018 with Dr A.H. Rosenstein as AFOSR project manager.

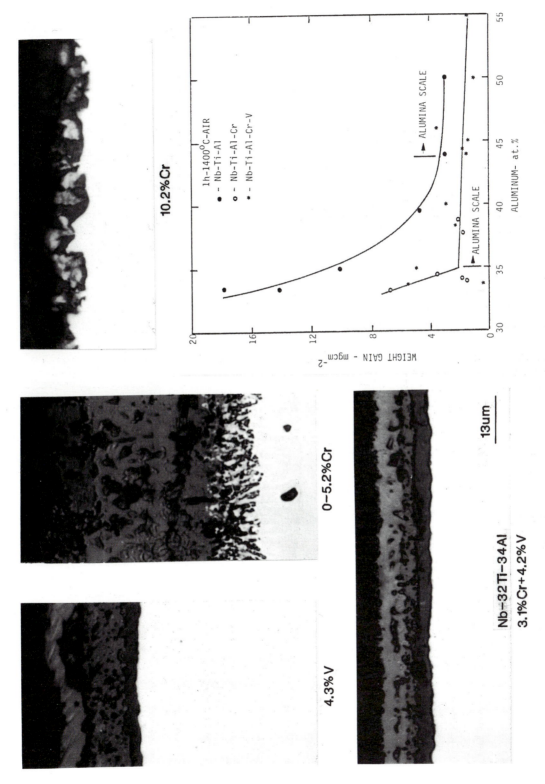

Figure 3: Effect of Cr and V on transition from internal to external oxidation of Nb-Ti-Al alloys in air at 1400°C.

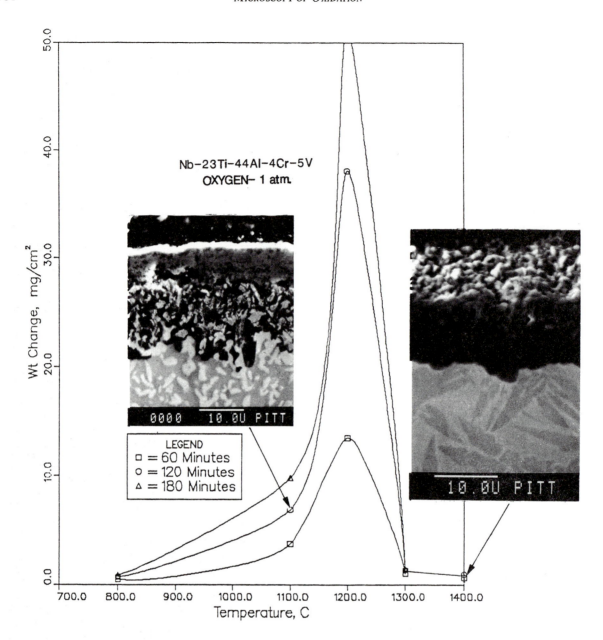

Figure 4: Effect of temperature on oxidation kinetics, oxides scales, and alloy microstructure of Nb-23Ti-44Al-4Cr-5V alloys.

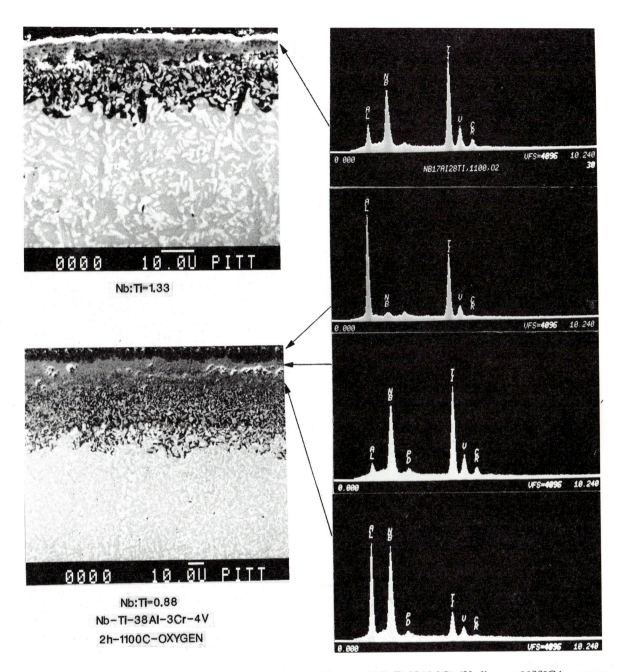

Figure 5: Effect of Nb:Ti ratio on oxide scales and internal oxidation of Nb-Ti-38Al-3Cr-4V alloys at 1100°C in oxygen.

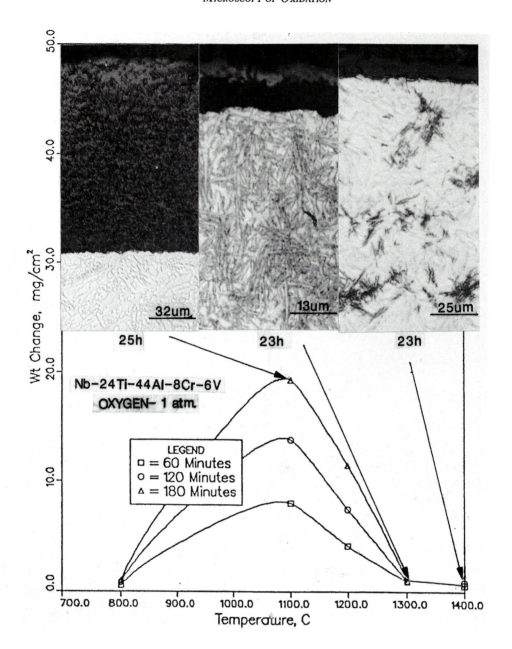

Figure 6: Effect of temperature on oxidation kinetics, oxide scales, and alloy microstructure of Nb-24Ti-44Al-8Cr-6V alloys.

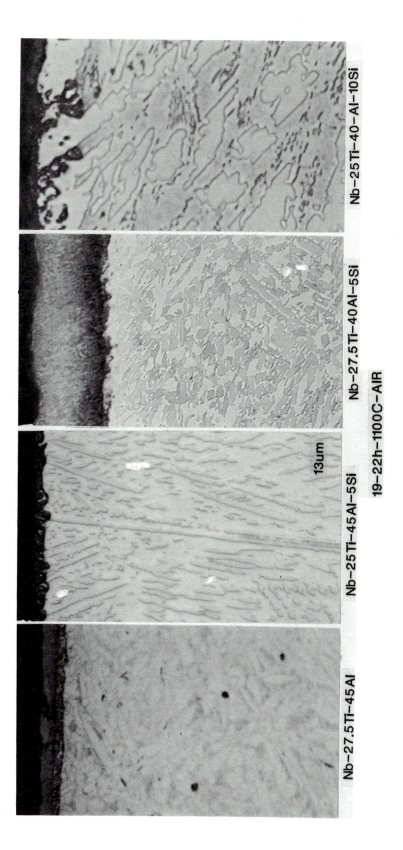

Figure 7: Effect of Si on oxide scales and internal oxidation of Nb-Ti-Al alloys in air at 1100°C.

REFERENCES

1. J.F. Stringer, High Temperature Corrosion of Aerospace Alloys, AGARD-AG200, 1975.

2. R.C. Svedberg, Properties of Hiqh Temperature Alloys, Eds. Z.A. Foroulis and F.S. Pettit, Electrochem Soc., 1976, p.331.

3. C.S. Wukusick, USAEC Contract AT(401) -2847, 1963.

4. R.A. Perkins, K.T. Chiang, G.H. Meier and R.A. Miller, AFOSR Contract F49620-86-C-0018, 1989.

5. R.A. Perkins, K.T. Chiang and G.H. Meier, Scripta Met, 22, 1988, p.419.

6. R.A. Perkins, K.T. Chiang, G.H. Meier and R. Miller, Oxidation of High Temperature Intermetallics, Eds. T. Grobstein and J. Doychak, TMS, 1988, p.157.

7. J.H. Perepezko, MRC Fall Meeting, Dec.1988.

8. C. Wagner, Z. Electrochem., 63, 1959, p.772.

24 *IN SITU* TEM INVESTIGATION OF THE OXIDATION OF TITANIUM ALUMINIDE ALLOYS

G. Welsch[1], S.L. Friedman* and A.I. Kahveci**

Department of Materials Science and Engineering, Case Western Reserve University, Cleveland, OH, USA.
[1]*Currently on leave at Deutsche Forschungsanstalt für Luft- und Raumfahrt (DLR), Köln, and at Universität Saarland, FRG.*
**On summer leave from School of Applied and Engineering Physics, Cornell University, Ithaca, NY, USA.*
***Now at Alcoa Research Laboratories, Alcoa Center, PA, USA.*

ABSTRACT

Transmission electron microscopy (TEM) of thin foil specimens during in-situ oxidation permits the study of oxide nucleation and growth. The results of such experiments on Ti_3Al (alpha-2), Ti_3Al/Nb (alpha-2 plus beta) and TiAl (gamma) alloys are presented. Direct images of the earliest observable oxide particles were obtained. The preferential sites of oxygen attack, e.g. on dislocations, grain boundaries and second phases were investigated. Diffraction ring patterns provide information on the nucleation and growth of the first oxide phases.

BACKGROUND

The intermetallic titanium aluminide compounds are strong, lightweight structural materials. They are of interest for structural components in hypersonic aircraft, space vehicles [1, 2] and gas turbine engines. The main problem areas are oxidation resistance at elevated temperature and brittleness at low temperature. Brittleness is in part due to low intrinsic dislocation mobility/multiplication and is severely aggravated by interstitial oxygen impurities. Oxidation consumes load-bearing metal, and dissolved oxygen embrittles the metal substrate [3].

Investigations of the oxidation kinetics and of oxide microstructures [4, 5] have shown the lack of self-protection of bulk samples at elevated temperatures. Porosity of the multi-layered oxide scales and heterogeneity of mutually insoluble TiO_2 and Al_2O_3 oxides [6] are believed to be reasons for the lack of protection [5] (Fig. 1b). The oxidation resistance of these compounds increases with increasing Al concentration and also by alloying with Nb (Fig. 1a). The crystal structures of the alloy and possible oxide phases are summarized in Tables 1 and 2.

EXPERIMENTAL

Alloys were supplied by General Electric Co. with the following compositions in atom percent:

Cast Ti_3Al (alpha-2 phase):	26.1 Al,	0.20 O,	0.07 N,	bal. Ti
Forged Ti_3Al/Nb (alpha-2 + beta):	25.1 Al,	10.1 Nb,	- ,	bal. Ti
Cast TiAl (gamma + some alpha-2):	49.3 Al,	0.19 O,	0.09 N,	bal. Ti

The cast alloys were homogenization-heat-treated at 1050°C for three hours and air-cooled. The forged alloy was used in the as-received condition. TEM foils were prepared by grinding slices on SiC paper to a thickness of 0.15mm. From these discs of 3mm diameter were electrolytically thinned in a solution of perchloric acid, methanol and butyl cellusolve, using a twin jet apparatus [4].

The Argonne HVEM/Tandem facility [18] was used for the in-situ oxidation experiments. The TEM foils were heated in a resistance heated platinum ribbon stage inside an environmental cell (Fig. 2a). The cell could be charged with several torr of oxygen, while the usual vacuum of about 10^{-6} Torr was maintained in the column of the electron microscope. Observations were made at an acceleration voltage of 1MV with the specimen at temperature or after it had cooled down to room temperature. Depending on the specimen holder used, the specimen temperature was measured either with a thermocouple, or it was estimated from calibration curves of measured temperature versus heating current in various oxygen partial pressures[19]. In a typical experiment the TEM foil (Fig. 2b) was heated incrementally and held for some time at constant temperature to observe the development of oxide. Figure 3 shows a typical temperature versus time diagram. Thermally induced specimen drift made it difficult to obtain good photographic images, but the progress of oxidation could be observed visually or with a video recorder. It could also be monitored from the evolution of diffraction rings. The diffraction patterns were not sensitive to small specimen drifts. For brightfield and darkfield imaging The The

Table 1. Alloy phases and structures

Alloy	Phases	Structures
Ti$_3$Al	α_2	Hexagonal Ordered[7] Structure (DO$_{19}$) a=5.775Å c=4.638Å
Ti$_3$Al/Nb	β & α_2	α_2 (structure as above) β Body Centered Cubic a=3.283Å
TiAl	γ	Ordered Face Centered[8] Tetragonal (L1$_0$) a=3.987Å c=4.07Å

Table 2. Possible oxide phases and structures

Oxide	Phases	Structures
Titanium Oxide	TiO$_2$ (Rutile) anatase and brookite convert to rutile at 700°C	Tetragonal [9] a=4.5983Å c=2.9592
Aluminum Oxide	α-Al$_2$O$_3$	Trigonal[10] a=4.758Å c=12.991Å
	β-Al$_2$O$_3$	Hexagonal[11] a=5.64Å c=22.65Å
Niobium Oxide	Nb$_2$O$_5$	Orthorhombic[12] a=6.168Å b=29.312Å c=3.936Å
	NbO$_2$	Tetragonal[13] (P4$_{2/mnm}$) a=13.71Å c=5.985Å

*NbO$_2$ can also be described by a lattice structure similar to rutile with a'=4.861Å and c'=2.993Å.

Oxide	Phases	Structures
Titanium Niobium Oxide	Ti$_2$Nb$_{10}$O$_{29}$	Monoclinic [14] a=15.57Å b=3.814Å c=20.54Å β=113°41'
	Ti$_2$Nb$_{10}$O$_{29}$	Orthorhombic[15] a=28.50Å b=3.805Å c=20.51Å
Aluminum Niobium Oxide	AlNbO$_4$	Orthorhombic[16] a=6.17Å b=7.38Å c=8.61Å
	AlNb$_{11}$O$_{29}$	Monoclinic [17] a=15.57Å b=3.813Å c=20.55Å β=113.26°

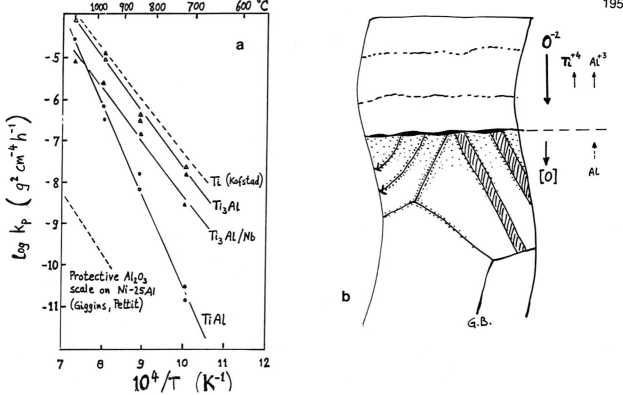

Figure 1: (a) Plot of the parabolic oxidation rate constants of titanium aluminide alloys in relation to those of pure Ti [25, 26] and a protective alumina-forming alloy, Ni-25%Al [27]. The data for Ti$_3$Al and TiAl are from Kahveci et al. [4, 5], and the data for Ti$_3$Al/Nb are from Mostafa and Welsch [28]. (b) Illustration of multilayered oxide scale and oxygen-enriched alloy surface layer on bulk titanium aluminide. Oxgen penetration is most likely via enhanced fast diffusion paths.

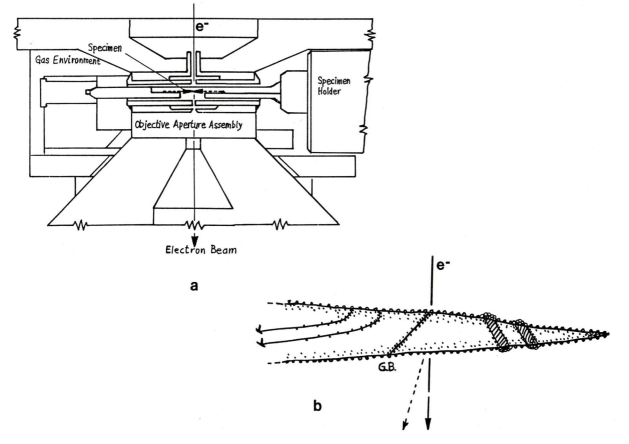

Figure 2: (a) The environmental specimen chamber of the Argonne HVEM-Tandem facility. (b) Illustration of the cross-section of a wedge-shaped TEM foil for *in situ* observation of oxide nucleation and oxygen penetration into the alloy.

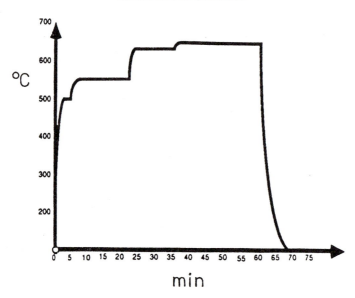

Figure 3: Typical temperature-time heating cycle during *in situ* oxidation. The example is for a Ti₃Al foil.

at temperature, short exposure times at corresponding high beam intensities were essential. A better method was to interrupt the heating cycle and to take photographs after the specimen had equilibrated at room temperature.

RESULTS

In Situ Oxidized Ti₃Al Foil

Figures 4, 5 and 6 show the earliest detectable oxide precipitates and their growth. The early oxide produces a spotty diffraction ring pattern, Fig. 4b, which indicates that an epitaxial relation with the alloy substrate exists. From the present preliminary data this relation could not be determined. However, in view of the fact that the alpha-2 intermetallic phase is isomorphic with the disordered alpha phase one might speculate that the epitaxy may be identical to that of TiO_2 on alpha titanium (Flower and Swann [20]):

$$(0001)_{alpha} \;//\; (010)_{rutile}$$
$$(1\bar{2}10)_{alpha} \;//\; (001)_{rutile}$$

Whatever the epitaxy between Ti₃Al and oxide is, it is quickly lost during further oxidation. This is evident from the continuous ring pattern in Fig. 5c.

There is only slight indication of preferential oxidation at a grain boundary, Fig. 4a, and it is similar with dislocations. The extremely small oxide grain size (few nanometers) and the movement and annihilation of dislocations in the thin parts of a TEM foil (visual observation) make it difficult to determine whether dislocations are preferred oxide nucleation sites. Nevertheless, an interaction of oxygen with dislocation lines is clearly visible in Fig. 6 in the form of splitting of dislocation images near their intersection sites with the surface. In other instances extremely narrow dislocation line contrasts are observed. These are not yet understood.

From diffraction ring patterns, such as the one in Fig. 5c, an analysis of the oxide phases was tried [21] - see Table 3. It shows that rutile is the dominant oxide during the nucleation stage.

In Situ Oxidized Ti₃Al/Nb Foil

Figures 7 and 8 show the results of oxidation of the alpha-2 and beta phase. The alloy element Nb is primarily concentrated in the beta phase. It is evident from both figures that the beta phase has oxidized faster than the surrounding alpha-2 phase. Another effect of Nb appears to be a stronger orientation relationship of the early TiO_2 oxide with the alloy substrate than was the case for the binary Ti₃Al. The limited diffraction data in Fig. 8 suggest the following orientation relation:

$$(011)_{rutile} \;//\; (11\bar{2}0)_{alpha-2}$$

Table 3. Oxide Nucleation on Ti₃Al
Comparison of lattice spacings determined from a diffraction ring pattern after *in situ* oxidation for 15 minutes at 655°C in 1 Torr oxygen. The numbers in parentheses are relative intensities of X-ray powder diffraction data. Underlined are those lattice spacings which could produce the observed rings

	$d_{observed}$ [Å]	d_{TiO_2}[9]	$d_{Al_2O_3-\alpha}$[10]	$d_{Al_2O_3-\beta}$[11]	$d_{Al_2O_3-\delta}$[22]	$d_{Al_2O_3-\iota}$[23]	$d_{Al_2O_3-\theta}$[24]
					11.9(70)		
				5.7(30)			
							5.45(100)
					4.57(12)		4.54(18)
					4.45(10)		
				4.07(10)	4.07(12)		
				3.78(10)			
					3.61(4)		
			3.479(75)			3.47(100)	
ring	3.25	**3.247**(100)				**3.23**(4)	
					3.05(4)		
						2.92(20)	
					2.881(8)		
				2.80(50)			2.837(80)
					2.728(30)	2.72(80)	2.730(65)
				2.68(70)			
					2.601(25)	2.59(80)	
			2.552(90)				2.566(14)
ring	2.47	**2.487**(50)		2.51(30)	**2.460**(60)	**2.46**(5)	2.444(60)
			2.379(40)	2.41(30)	2.402(16)		
						2.34(80)	
					2.314(8)		2.315(45)
		2.297(8)			2.279(40)		
ring	2.24	**2.188**(25)		2.24(30)		**2.24**(80)	**2.257**(35)
			2.165(<1)		2.160(4)		
				2.14(30)		2.15(40)	
			2.085(100)				
ring	2.06	**2.054**(10)		**2.03**(50)			
							2.019(45)
					1.986(75)		
			1.964(2)		1.953(40)		1.9544(8)
				1.93(30)	1.914(12)		1.9094(30)
						1.88(20)	
				1.845(30)	1.827(4)		
					1.816(4)		
			1.740(45)	1.742(30)	1.73(40)		
ring	1.70	**1.687**(60)					
ring	1.64						
		1.6237(20)				**1.628**(4)	
						1.61(20)	
			501(80)	1.591(70)	1.604(4)		
					1.563(30)		1.55(60)
			.46(4)		1.538(4)		1.5426(25)
			1.514(6)		1.517(16)		
			1.510(10)				
ring	1.48		ϑ)		**1.483**(30)		**1.492**(10)
		1.4528(10)			1.456	1.461(10)	1.4526(25)
		1.4243(2)					
ring	1.38		1.404(30)	1.403(100)	1.407(50)	1.41(5)	
					1.396(100)		**1.3883**(100)
			1.374(50)	**1.369**(50)			
		1.3598(20)					

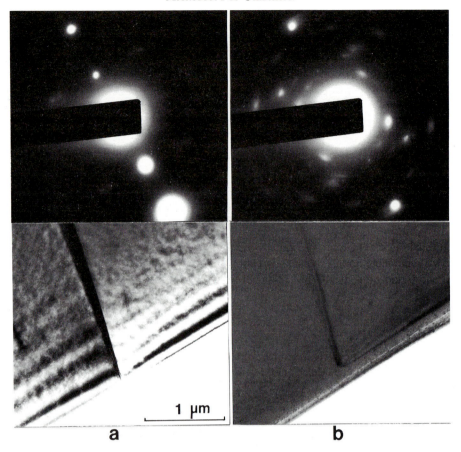

Figure 4: TEM images and selected area diffraction patterns of a wedge-shaped Ti₃Al foil that contains a grain boundary. (a) Before oxidation. (b) Oxidized *in situ*, 5 minutes, 625°C, 1 Torr oxygen. The thinnest part of the foil, a strip along the foil edge, has been oxidized through and consists of nanometer-size oxide grains. Spotty oxide diffraction rings have developed and are superimposed on the alloy pattern.

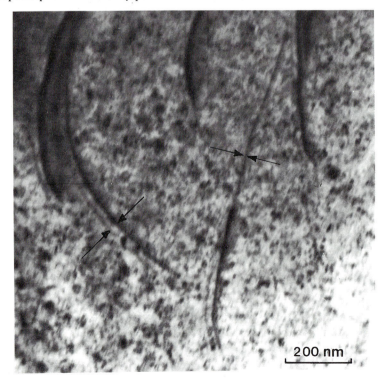

Figure 6: Ti₃Al foil, oxidized 15 minutes, 655°, 1 Torr oxygen. The oxide grains are less than 10nm in size. During oxidation the dislocation segments near the surface of the foil developed the peculiar split images shown (approximately 20nm width). This reaction is not understood.

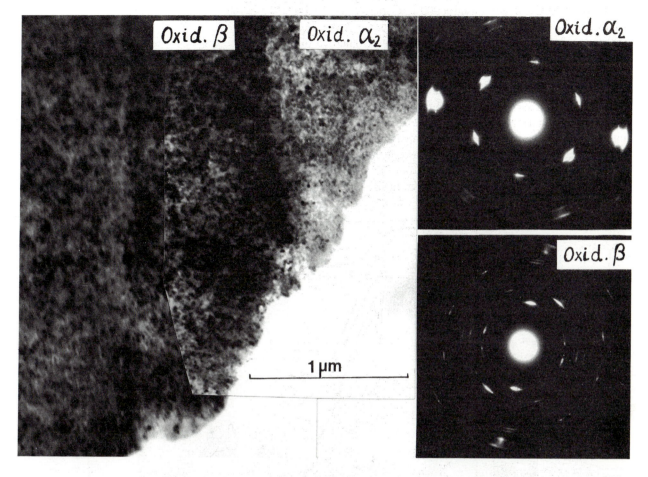

Figure 8: Ti$_3$Al/Nb foil, oxidized for 1 hour between 600 and 700°C in 1 Torr oxygen. The area is the same as in Fig. 7. The brightfield image shows a completely ozidized foil. The diffraction patterns are from former alpha-2 (top) and from beta phase (bottom). The pattern of oxidized alpha-2 phase contains only faint (110)$_{rutile}$ spots, but strong arced (011)$_{rutile}$ spots are aligned in <11$\bar{2}$0> directions of the former alpha-2 phase, see diffraction pattern in Fig. 7.

Ti$_3$Al/Nb:

Rutile, variants of alumina and mixed titanium- and aluminum-niobium oxides grow on alpha-2 and on beta phase. The oxide nucleation and early growth is accelerated on the Nb-rich beta phase. The nucleation and growth of more highly orientation-related oxide is

$$(011)_{rutile} \ // \ (11\bar{2}0)_{alpha-2}$$

favoured by Nb. It indicates that the oxide grains on Ti$_3$Al/Nb are more intimately bonded to each other than is the case on binary Ti$_3$Al alloy.

TiAl:

Variants of alumina and rutile are the early oxides on TiAl. They develop little or no orientation relation with the alloy substrate. The gamma phase is more oxidation resistant than alpha-2 laths.

ACKNOWLEDGEMENTS

Thanks are due to Dr. C. Allen, Dr. E. Ryan and Mr. S. Okers for their support at the HVEM-Tandem facility of Argonne National Laboratory, Argonne, IL. The research was part of a project supported by General Electric Corporation, Aircraft Engine Division, Cincinnati, OH, by courtesy of Dr. R. McCarron. One of us (GW) acknowledges the support during a sabbatical leave by DLR, Institute duer Werkstoff-Forschung, Prof. W. Bunk.

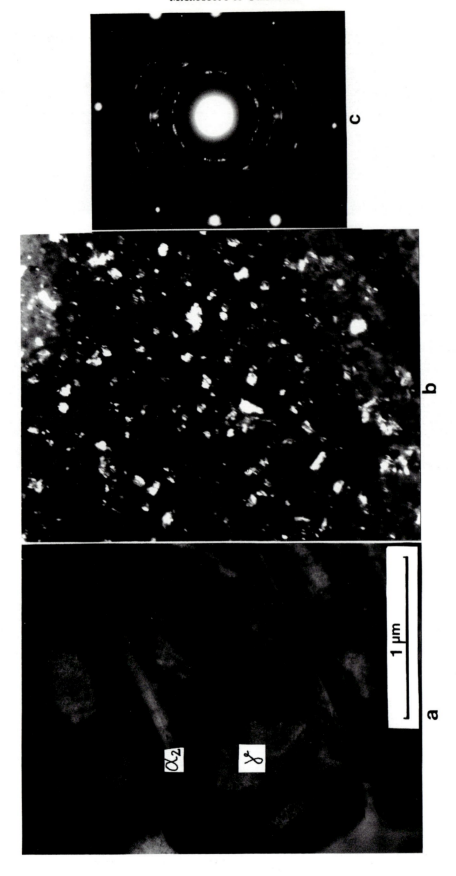

Figure 9: Ti-Al foil, consisting of gamma phase with some interspersed alpha-2 laths. (a) Oxidized for 1 minute at 600°C in 1 atmosphere oxygen: brightfield. (b) The same area after oxidation for 55 minutes between 640 and 720°C in 1 Torr oxygen: darkfield. (c) Diffraction pattern of condition (b). The inner ring is of $(110)_{rutile}$ (see also Table 5).

Table 4. Oxide nucleation on Ti₃Al/Nb

Comparison of lattice spacings determined from diffraction ring patterns taken from oxidised alpha-2 and beta phase regions. In parentheses are relative intensities (per cent) of x-ray powder diffraction data. Underlined are those lattices spacings which could produce the observed rings.

d_observed [Å]	d_{TiO_2}	$d_{Al_2O_3-\alpha}$	$d_{Al_2O_3-\beta}$	$d_{Al_2O_3-\delta}$	$d_{NbO_{25}}$ [12]	d_{AlNbO_4} [16]	$d_{Ti_2Nb_{10}O_{29}}$ [14,15]
			11.9(70)			10.3(10)	
					8.86(1)	9.62(40)	
					7.37(3)	7.13(30)	
					6.15(25)		
		5.7(30)			5.70(12)		
					5.22(5)		5.13(40)
					4.89(1)	5.04(35)	5.04(30)
							4.75(40)
				4.57(12)			
				4.45(10)		4.33(5)	
			4.07(10)	4.07(12)	3.93(90)	4.16(3)	
			3.78(10)				
							3.74(80)
				3.61(4)		3.69(30)	3.71(80)
faint 3.50		3.479(75)				3.53(100)	3.56(100)
							3.48(20)
							3.42(80)
							3.39(30)
							3.32(40)
bright ring 3.25	3.247(100)			3.23(4)		3.22(10)	3.22(10)
							3.19(10)
					3.15(100)		
					3.09(45)	3.09(55)	3.13(10)
faint 2.92				3.05(4)	3.07(15)	2.98(30)	2.92(20)
				2.881(8)		2.88(20)	2.85(20)
faint 2.75			2.80(50)				2.78(60)
				2.728(30)	2.73(5)		2.75(40)
faint 2.71			2.68(70)			2.673(40)	2.67(20)
					2.608(3)		2.62(40)
				2.601(25)	2.600(1)		2.57(30)
							2.55(10)
bright ring 2.49	2.487(50)	2.552(90)	2.51(30)			2.514(16)	2.51(20)
faint 2.47				2.460(60)	2.460(50)		2.47(40)
					2.428(20)	2.448(18)	2.43(10)
					2.420(10)		
faint 2.34		2.379(40)	2.41(30)	2.402(16)		2.361(7)	2.37(10)
				2.314(8)			2.30(40)
weak 2.30	2.297(8)			2.279(40)		2.279(12)	
ring 2.20			2.24(30)				
						2.216(12)	
faint 2.18	2.188(25)						
				2.160(4)		2.153(7)	
		2.14(30)			2.125(3)		
				2.118(4)		2.09(10)	
		2.085(100)				2.056(18)	2.05(40)
	2.054(10)						2.04(40)
			2.03(50)				
					2.017(6)	2.018(25)	
					2.013(6)		
				1.986(75)	1.983(1)		
	1.964(2)			1.953(40)	1.969(25)	1.960(12)	
			1.93(30)	1.914(12)		1.923(5)	
						1.876(4)	1.90(80)
					1.865(6)	1.857(20)	
			1.845(30)	1.827(4)	1.832(16)	1.826(12)	
				1.816(4)	1.808(1)		
					1.793(25)	1.767(8)	
						1.755(5)	
						1.736(5)	
		1.740(45)	1.742(30)				
bright ring 1.69	1.687(60)				1.669(25)	1.696(6)	
ring 1.68						1.672(6)	
					1.660(20)	1.670(7)	
					1.659(20)		
					1.651(3)		
faint 1.62	1.6237(20)			1.628(4)	1.632(15)		
		1.601(80)	1.591(70)	1.604(4)		1.597(18)	
				1.563(30)	1.575(10)	1.567(40)	
		1.546(4)		1.538(4)	1.542(4)		
		1.514(6)		1.517(16)	1.520(3)		
		1.510(10)					
					1.491(1)	1.487(10)	
ring 1.47	1.4797(10)		1.483(30)			1.484(9)	
	1.4528(10)			1.456	1.462(9)	1.458(8)	
						1.451(5)	
	1.4243(2)				1.435(4)	1.422(10)	
		1.404(30)	1.403(100)	1.407(50)	1.399(6)		
				1.396(100)			
						1.385(10)	
ring 1.37	1.374(50)		1.369(50)			1.373(10)	
ring 1.36	1.3598(20)						

Table 5. Oxide nucleation on TiAl

Comparison of lattice spacings determined from a diffraction ring pattern after *in situ* oxidation for 1 hour between 600 and 720°C in 1 Torr oxygen. In parentheses are relative intensities (per cent) of X-ray powder diffraction data. Underlined are those lattice spacings which could produce the observed rings. Additional references are given in Table 3

$d_{observed}$ [Å]		d_{TiO_2}	$d_{Al_2O_3-\alpha}$	$d_{Al_2O_3-\beta}$	$d_{Al_2O_3-\delta}$	$d_{Al_2O_3-\iota}$	$d_{Al_2O_3-\theta}$
					11.9(70)		
				5.7(30)			
							5.45(100)
very faint	4.4				4.57(12)		4.54(18)
					4.45(10)		
faint	3.75			4.07(10)	4.07(12)		
				3.78(10)			
					3.61(4)		
good ring	3.25		3.479(75)			3.47(100)	
		3.247(100)				3.23(4)	
					3.05(4)		
faint	2.89					2.92(20)	
					2.881(8)		
				2.80(50)			2.837(80)
					2.728(30)	2.72(80)	2.730(65)
				2.68(70)			
					2.601(25)	2.59(80)	
			2.552(90)				2.566(14)
		2.487(50)		2.51(30)			
ring	2.47				2.460(60)	2.46(5)	2.444(60)
			2.379(40)	2.41(30)	2.402(16)		
ring	2.33					2.34(80)	
					2.314(8)		2.315(45)
		2.297(8)			2.279(40)		
				2.24(30)		2.24(80)	2.257(35)
		2.188(25)					
good ring	2.14				2.160(4)		
				2.14(30)		2.15(40)	
			2.085(100)				
		2.054(10)		2.03(50)			
							2.019(45)
					1.986(75)		
			1.964(2)		1.953(40)		1.9544(8)
				1.93(30)	1.914(12)		1.9094(30)
						1.88(20)	
				1.845(30)	1.827(4)		
					1.816(4)		
			1.740(45)	1.742(30)		1.73(40)	
ring	1.68	1.687(60)					
		1.6237(20)			1.628(4)		
						1.61(20)	
			1.601(80)	1.591(70)	1.604(4)		
					1.563(30)		1.55(60)
			1.546(4)		1.538(4)		1.5426(25)
			1.514(6)		1.517(16)		
spots	1.51		1.510(10)				
spots	1.47	1.4797(10)		1.483(30)		1.492(10)	1.4883(25)
	1.45	1.4528(10)			1.456	1.461(10)	1.4526(25)

REFERENCES

1. H.A. Lipsitt, Titanium Aluminides - An Overview. MRS Symp., 39, 1985, p.351.

2. The X-30, Giving Wing to New Materials. Article in National Geographic, Dec. 1989.

3. N.W. Kearns and J.E. Restall, Oxidation Resistant Coatings for Titanium Alloys, Eds. P. Lacombe, R. Tricot and G. Beranger, Sixth World Conf. on Titanium, Les Editions de Physique, 1988, p.1753.

4. A.I. Kahveci, G. Welsch and G.E. Wasielewski, Oxidation of Titanium Aluminides, Eds. P. Lacombe, R. Tricot and G. Beranser, Sixth World Conf. on Titanium, Les Editions de Physique, 2, 1988, p.1015.

5. G. Welsch and A.I. Kahveci, Oxidation Behavior of Titanium Aluminide Alloys, Eds. T. Grobstein and J. Doychak, Oxidation of High Temperature Intermetallics, TMS, 1989, p. 207.

6. P. Kofstad, Nonstoichiometry, Diffusion and Electrical Conductivity in Binary Metal Oxides, R.E. Krieger Publ. Co., 1983, p.142.

7. E.S. Bumps, H.D. Kessler and M. Hansen, Trans. Am. Inst. Min. Engr., vol. 194, 1952, p.609.

8. P. Duwez and J.L. Taylor, Trans. AIME, J. Metals, 1952, p.71.

9. X-Ray Powder Diffraction File, JCPDS International Centre for Diffraction Data, Inorganic Vol., 1980, Set 21-1276.

10. X-Ray Powder Diffraction File, JCPDS International Centre for Diffraction Data, Inorganic Vol., 1980, Second Printing, Set 10-173.

11. X-Ray Powder Diffraction File, JCPDS International Centre for Diffraction Data, Inorganic Vol., 1980, Second Printing, Set 10-414.

12. X-Ray Powder Diffraction File, JCPDS International Centre for Diffraction Data, Inorganic Vol., 1986, Set 27-1003.

13. X-Ray Powder Diffraction File, JCPDS International Centre for Diffraction Data, Inorganic Vol., 1980, Second Printing, Set 9-235.

14. X-Ray Powder Diffraction File, JCPDS International Centre for Diffraction Data, Inorganic Vol., 1972, Set 13-317.

15. X-Ray Powder Diffraction File, JCPDS International Centre for Diffraction Data, Inorganic Vol., 1972, Set 13-316.

16. X-Ray Powder Diffraction File, JCPDS International Centre for Diffraction Data, Inorganic Vol., 1984, Set 26-30.

17. X-Ray Powder Diffraction File, JCPDS International Centre for Diffraction Data, Inorganic Vol., 1980, Set 22-9.

18. Argonne National Laboratory, Center for Electron Microscopy, HVEM/Tandem Facility, Chicago, Illinois. Private communication.

19. E.A. Ryan, Argonne HVEM/Tandem Eacility, private communication, calibration curves (1989).

20. H.M. Flower and P.R. Swann, Acta Metall., 22, 1974, p. 1339.

21. S.L. Friedman, Unpublished research (1989), Case Western Reserve University, Cleveland, OH.

22. X-Ray Powder Diffraction File, JCPDS International Centre for Diffraction Data, Inorganic Vol., 1974, Set 16-394.

23. X-Ray Powder Diffraction File, JCPDS International Centre for Diffraction Data, Ioorganic Vol., 1972, Set 12-539.

24. X-Ray Powder Diffraction File, JCPDS International Centre for Diffraction Data, Inorganic Vol., 1983, Set 23-1009.

25. P. Kofstad, High Temperature Oxidation of Metals, J. Wiley, N.Y., 1966.

26. P. Kofstad, K. Hauffe and H. Kjöllesdal, Acta Chem. Scand., 12, 1958, p. 239.

27. C.S. Giggins and E.S. Pettit, J. Electrochem. Soc., 118, 1971, p.1782.

28. I. Mostafa and G. Welsch, Unpublished research, Case Western Reserve University, 1990.

25 PROTECTIVE Al$_2$O$_3$ SCALE FORMATION ON NbAl$_3$-BASE ALLOYS

J. Doychak and M.G. Hebsur

Sverdrup Technology, Inc., NASA Lewis Research Center Group,
Brookpark OH 44142, USA.

ABSTRACT

The oxidation of NbAl$_3$ with additions of Cr and Y was studied to determine the mechanisms of the beneficial effects of these elements upon oxidation. Cr additions to the binary NbAl$_3$ alloy of up to 6.8 at% reduced scale growth rates and promoted α-Al$_2$O$_3$ formation over much longer times relative to binary NbAl$_3$. A major effect of Cr is to form a layer of AlNbCr at the metal/scale interface, which appears to be inherently more oxidation resistant than the matrix alloy. Yttrium additions to a Cr-containing alloy improvod the scale growth rate and adherence, and changed the scale microstructure to mimic that of a typical protective Al$_2$O$_3$ scale.

INTRODUCTION

The niobium aluminide, NbAl$_3$, is a potential matrix material in fiber reinforced composites being developed for applications in advanced turbine engines. Advantages of NbAl$_3$ are its relatively low density ($\rho = 4.5$ g/cc), high melting temperature (solidus T$_m$ $\approx$ 1650°C), and good chemical and thermal expansion compatibility with potential Al$_2$O$_3$ reinforcing fibers. The major disadvantages of NbAl$_3$ are its lack of ductility, poor toughness and difficulty of processing. In terms of oxidation resistance, NbAl$_3$ possesses the slowest scale growth kinetics of the niobium aluminides, however the growth rates are still at least two orders of magnitude higher the conventional alumina forming alloys[1]. In addition, NbAl$_3$ has been found to undergo post attack during oxidation at intermediate (700°C- 900°C) temperatures[2].

A study by Hebsur et al. was performed to improve the oxidation behavior of NbAl$_3$-base alloys containing ternary, quaternary, and quinary additions [3]. The intent was to expand the NbAl$_3$ phase field to lower aluminum levels such that oxidation to form Al$_2$O$_3$ would delay the eventual formation of Nb$_2$Al in the alloy (Nb$_2$Al does not form a protective Al$_2$O$_3$ scale). The ternary alloying scheme for NbAl$_3$ resembled that developed by Perkins et al. on the oxidation of Nb-Ti-Al alloys to stabilize the β phase which has a BCC structure, thus raising the alloy aluminum diffusivity [4]. Hebsur et al.'s study was successful in that the oxidation behaviour of the NbAl$_3$-base alloys was improved. Indeed, the isothermal scale growth kinetics on a patented NbAlC alloy are the same as conventional alumina forming alloys [5]. However, the protective scale formation appeared to be related to Cr-rich intermetallic grain boundary phases rather than the NbAl$_3$-base matrix [6].

The purpose of the present study is to determine the oxidation mechanisms of ternary NbAl$_3$+Cr and quaternary NbAl$_3$+Cr,Y alloys. An understanding of the processes involved in protective scale formation is required for future alloy development.

EXPERIMENTAL

Alloys of nominal compositions shown in Table 1, were induction melted and drop cast into slabs approximately 10 x 4 x 0.5 cm. Oxidation coupons and transmission electron microscopy (TEM) samples were machined from the slabs. Specimen surfaces were abraded using successively finer grits of SiC paper, then polished to a final surface finish using 1μm diamond paste. Specimens were washed in ethanol then acetone prior to oxidation tests.

TEM samples were oxidized at 1200°C in air for 0.1 or 1.0 h. Isothermal gravimetric (TGA) tests were performed at 1200°C in flowing oxygen for 50,100 or 500 h.

Characterization of unoxidized specimens using analytical TEM required electrolytic thinning in a 2:1 methanol:nitric acid solution at -20°C using a potential of 30 V. Oxidized specimens were backthinned by first carefully removing the

Table 1. Alloy designations and nominal compositions

Alloy Designation	Composition (at%)			
	Nb	Al	Cr	Y
1.2Cr	24.7	74.1	1.2	
2.4Cr	24.4	73.2	2.4	
4.8Cr	23.8	71.4	4.8	
5.1Cr.5Y	23.6	70.8	5.1	0.5
6.8Cr	23.3	69.9	6.8	
6.8Cr.25Y	23.2	69.7	6.8	0.2

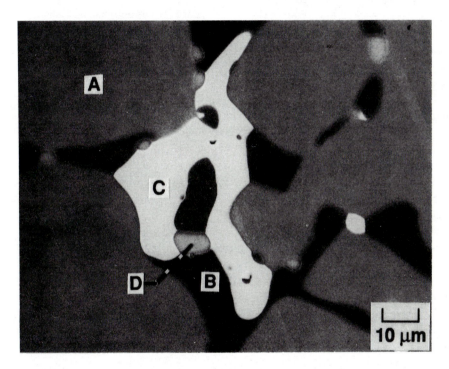

Figure 1: Secondary electron image of the as-cast microstructure from a 6.8Cr.25Y alloy; (A) NbAl$_3$+Cr matrix, (B) h-Cr$_2$Al, (C) AlNbCr, (D) Y-rich phase.

scale with SiC paper from one side of the specimen, then electropolishing from the exposed metal side. Ion thinning of oxide scales protruding over the hole formed during electropolishing was sometimes necessary.

Additional characterization of all specimens was performed using analytical scanning electron microcopy (SEM). Quantitative energy dispersive spectroscopy (EDS) was performed in some cases, using a binary NbAl$_3$ specimen as standards for niobium and aluminum, and elemental Cr as a standard for chromium.

RESULTS

The experimental results consist of oxidation weight gains from thermal gravimetric analysis (TGA), and electron microscope (scanning and transmission) images and EDS data. The results are divided into alloy microstructure from as-cast and oxidized specimens, microstructures and morphologies of oxide scales, and TGA results.

Alloy Microstructures

An SEM secondary electron image of a typical alloy microstructure is shown in Fig. 1. The specimen in this case is a 6.8Cr.25Y alloy in the as-cast condition. Four different phases are indicated as A-D. Phases A-C contain Nb, Al and Cr as determined by EDS, while phase D also contains Y. Phases A-C were observed in all alloys. The compositional results of individual analyses determined by quantitative EDS for these and other phases are shown in Table 2 for all specimens. The quantitative results are indicated in wt%, and include the total wt% based on the sum of all constituents. A total wt% of 100 is ideal, with ± 5 wt% generally indicating good analyses. Normalized concentrations are also included in at% for better comparison. Phase A is the $NbAl_3$+Cr matrix. As listed in Table 2, Cr levels up to 2 at% were observed in the matrix phases, while up to 4.1 at%Cr was observed in precipitates of $NbAl_3$+Cr trapped within the Cr-rich grain boundary phases. The amount of Cr in the matrix is observed to increase with increasing Cr content in the alloy. The darkest phase (B) in Fig. 1 contains mostly Cr and Al in the ratio of 2:1 or 3:2 with about 1-2% Nb. The Cr-Al binary phase diagram reveals η-Cr_2Al to be the only phase in this composition region. [7] Selected area electron diffraction (SAD) patterns revealed this phase to have the hexagonal crystal structure of $Al_2Cr_4C_2$($P6_3$/mmm) rather than that of tetragonal Cr_2Al (Figs. 2a and 2b). Attempting to identify carbon in this phase using EDS in the TEM and SEM modes revealed no carbon enrichment. Electron energy loss spectroscopy (EELS) has not yet been attempted. Further evidence for this phase not being a carbide is that the microstructure of this phase is a semicontinuous grain boundary phase rather than discrete precipitates which is more typical for carbides with an alloy matrix. It is believed that the small amount of Nb in this phase has stabilized the $P6_3$/mmm structure. The lightest phase (C) shown in Fig. 1 exists in appreciable amounts within the grain boundary regions. The compositions of this phase from many different alloys reveal the stoichiometry to be approximately 4:3:3 of Al:Nb:Cr. SAD revealed this phase to have a $P6_3$/mmm structure, as well. Although no powder diffraction file card has been issued for this phase, an intermetallic phase with the formula AlNbCr and having the structure reported here has been identified elsewhere[8]. This intermetallic phase is therefore believed to have a relative large phase field with respect to aluminum solubility. The phase indicated as D in Fig. 1 contains Y as well as Al, Cr and Nb. Quantitative analysis was not performed because of the lack of appropriate standards for Y at the time. This phase, which has the approximate compositions indicated in Table 2, was observed in both Y-containing alloys.

An additional grain boundary phase was observed in the 1.2Cr alloy during TEM characterization. SAD revealed this phase to have a structure corresponding to Cr_5Al_8 (Fig. 2c). The composition of this phase (Table 2) combined with the crystal structure indicates that it is either the ϵ or the ζ_1 phase.

Secondary images of an as-cast 4.8Cr alloy cross section are shown in Figs. 3a and 3b. Towards the interior of the specimen, appreciable amounts of η-Cr_2Al (darkest phase) are observed at the grain boundaries along with small amounts of AlNbCr (lightest phase). More AlNbCr is formed near the outer specimen surface during the casting process due to its higher melting temperature (solidus $T_m \approx 1580°C$) than for η-Cr_2Al (solidus $T_m = 911°C$). The effects of oxidation on the alloy's microstructure are shown in Figs. 3c and 3d. AlCrNb increased significantly both near the middle of the specimen and especially near the metal/oxide scale surface. In addition, porosity is observed near the metal/oxide scale surface. No internal oxidation was observed.

Oxidation Kinetics

The isothermal weight changes per unit area for 1200°C oxidation are plotted in Fig. 4 as a function of square root of time. Results for true parabolic kinetics would appear as straight lines, as depicted from dotted lines in Fig. 4 which correspond to decades of parabolic growth rates constants, k_p. A k_p for typical alumina scales, such as those formed on NiCrAl alloys, is approximately 0.02 $mg^2/cm^4/h$.

$NbAl_3$ has been observed to oxidize according to linear kinetics. When plotted parabolically, the kinetics of scale growth on $NbAl_3$ are extremely fast having a k_p approximately 2-3 orders of magnitude greater than typical α-Al_2O_3 scales.

Scales formed on the 2.4Cr and 4.8Cr alloys showed much slower kinetics than for binary $NbAl_3$, but approximately one order of magnitude greater than typical α-Al_2O_3 scales. The scallops in the kinetic curves appear to be related to scale cracking and subsequent rapid oxidation as will be shown later when discussing scale morphologies and microstructures. After 400 h at 1200°C for the 4.8Cr alloy, oxidation kinetics are increasing and breakaway oxidation is imminent.

The initial testing of the 6.8Cr alloy was performed using a very porous and cracked casting, which is believed to be responsible for the initial large weight gain. After initial oxidation, the slope of the weight change kinetic curve for the 6.8Cr alloy (and for a 6.8Cr alloy from a previous study (6.8Cr(b)) [4]), approaches the slopes of the 2.4Cr and 4.8Cr alloys.

Table 2. Compositions of alloy phases determined using EDS

Phase	Alloy		Nb	Al	Cr	Total
ϵ, ζ_1-Cr_5Al_8	1.2Cr	wt%	---	--	---	
		at%	2.0	54.4	43.6	
		wt%	0.6	26.2	73.9	100.7
		at%	0.3	59.3	40.4	
			2.7	42.4	55.6	100.7
			1.1	58.9	40.0	
η-Cr_2Al	1.2Cr	wt%	---	---	---	
		at%	1.1	36.1	62.8	
	2.4Cr	wt%	1.6	19.1	76.4	97.1
		at%	0.8	32.2	67.0	
	4.8Cr		3.6	20.2	70.6	94.4
			1.8	34.9	63.3	
	6.8Cr		2.4	21.5	78.5	102.4
			1.1	34.2	64.7	
	6.8Cr.25Y		3.4	20.3	73.3	97.0
			1.7	34.2	64.1	
AlNbCr	2.4Cr		44.8	23.3	27.4	95.5
			25.8	46.1	28.1	
	4.8Cr		49.2	18.1	24.5	91.8
			31.7	40.1	28.2	
	5.1Cr.5Y		58.7	22.3	23.5	104.5
			33.1	43.2	23.7	
	6.8Cr		54.7	21.8	28.7	105.2
			30.2	41.4	28.4	
	6.8Cr.25Y		51.6	20.7	27.2	99.5
			30.1	41.6	28.3	
$NbAl_3$+Cr matrix	2.4Cr		53.3	46.0	1.9	101.2
			24.8	73.7	1.5	
	4.8Cr		52.5	45.3	1.9	99.7
			24.8	73.6	1.6	
	5.1Cr.5Y		56.5	47.7	2.0	106.2
			25.2	73.2	1.6	
			52.6	46.1	5.1	103.8
			23.9	72.0	4.1	
	6.8Cr		56.0	48.8	2.3	107.1
			24.5	73.7	1.8	
			55.8	49.1	3.5	108.4
			24.1	73.1	2.8	
	6.8C.25Y		53.8	45.8	2.4	102.0
			24.9	73.1	2.0	

Phase	Alloy		Nb	Al	Cr	Y
Y-rich Phase	6.8Cr.25Y	at%	7.4	57.5	22.2	12.9

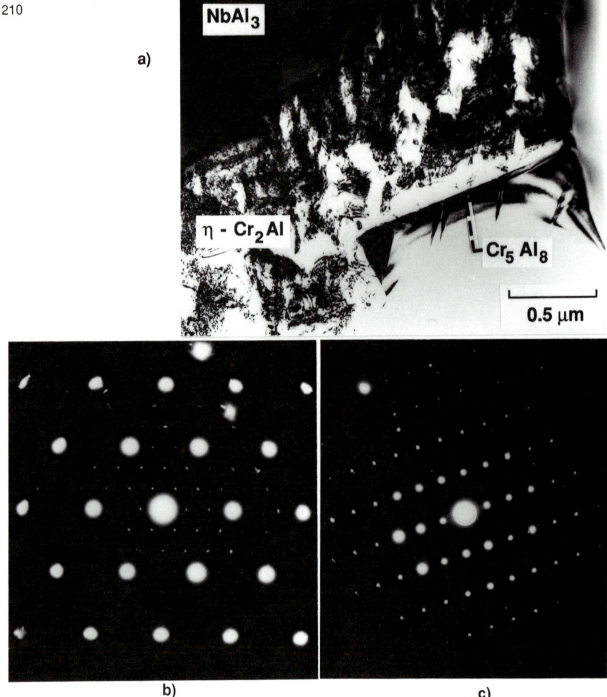

Figure 2: TEM images from an as-cast 1.2Cr alloy; (a) BF image showing grain boundary phases, (b) SAD of Cr_2Al having the $Al_2Cr_4C_2$ structure, (c) SAD of Cr_5Al_8.

The oxidation kinetics for the 6.8Cr.25Y alloy are significantly reduced relative to the Cr-containing alloys without yttrium. The oxide scale growth rates after 100 h are comparable to the growth rates of typical alumina scales.

Oxide Scale Microstructures and Morphologies

TEM bright field images of scales formed after 0.1 h at 1200°C on binary $NbAl_3$, ternary 1.2Cr, a quaternary 6.8Cr.25Y are shown in Figs 5a, 5b, and 5c, respectively. On the binary $NbAl_3$ alloy (Fig. 5a), the oxide scale consists of fine-grained (grain size ≈ 0.2 μm) α-Al_2O_3. Thicker scale formed over scratches the substrate, while a network of scale ridges, still α-Al_2O_3, is observed, and is independent of any substrate microstructure. The ridges are not convolutions within the scale layer, but appear to coincide with an even thicker scale.

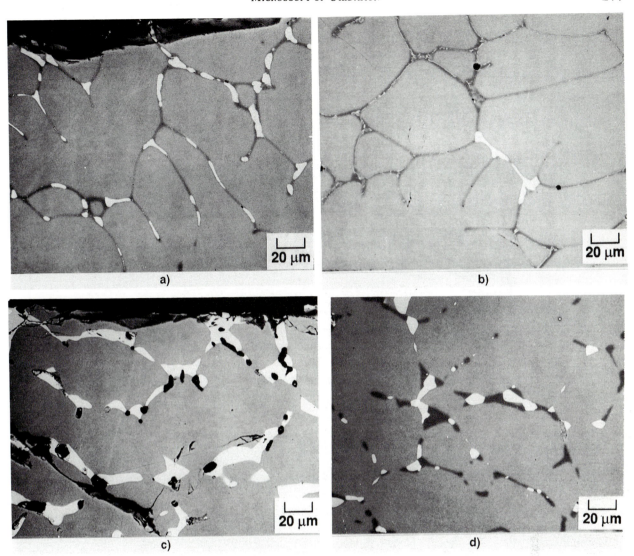

Figure 3: Secondary electron images of a 4.8Cr alloy in the as-cast condition near (a) the cast outer surface, near the middle of the specimen, and after oxidation for 500 h at 1200°C near (c) the metal/oxide interface, and (d) near the middle of the specimen.

The scale that forms on a 1.2Cr alloy (Fig. 5b) is definitely affected by the alloy microstructure. The scale that formed over the NbAl$_3$+Cr matrix is similar in all respects to the majority scale formed on binary NbAl$_3$. However, a thinner, slightly larger-grain α-Al$_2$O$_3$ formed over the grain boundary phases. The grain boundary phases are presumed to be predominantly η -Cr$_2$Al with some ε- or ζ$_1$-Cr$_5$Al$_8$. No distinct Cr-O phases were observed, but EDS indicated that the scale formed over the alloy grain bounadaries contained about 2% Cr solution.

Oxidation of 6.8Cr.25Y matrix was again similar to binary NbAl$_3$. Likewise, oxidation of the majority of grain boundary phases was similar to a 1.2Cr alloy. However, platelets or rods of an oxide phase similar to Al$_2$O$_3$ formed over the AlNbCr grain boundary phase in addition to larger-grain α-Al$_2$O$_3$ (Fig. 5c). An SAD of the AlNbCr substrate and overlapping oxide scale is shown in Fig. 6. Microdiffraction of individual platelets revealed an oxide phase having the same crystal structure as θ-Al$_2$O$_3$, but lattice parameters wc slightly larger than quoted for θ-Al$_2$O$_3$. The platelets or rods were often twinned, and had the same appearance as θ-Al$_2$O$_3$ platelets formed on NiAl [9]. Larger concentrations of impurities such as Mo, Fe, and Cr were also observed in the oxide platelets or rods, again similar to the NiAll case. Although no specific orientation relationship was determined, it is obvious from Fig. 6 that there is some preferential orientation of the θ-Al$_2$O$_3$ with respect to the AlNbCr substrate.

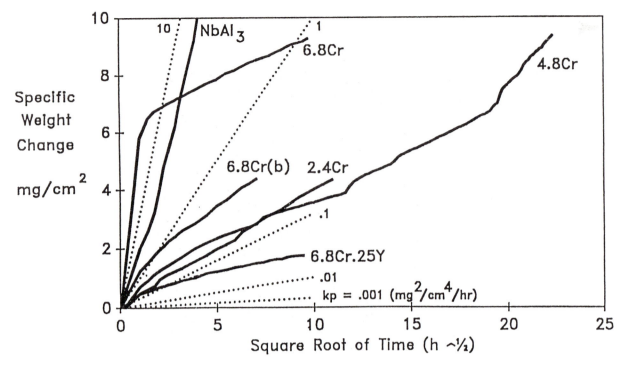

Figure 4: Oxidation kinetics at 1200°C for NbAl$_3$+Cr,Y alloys plotted as specific weight change versus the square root of time. Dotted lines represent decades of parabolic rate constants, k$_p$, in mg^2/cm^4/h.

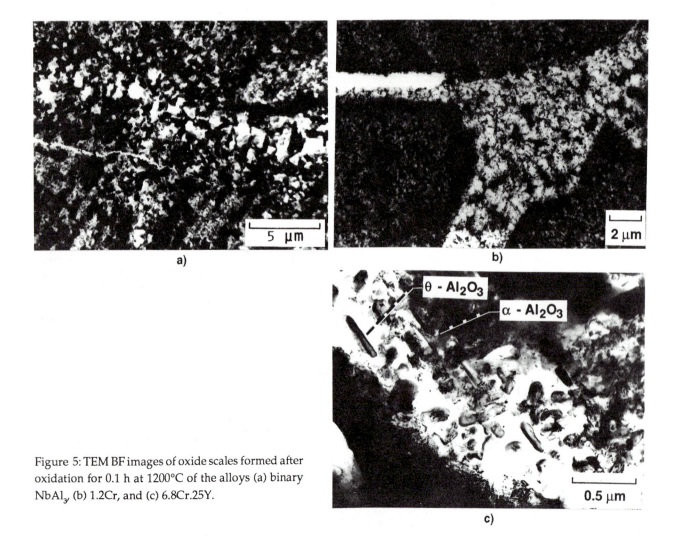

Figure 5: TEM BF images of oxide scales formed after oxidation for 0.1 h at 1200°C of the alloys (a) binary NbAl$_y$ (b) 1.2Cr, and (c) 6.8Cr.25Y.

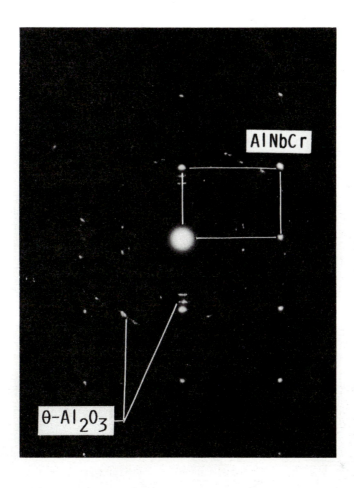

Figure 6: SAD of textured θ-Al$_2$O$_3$ and the underlying AlNbCr substrate from a 6.8Cr.25Y specimen oxidized for 0.1 h at 1200°C.

Figure 7a is a secondary electron image of a 2.4Cr specimen oxidized for 1.0 h at 1200°C then fractured. EDS analysis of the scales revealed only aluminum and oxygen suggesting the scale consists entirely of Al$_2$O$_3$. The scale is nonuniform and appears to form distinct thin regions a few microns thick, or thicker regions in upwards of 5-10 μm thick. The two distinct regions are observed in backscatter (BS) mode (Fig. 7b) as bright and dark regions corresponding to thin and thick scale, respectively. Image analysis was performed on BS images for oxidized NbAl$_3$ containing up to 4.8% Cr. (The ternary 6.8Cr specimen was nonuniform due to poor alloy integrity.) The relative area fraction of thin scale formed after oxidation for 1.0 h increased with increasing alloy Cr content, as shown in Fig. 8.

Beyond 1 h of oxidation at 1200°C, the oxides scales formed on Cr-containing alloys were essentially α-Al$_2$O$_3$ revealed by X-ray diffraction. Secondary electron images of scale surfaces from alloys oxidized for 50 or 100 h are shown in Figs. 9a-9f. Scales formed after 100 h on ternary NbAl$_3$+Cr alloys containing 4.8Cr or less are similar in appearance to Figs. 9a and 9b which are for a 2.4Cr alloy. The major features are gross convolutions of the α-Al$_2$O$_3$ scales. However, the scale formed on a 6.8Cr.25Y alloy is uniformly thin except for Y-rich oxides which form above alloy grain boundaries (Figs. 9c and 9d). The scale that formed after 50 h, a rapidly solidified 6.8Cr alloy (Figs. 9e and 9f) appears similar to the 6.8Cr.25Y alloy except for the absence of Y-rich oxide phases.

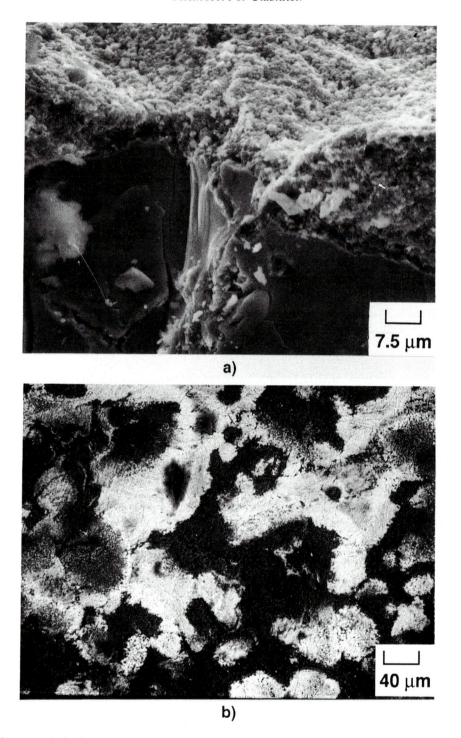

a)

b)

Figure 7: SEM images of a 2.4Cr alloy oxidized for 1.0 h at 1200°C; (a) secondary electron image from fracture cross section, (b) backscatter electron image of the scale surface.

The features described above are often better demonstrated by observing cross sections of the substrates with attached oxide scales (Figs. 10a-10d). Convolutions in the scale after 100 h on the lower-Cr alloys appear be regions where the scale cracks resulting in new scale formation on the exposed alloy substrate (Fig. 10a). The scale that formed after 100 h on the 6.8Cr alloy was only partially adherent to the alloy, and showed indications of cracking and porosity resembling scales formed on the lower Cr-containing alloys (Fig. 10b). Note also a complete layer of AlNbCr is present beneath the oxide layer. The scale that formed after 50 h on rapidly solidified 6.8Cr alloy is the same as on a conventionally cast alloy (Fig. 10c). The substrate layer directly beneath the oxide scale is AlNbCr, as is the lighter alloy phase. An optimum scale microstructure appears to be associated with the presence of Y in the alloy, as shown by the uniformly thin, compact, flat,

Relative Fraction of Thin Scale

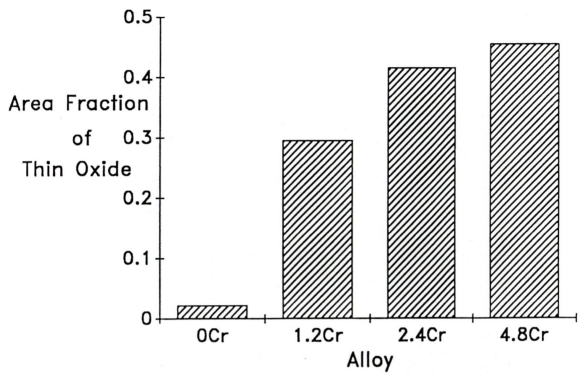

Figure 8: Plot of the relative area fraction of thin scale (white regions in Fig. 7b) for alloys containing up 4.8% Cr oxidized for 1.0 h at 1200°C.

adherent α-Al_2O_3 scale formed on the 6.8Cr.25Y alloy after 100 h (Fig. 10d). Some Y-rich oxide phases are indicated within the scale.

Longer oxidation exposures of low Cr-containing alloys, especially in regions where little Cr-rich grain boundary phases occurred, resulted in the formation of thick oxide nodules (Fig. 11a). The oxidation mechanism for nodule formation is the same as that for oxidation of binary $NbAl_3$ [4]. The process of mixed $AlNbO_4$/α-Al_2O_3 oxidation is shown in Fig. 11b.

DISCUSSION

Alloy Microstructures

Oxidation of $NbAl_3$+Cr results in the formation of predominantly α-Al_2O_3 scales. α-Al_2O_3 forms only transient conditions on AlNbCr, and $AlNbO_4$ forms only during non-protective scale formation. Therefore, as the alloy is oxidized, aluminum is removed from the alloy. The presence of increased amounts of AlNbCr with alloy grain boundaries (e.g. Figs. 3a and 3b) and at the metal/scale interface (e.g. Figs. 10b-10d) following oxidation, indicates that removal of aluminum by oxidation leads to formation of AlNbCr. In addition, it was observed that the higher the alloy Cr content, the higher was the fraction of AlNbCr as a constituent phase. This indicates that, with the experimental result of Al removal by oxidation, AlNbCr is favored by increased Cr:Al ratios in an otherwise binary $NbAl_3$ system, as would be expected from stoichiometry.

The parent phases are the $NbAl_3$+Cr matrix and typically Cr_5Al_8. The possible diffusion paths that could be taken to result in AlNbCr can be surmised from a ternary Al-Nb-Cr phase diagram. Although a ternary Al-Nb-Cr phase diagram is not available in the literature, an estimate of the pertinent phase fields can be obtained using information from the binary diagrams, from room temperature phase compositions in the present study and from Cr solubilities in Nb-Al phases provided by Argent [10]. Figure 12 shows a possible ternary phase diagram at 1200°C for the Cr-lean region of the Al-Nb-Cr system.

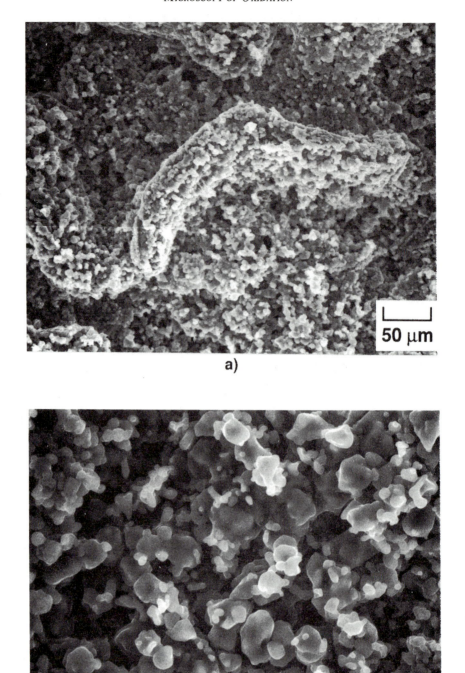

Figure 9: Secondary electron images of oxide scales formed at 1200°C on (a), (b) 2.4Cr oxidized for 100 h.

Since formation of AlNbCr appears to be occurring predominantly at the metal/scale interface and at alloy grain boundaries, Al is being depleted primarily from the grain boundaries. Further evidence for this is the large amount of grain boundary porosity (Fig. 4b) due to oxidation, which most likely is Kirkendall porosity from the selective removal of Al. Therefore, in the grain boundary regions, the relative ratio of Cr:Al increasing with the Nb contribution being controlled by the surrounding matrix. Since no formation of Nb was observed under normal circumstances and the volume fraction of Cr-Al phases does not appear to increase it is believed that the diffusion path for formation of AlNbCr from $NbAl_3$ follows directly from $NbAl_3$ into two phase $NbAl_3$-AlNbCr field. Similarly, formation of AlNbCr from Cr_5Al_8 moves directly from Cr_5Al_8 into two phase Cr_5Al_8-AlNbCr field.

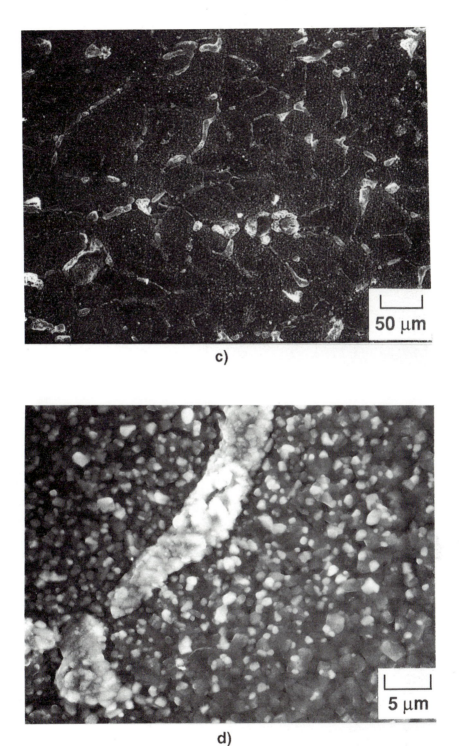

Figure 9 (cont'd): Secondary electron images of oxide scales formed at 1200°C on (c), (d) 6.8Cr.25Y oxidized for 100 h.

Another problem that is a concern for the use of $NbAl_3$+Cr alloys can be understood and resolved through the use of the ternary phase diagram in Fig. 12. The problem is that Cr_5Al_8 (and η-Cr_2Al) has a low melting temperature relative to $NbAl_3$ and AlNbCr which would be a limiting factor for the use of these alloys. Argent reports the solubility of Cr in $NbAl_3$ to be 5 at% at 1000°C [10]. At 1200°C, the solubility should be slightly higher (7 at% was used in Fig. 12) as surmised from binary diagrams. Therefore, as the alloys are heated less Cr_5Al_8 (and η-Cr_2Al) should be present because more Cr goes into solution. Finally, as oxidation continues, Al is removed from the system moving the equilibrium phase volume fractions away from the lower melting temperature Cr_5Al_8 and (and η-Cr_2Al) phase fields.

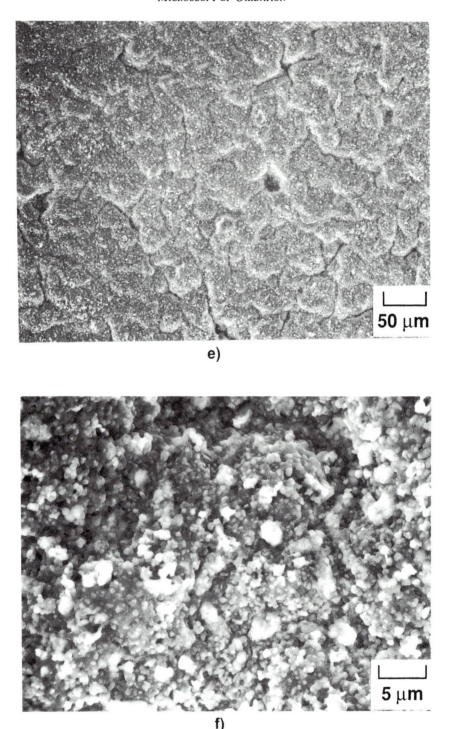

Figure 9 (cont'd): Secondary electron images of oxide scales formed at 1200°C on (e), (f) rapidly solidified 6.8Cr alloy oxidized for 50 h.

Oxidation Mechanisms

The initial oxidation mechanisms (through 0.1 h at 1200°C) for the binary alloys, and the matrices of the ternary and quaternary alloys appear to be similar. This would indicate that the Al activity at the alloy surface is sufficient to form Al_2O_3, and that Cr additions in $NbAl_3$ up to 6.8 at% have little affect on the initial scale formation. Above the grain boundary phases in the ternary and quaternary alloys, the scales are thin possibly due to the same beneficial effect of Cr on transient oxidation as was observed for NiCrAl alloys especially during the specimen heating cycle [11]. The oxidation mechanisms are shown schematically in Fig. 13. The scale thicknesses for the different oxidation times are only approximated.

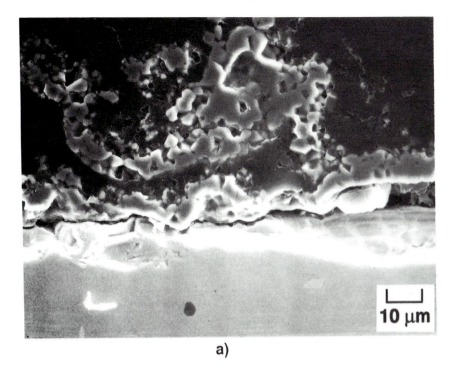

a)

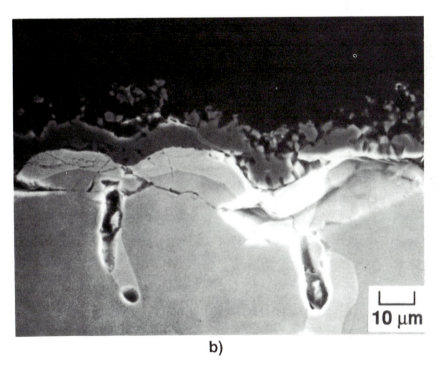

b)

Figure 10: Secondary electron images of specimen cross sections after oxidation at 1200°C of (a) 2.4Cr for 1 h, (b) 6.8Cr oxidized for 100 h.

After 1.0 h of oxidation at 1200°C, the scales formed on binary alloys are more uniformly thick relative to the NbAl₃+Cr ternary alloys (see Fig. 13). The fraction of the thinner scale increases with alloy Cr content indicating that Cr plays a dominant role in reducing overall weight gains due to oxidation. Recalling that the effect is observed during transient oxidation and a complete layer of AlNbCr is observed at the metal/scale interface after longer oxidation times, the effect of Cr is possibly to promote formation of AlNbCr. transient oxidation stage in the Cr-containing alloys can then be defined as the time necessary to form a complete layer of AlNbCr beneath the oxide scale. This phase appears to be inherently more oxidation resistant than the NbAl₃+Cr matrix. Although oxidation growth rates for alumina scales are typically controlled diffusion through the scale (a parabolic process), the growth rates of scales that form on binary NbAl₃ or NbAl₃

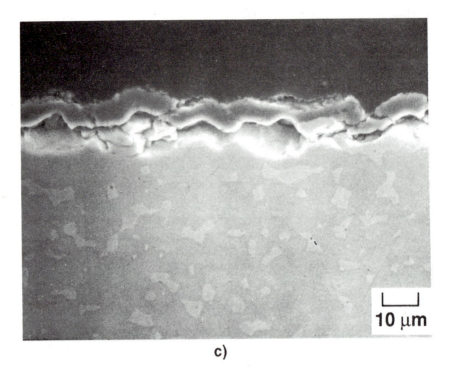

c)

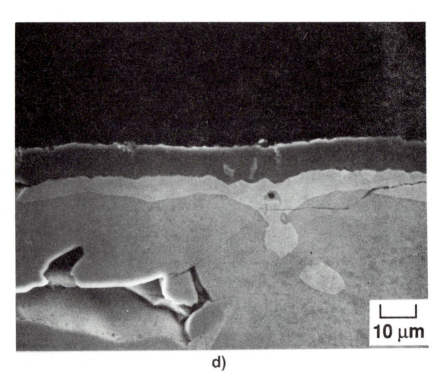

d)

Figure 10 (cont'd): Secondary electron images of specimen cross sections after oxidation at 1200°C of (c) rapidly solidified 6.8Cr oxidized for 50 h, and (d) 6.8Cr.25Y oxidized for 100 h.

with Cr in solution are probably controlled by fast diffusion through physical defects in the scales such as cracks. Evidence for this is provided in the microstructures and in Fig. 4 where scallops form in the kinetic curves on low Cr-containing alloys, and long term scale growth rates (4.8Cr) are faster than for parabolic kinetics.

Although a complete layer of AlNbCr was not observed using SEM at the metal/scale interface after oxidizing the 2.4Cr and 4.8Cr alloys, this is not conclusive evidence that a thin layer of a second phase does not exist. Further studies need

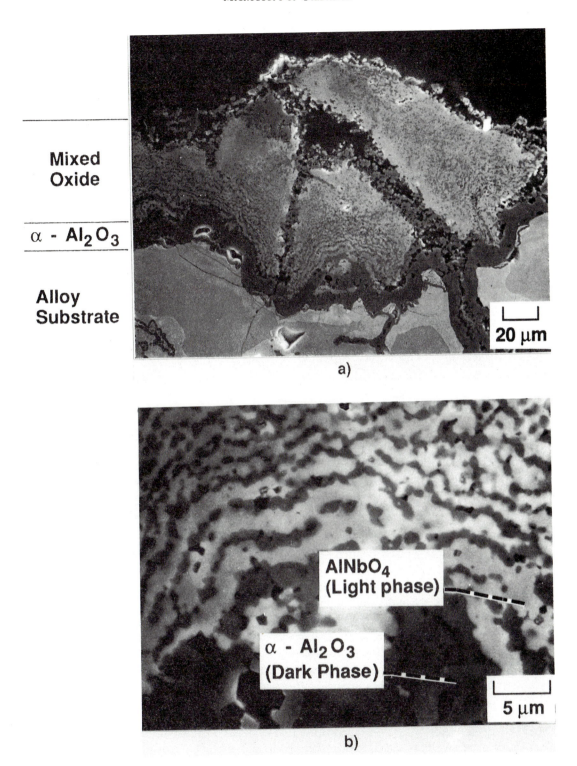

Mixed Oxide

α - Al_2O_3

Alloy Substrate

20 µm

a)

$AlNbO_4$
(Light phase)

α - Al_2O_3
(Dark Phase)

5 µm

b)

Figure 11: Secondary electron cross section images from a 4.8Cr specimen oxidized for 500 h at 1200°C showing (a) oxide nodule, (b) products from mixed oxidation behavior.

to be performed to determine the composition of the alloy near the metal/scale interface and if $NbAl_3$ with Cr in solution is sufficient to promote continued Al_2O_3 formation over the long term. Based on the 1200°C, 500 h kinetic study of the 4.8Cr alloy and not observing a continuous AlNbCr layer, it appears that Cr in solution does improve oxidation resistance only over the short term. These improvements would occur by gettering or an increase in the flux of aluminum to the metal/scale interface as might arise from an increase in the alloy aluminum diffusivity or a decrease in the alloy oxygen solubility [15].

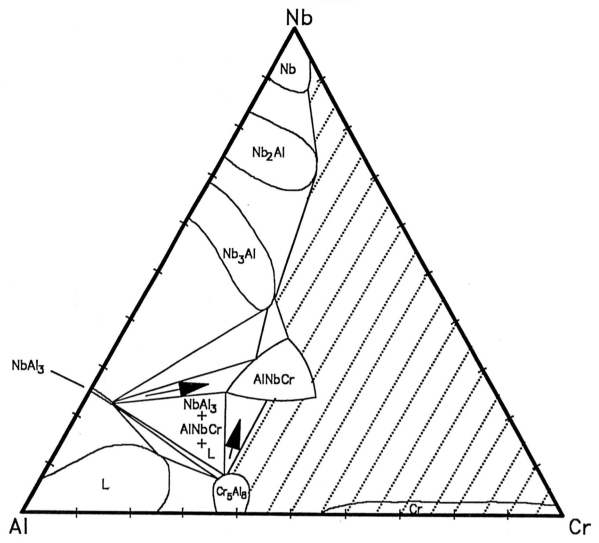

Figure 12: Possible Nb-Al-Cr ternary phase diagram at 1200°C.

Yttrium additions to the 6.8Cr alloy significantly improved the scale growth kinetics to the level of typical alumina formers. Not only was the adherence of the scale improved, but the scale microstructure was completely compact and uniformly thin. In contrast to scales on the Y-free alloys, the scale formed on the Y-containing alloy was not convoluted (Fig. 13). Therefore, yttrium appears to promote effects of improved scale adherence, reductions in the scale growth rate, and prevention of convolutions in the scale. These effects have been observed previously in other systems, and are termed "reactive element effects" [12]. Although many of the reactive element effects are observed in these alloys, more study is necessary to determine whether yttrium has a primary effect and to optimize its concentration.

CONCLUSIONS

(1) Chromium additions improve the isothermal oxidation resistance over the binary alloy by promoting continuous Al_2O_3 formation. This could occur in the short term and in low Cr-containing alloys by an increase in the aluminum flux to the metal/scale interface. In the higher Cr-containing alloys and over the long term, a complete layer of AlNbCr forms beneath the scale which appears to be inherently more oxidation resistant than the alloy matrix.
(2) Yttrium additions reduce the scale growth rate, improve the scale adherence and reduce convolutions with the scale. All these effects have been observed in prior studies on other systems, and are characteristic of reactive element effect.

ACKNOWLEDGEMENTS

This work was supprted under the HITEMP program at NASA Lewis Research Center. Helpful discussions with Dr. J.L. Smialek and Dr. J.A. Nesbitt are appreciated.

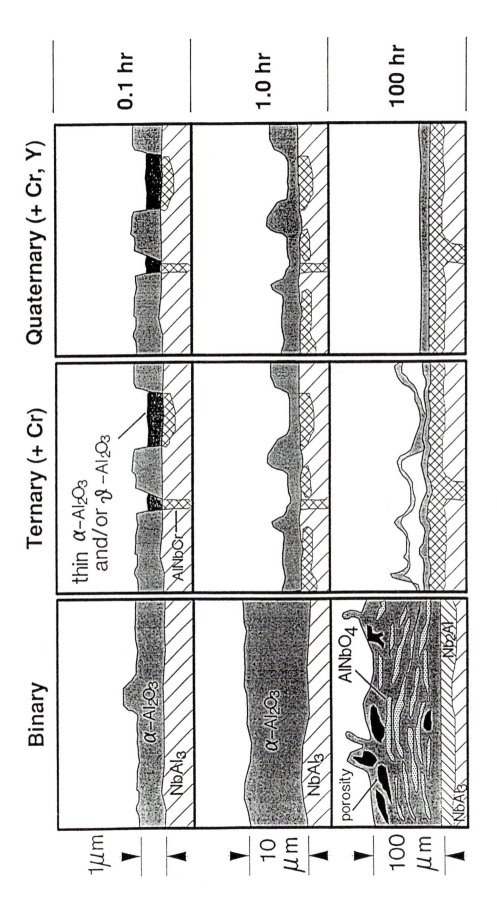

Figure 13: Schematics of oxide scales formed on NbAl$_3$+Cr,Y alloys oxidized at 1200°C.

REFERENCES

1. R.C. Svedburg, Properties of High Temperature Alloys, Eds. Z.A. Foroulis and F.S. Pettit, The Electrochemical Society, 1976, p. 331.

2. J. Berkowitz-Mattuck and M. Rossetti, NASA CR-125071, 1971.

3. M.G. Hebsur, J.R. Stephens, J.L. Smialek, C.A. Barrett and D.S. Fox, Oxidation of High-Temperature Intermetallics, Eds. T. Grobstein and J. Doychak, TMS:Warrendale, PA, 1989, p.171.

4. R.A. Perkins, K.T. Chiang and G.H. Meier, Scripta Met., 22,1988, p. 419.

5. M.G. Hebsur and J.R. Stephens, 1990, Approved U.S. Patent Serial # 07/406,700.

6. J. Doychak, C.A. Barrett, J.L. Smialek, M.G. Hebsur and D.L. Humphrey, NASA CP-10039, 1989, 23-1.

7. T.B. Massalski, Ed.-in-chief, Binary Phase Diagrams, ASM, vol. 1,1986.

8. P. Villars and L.D. Calvert, Pearson's Handbook of Crystallographic Data for Intermetallic Phases, ASM, 2, 1985, p. 939.

9. J. Doychak, J.L. Smialek and T.E. Mitchell, Met. Trans., 20A, 1989, p. 499.

10. B.B. Argent, Niobium: Proceedings of the International Symposium, Ed. H. Stuart , TMS:Warrendale, PA, 1984, p.325.

11. B.H. Kear, F.S. Pettit, D.E. Fornwalt and L.P. Lemaire, Oxid. Met., 3,1971, p. 557.

12. D.P. Whittle and J. Stringer, Phil. Trans. R. Soc. Lond., A295, 1980, p. 309.

26 MICROSCOPY OF THE CORROSION OF HIGH-TEMPERATURE COATINGS

G. H. Meier and F. S. Pettit

*Department of Materials Science and Engineering,
University of Pittsburgh*

ABSTRACT

An important aspect of minimizing the effects of high temperature corrosion involves understanding the mechanisms of degradation. The development of this understanding involves the ability to reproduce the corrosion morphologies found in service by conducting well controlled laboratory experiments. This paper describes experiments in which this has been attempted. The results of morphological studies of the oxidation and hot corrosion of uncoated superalloys as well as diffusion aluminide, CoCrAlY overlay, and zirconia thermal-barrier coatings are presented. The influence of Pt additions on the cyclic oxidation and hot corrosion behavior of diffusion aluminides is described in detail. It will be shown that the composition of the substrate has a substantial influence on the corrosion rates and morphologies. The development of the hot corrosion morphologies of CoCrAlY coatings on Ni-base superalloys is described and the use of transmission electron microscopy in characterizing the fine details of these morphologies discussed. Finally, the development of the hot corrosion morphologies of stabilized zirconia thermal-barrier materials is described.

INTRODUCTION

The documentation and characterization of corrosion microstructures via optical, scanning, and transmission electron microscopy is one of the most effective means to correlate field experience and laboratory test results. The capability to reproduce certain corrosion morphologies in the laboratory permits the formulation of models to describe the important degradation processes as well as the development of more corrosion resistant coatings-alloys systems. In the following examples will be presented to illustrate the use of optical and electron microscopy in describing alloy and coating corrosion degradation.

RESULTS AND DISCUSSION

Uncoated Superalloys

The examination and documentation of uncoated alloy oxidation or hot corrosion is often a prerequisite for coating selection. In Fig. 1 the depleted zone formed upon oxidation of alloys such as PWA 1480 and CMSX-3(Compositions given in Table 1) at 1100°C is shown. An aluminum-depleted zone has developed as a result of oxidation but a recrystallized zone, consisting of γ' lamellae in γ, has also been formed. This recrystallized zone formed as a result of residual stresses established at the surface of the alloy upon removing mold material by grit blasting. An acicular phase was also present in the depleted zone on numerous specimens, Fig. 2. Such acicular phases were rich in the refractory metals tantalum and tungsten. Moreover, the acicular phases were rapidly oxidized, as shown in Fig. 3. Metallographic results, such as those illustrated in Figs. 1 to 3, show that alloys such as PWA 1480 and CMSX-3 must be coated if they are to be used in oxidizing gases at temperatures as high as 1100°C.

Diffusion Aluminide Coatings

Aluminide coatings applied by pack cementation are commonly used to extend the lives of Ni-base superalloys by producing an Al-rich surface which forms a protective alumina film in service. Aluminide coatings are degraded by reaction with the environment(corrosion) and by interdiffusion with the substrate, both of which deplete Al from the coating. A significant advance in the technology of diffusion aluminides was made when Pt was incorporated into aluminide coatings [1, 2] which greatly improved the degradation resistance of the coatings [2-8]. This section of the paper is concerned with defining the types of degradation for which Pt additions are effective means to extend coating lives and describing the mechanisms by which Pt produces beneficial effects.

Several straight aluminide and platinum aluminide coatings were studied on the superalloy substrates whose compositions are indicated in Table 1. Micrographs showing the microstructures of typical as-processed coatings are presented

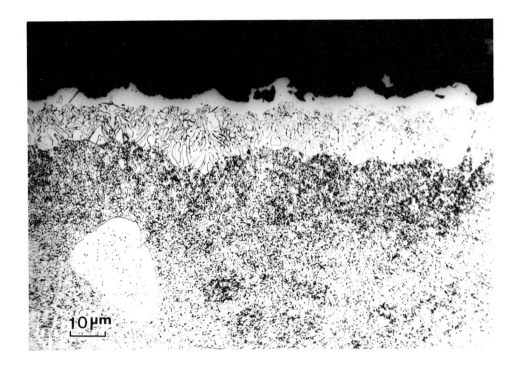

Figure 1: Depleted zones typical of those formed on PWA 1480 and CMSX-3 during oxidation in air at 1100°C.

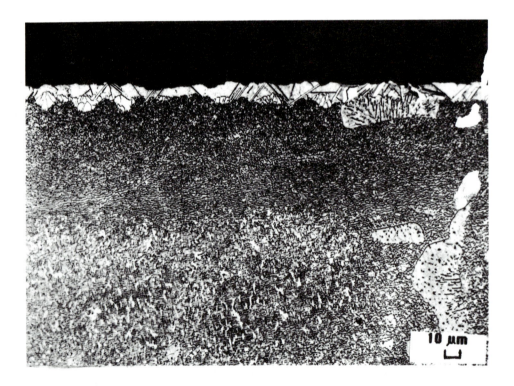

Figure 2: Oxidation-affected zone containing plates and coarsened γ′. Structures typical of those formed on PWA 1480 and CMSX-3 for 20 hours at 1100°C.

Table 1. Compositions of alloy substrates (wt%)

Sample	Ni	Cr	Al	Ti	Co	Ta	Mo	W	Hf	B	Zr	C	Nb
IN738	Bal.	16	3.4	8.5	1.7	1.7	1.7	2.6	–	.01	–	.17	.9
MAR M200	Bal.	9	5	2	10	–	–	12	–	.015	.05	.15	1
RENE 80	Bal.	14	3	5	9.5	–	4	4	*	.015	.32	.17	–
CMSX-3	Bal.	7.7	5.5	.9	4.9	5.8	.6	7.8	.1	–	–	–	–
PWA 1480	Bal.	9.8	5.1	1.5	4.7	11.9	–	4	–	–	–	–	–

*EDS indicated the presence of Hafnium.

in Fig. 4. The coatings are a straight aluminide (PWA 73) and a Pt-aluminide on IN738. The light discontinuous phase evident at the surface of the Pt-modified coating is $PtAl_2$ which lies in a matrix of (Pt,Ni)Al. Carbides exist in the inner zone. Alumina particles from grit blasting are also evident in the Pt-aluminide coating. These particles are in the inner zone and appear to be from 1-4 µm in diameter. Alpha-refractory metal precipitates can be observed beneath the $PtAl_2$ layer. The PWA 73 coating has a three zone structure.

Isothermal oxidation tests were performed on a number of the coating systems at 1100°C in air. The platinum modified coatings usually had smaller weight increases than the conventional aluminide coatings, however, some of the conventional aluminides also exhibited very small weight increases. This test was further complicated by the fact that some weight losses may have occurred from the platinum modified samples as a result of PtO_2 vaporization from the platinum tabs used to support the coupons for electroplating. Cross sections of the scales of both coating types on MAR M200 are presented in Fig. 5. The scale upon the straight aluminide coating is about 4 times thicker than that on the Pt-modified coating. These thicknesses were used to calculate parabolic rate constants. Values of 2.8×10^{-12} and 2×10^{-13}g^2/cm^4-sec were obtained for the oxides formed on the conventional and platinum aluminides, respectively. The value of 2×10^{-13} is in good agreement with alumina growth on platinum-aluminum alloys. The coating microstructures beneath the scales in Fig. 5 (not shown) were very different for the two coatings. The straight aluminide had formed an appreciable amount of 'γ', and in some areas γ, next to the scale. The platinum aluminide on the other hand displayed a continuous layer of β phase adjacent to the scale with γ' appearing only at the grain boundaries.

Cyclic oxidation tests were performed at 1100, 1135, and 1200°C and typical weight change versus time measurements from these tests are presented in Fig. 6 for different coating systems. In all of these tests the platinum modified coatings performed better than the conventional aluminides. In particular, protective alumina scales were maintained for longer periods of time on the platinum modified coatings compared to the conventional coatings. Some platinum modified coatings exhibited lives greater than others because substrate elements influenced the coating lives. Figure 7 shows micrographs of the surface of coated IN738 after 322 cycles of testing. The Pt-aluminide coating displays alumina whiskers and outgrowths upon a web or lacy structure. The EDS of the general scale on a Pt-aluminide indicates a scale rich in Al and Cr. The surface of the straight aluminide shows a considerable amount of relief and no one oxide morphology dominates. The EDS analyses of these scales are similar to the Pt modified coating but indicate a higher amount of Ti. Figure 7 indicates substantial spalling of the alumina from both coatings. Therefore, it does not appear that the beneficial effects of Pt are associated with improved scale adherence. Results obtained from acoustic emission tests with several coating systems are also consistent with the proposal that the platinum modified coatings do not exhibit

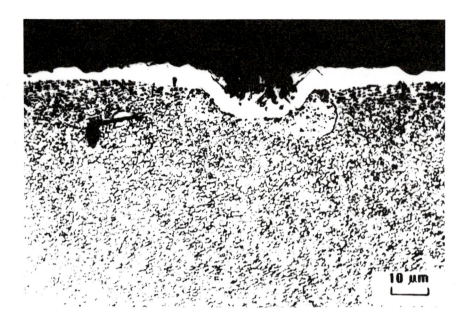

Figure 3: Photograph showing the preferential oxidation of the plate phase during cyclic oxidation of a specimen. Microstructures typical of those formed in PWA 1480 and CMSX-3 after 24 hours of oxidation in air at 900°C.

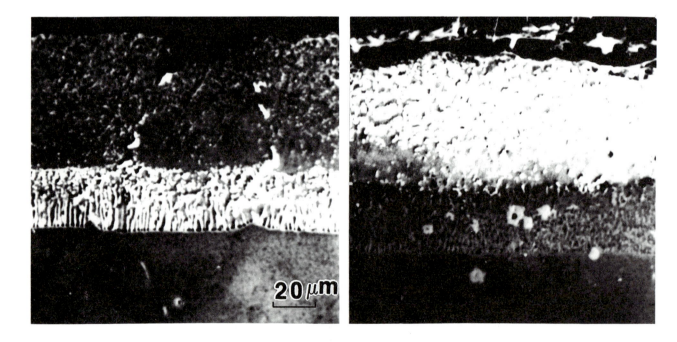

Figure 4: SEM micrographs of typical diffusion aluminide coating microstructures. Left is PWA 73 on IN738. Right is Pt-aluminide on IN738.

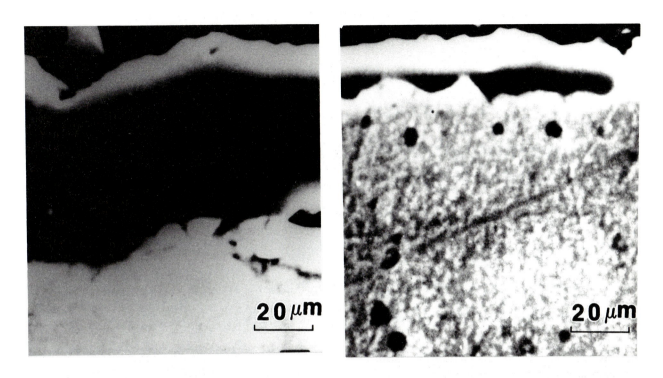

Figure 5: Cross-sections of scales after isothermal oxidation in air for 1 week at 1100°C. The left is PWA 73 on MAR M200 and the right is Pt-aluminide on MAR M200.

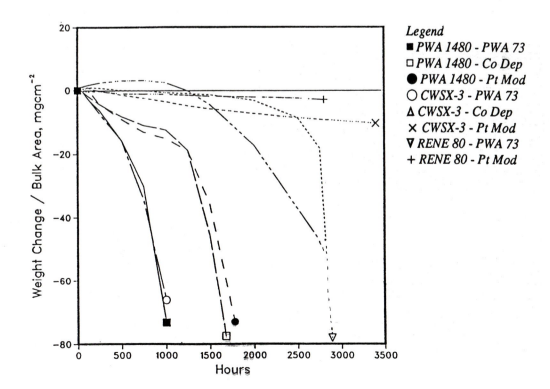

Figure 6: Cyclic oxidation of various systems at 1135°C in air.

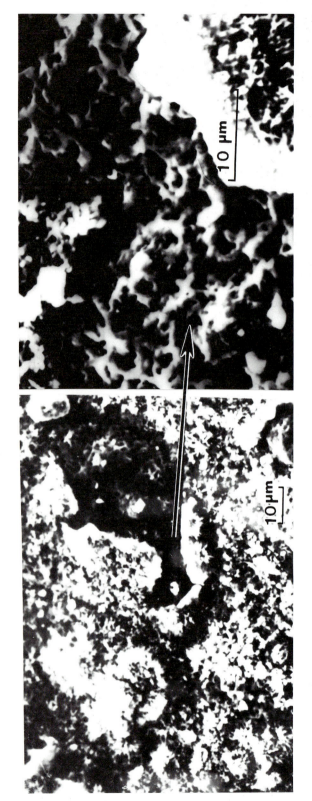

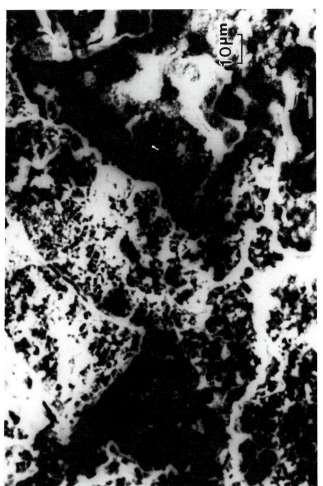

Figure 7: Surface micrographs of PWA 73 (top) and Pt-aluminide (bottom) on IN738 after 322 hours of cyclic oxidation at 1100°C. Note that there is evidence of spalling from both coatings.

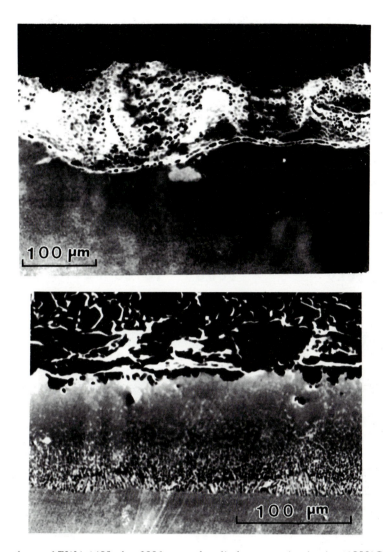

Figure 8: Cross-sections of coated PWA 1480 after 300 hours of cyclic hot corrosion in air at 1000°C. Coating (top) is PWA 73 while below is Pt-aluminide.

substantially better oxide scale adherence than the straight aluminide coatings. Therefore, it is proposed that Pt improves the ability of the coatings to reform alumina scales after spalling.

The various aluminized alloys were exposed to high temperature hot corrosion conditions which consisted of coating with Na_2SO_4 deposits and exposing to air at 1 000°C under thermal cycling. Weight change results obtained from these tests show conclusively that the platinum modified coatings are degraded less than coatings that do not contain platinum. Typical microstructures of specimens exposed to cyclic hot corrosion conditions are presented in Fig. 8. The amount of degradation is much more extensive in the coating that does not contain any platinum. Examination of the weight change and microstructural data also show that there is an effect of substrate composition. Coatings have shorter lives on Rene 80 compared to CMSX-3 which in turn are not as good as coatings on PWA1480. The significant difference between these substrates is the amount and type of refractory elements. Rene 80 contains both Mo and W. CMSX-3 contains mainly W where as PWA 1480 contains predominantly Ta. In the hot corrosion literature [9] it is well documented that refractory elements such as Mo and W can have a profound effect upon hot corrosion behavior. The results obtained from the high temperature hot corrosion tests show that platinum extends the lives of diffusion aluminide coatings. Optimum lives, however, depend on the composition of the superalloy substrate. These results are interpreted in terms of the Pt additions excluding refractory metals from the outer regions of the coatings and in improving the ability to reform alumina scales under cyclic conditions as described above.

Low temperature hot corrosion experiments were conducted by coating the aluminized alloys with Na_2SO_4 and exposing them to oxygen plus SO_3 under isothermal conditions at 700°C. Examination of these results showed that the weight

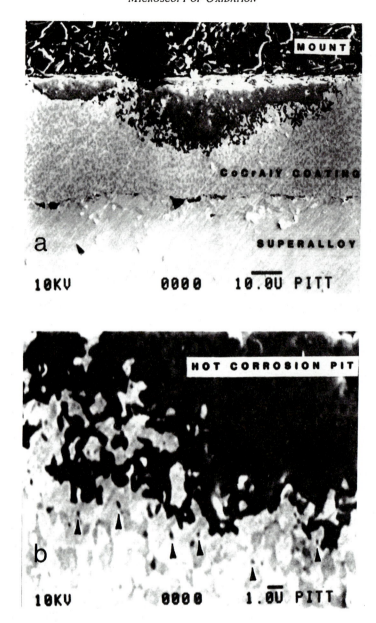

Figure 9: (a) Overall cross-sectional SEM view of a hot corrosion pit in a CoCrAlY overlay coating after 2 weeks at 700°C. (b) High magnification micrograph of the reaction front with arrows showing titanium sulfides in the coating.

changes for the coatings which contain platinum are usually significantly less than those which do not contain any platinum. This observation was also supported by metallographic examination of the exposed specimens. The improvement afforded by the platinum was, however, far less dramatic than that observed in high temperature hot corrosion and no obvious influence of substrate composition was evident in the low temperature hot corrosion tests.

Hot Corrosion of CoCrAlY Overlay Coatings

The exposure of CoCrAlY overlay coatings to the low temperature hot corrosion conditions described above, for example in marine turbines, is known to cause even more severe degradation than is experienced by the aluminide coatings. The typical morphology is a pitting type as illustrated in Fig. 9 (a). Higher magnification SEM observation of the corrosion front, Fig. 9(b), indicates some preferential attack of the Al-rich β-CoAl phase and precipitation of sulfides ahead of the corrosion front. EDS analysis of these sulfides shows them to be Ti-sulfides. The Ti has diffused up from the IN738 substrate. Combination of SEM, cross-sectional TEM, and selected area electron diffraction, Fig. 10, indicates that the bulk of the pit contains Cr_2O_3 and a mixed spinel of Co, Cr, and Al. Combination of bright-field and dark-field microscopy, Fig. 11, indicated that the metal islands in the lower part of the pit were virtually pure Co. The combination of the above morphological observations with thermodynamic and kinetic analyses, which have been eliminated for the sake of

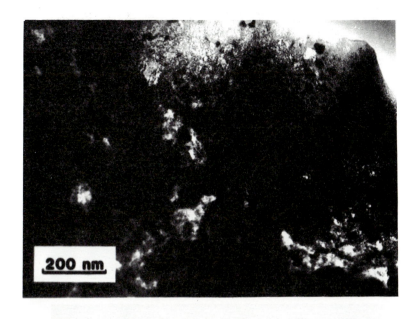

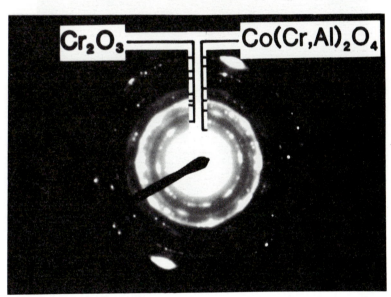

Figure 10: (Top) TEM image of the corrosion product in the center of a typical low temperature hot corrosion pit. (Bottom) Selected area diffraction pattern from the reaction product in the pit.

brevity, have lead to a model for this form of degradation. This model involves the preferential removal and reprecipitation of Al at the corrosion front, which prevents a protective oxide from forming, followed by the dissolution of Co into the sulfate melt.

Vanadium-Induced Corrosion of Thermal Barrier Coatings

Thermal barrier coatings, such as stabilized zirconia, are now being used to reduce metal surface temperature. It is well established that deposition of vanadium-containing compounds on these coatings can cause severe corrosive attack. Under some conditions this attack is associated with removal of the stabilizer from the coating. The experimental procedure to examine the effects of vanadium was similar to that for sulfate hot corrosion except that $Na_2SO_4 + NaVO_3$ deposits were used. The activity of V_2O_5 in the deposit was fixed by the SO_3 pressure in O_2 - SO_3 mixtures. Polycrystalline zirconia specimens stabilized with 6 and 12% Y_2O_3 were examined. A single crystal of ZrO_2-6% Y_2O_3 was also studied. At 900°C it was found that the corrosion attack could be described by measuring the thickness of a porous zone which developed at the surface of the specimens, Fig. 12. The thickness of this zone as a function of time is presented in Fig.

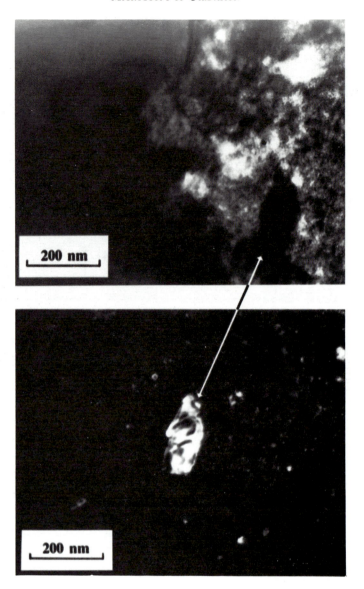

Figure 11: (Top) TEM image of the corrosion product at the base of a low temperature hot corrosion pit showing a metallic particle at arrow. (Bottom) Dark-field image of the particle using an fcc-Co reflection.

13 for 6 wt% Y_2O_3 (3.3 mole %). The kinetics conform to a parabolic rate and the rates become greater as the activity of V_2O_5 in the deposits is increased. Examination of degraded specimens shows that Y_2O_3 is removed from the ZrO_2 and precipitated at the surface of the specimen as YVO_4, Fig. 12. Some zirconium is also removed from the samples since a small amount of a phase containing zirconium and vanadium was detected upon the surfaces of specimens, Fig. 12. The porosity is concentrated at grain boundaries of the specimens. The depths of the porous zones formed upon the 12% Y_2O_3 specimens were not as great as those formed upon the 6% specimens.

The morphological observations have lead to a model for the development of the porous zone and compound formation on the specimen surfaces which is presented in Fig. 14. The model involves the dissolution of Y_2O_3 and ZrO_2 concomitantly by reactions of the type,

$$Y_2O_3 + 3ZrO_2 = 2Y^{3+} + 3ZrO_3^{2-} \quad (1)$$

At sufficiently high VO_3^{1-} activities precipitation of YVO_4 occurs on the specimen surface and the zirconium remains in solution but does precipitate on cooling. The melt gradually becomes enriched in Na_2SO_4.

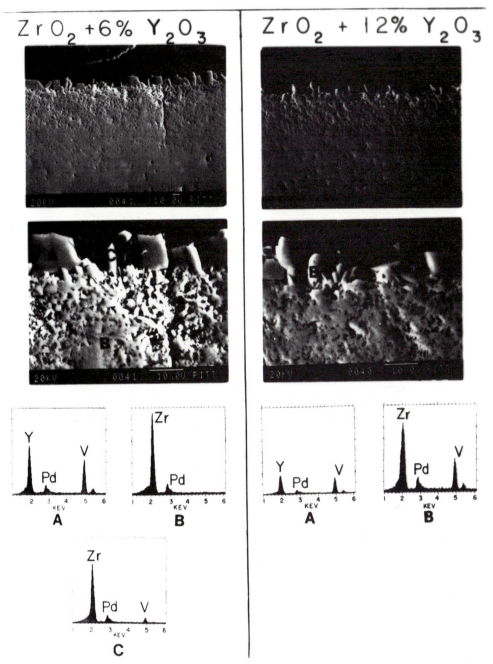

Figure 12: SEM micrographs and EDS spectra from the cross-sections of ZrO_2 + 6wt% Y_2O_3 and 12wt% Y_2O_3 after cyclic hot corrosion at 900°C.

SUMMARY

The above examples have illustrated the use of various forms of microscopy in the evaluation of high temperature corrosion morphologies. They have, hopefully, illustrated the necessity for utilizing this tool in developing mechanisms for high temperature corrosion processes.

ACKNOWLEDGEMENTS

The authors gratefully acknowledge the financial support of this work by the Office of Naval Research under Contract No. N00014-81-K-0355/P00007 and the Materials Technology Laboratory (Department of the Army). They also acknowledge the experimental contributions of J. Caola, G. Kim, J. Schaeffer and B. Warnes.

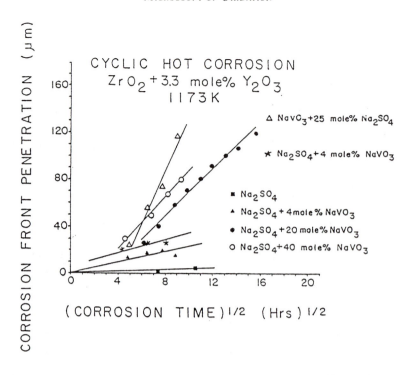

Figure 13: Kinetics of cyclic hot corrosion of $ZrO_2 + 6wt\%$ Y_2O_3 at 900°C in three different deposit-gas combinations.

FLUXING

$$Y_2O_3 + 3ZrO_2 = 2Y^{3+} + 3ZrO_3^{2-}$$

PRECIPITATION

$$2Y^{3+} + 3ZrO_3^{2-} + 12VO_3^{1-} = 2YVO_4 + 3Zr^{4+} + 3V_2O_7^{4-} + 4VO_4^{3-}$$

GAS-MELT INTERFACE

$$4VO_4^{3-} + 4SO_3 = 4VO_3^{1-} + 4SO_4^{2-}$$

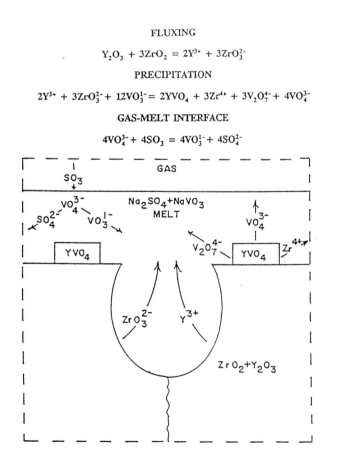

Figure 14: Important chemical reactions and model for hot corrosion of Yttria-stabilized Zirconia.

REFERENCES

1. D. A. Joshep, U. S. Patent No. 3,495,748, 1970.

2. H. H. Todd, U. S. Patent No. 33,494,748.

3. M. J. Fleetwood, Metals Sci., 98, 1970, p. 503.

4. P. Deb and D. H. Boone, "Microstructural Formation and Effects on the Performance of Platinum Modified Aluminide Coatings, Tech. Report, Naval Postgraduate School,1985.

5. K. Bungardt et al., U. S. Patent No. 33,677,789, 1972.

6. G. J. Tatlock, T. J. Hurd and J. S. Punni, Platinum Metals Rev., 31, 1987, 26, p. 26.

7. G. R. Johnston and P. G. Richards, "The Relative Durabilities of Conventional Aluminide and Platinum-Modified Aluminide Coatings in an Operational Gas Turbine Engine" in Corrosion in Fossil Fuel Systems, Ed. I. G. Wright , The Electrochem. Soc., 1983, p. 456.

8. R. Bauer, K. Schneider and H. W. Grunling, High Temp. Tech., 3, 1985, p. 59.

9. F. S. Pettit and C. S. Giggins, "Hot Corrosion" in Superalloys 11, Eds. C. T. Sims, N. S. Stoloff and W. C. Hagel, Wiley, 1987, p. 327.

27 A STUDY OF THE EFFECT OF MAGNESIUM ADDITIONS ON THE OXIDE GROWTH MORPHOLOGIES ON LIQUID ALUMINIUM ALLOYS

S. Impey, D. J. Stephenson and J. R. Nicholls

School of Industrial Science,
Cranfield Institute of Technology, Cranfield, Bedford MK43 0AL, UK.

ABSTRACT

This paper examines the effect of magnesium additions (0-5wt%) on the morphology of surface oxides formed on metals at 750°C, and compares it to scale formation on pure aluminium. It is shown that on pure aluminium dross formation proceeds by the rapid oxidation associated with the development of nodular growths. These are produced by the exudation of liquid metal from sites at which local breakdown of the thin first-formed protective oxide occurs. The number and growth of exudations increases with time, with nodules consisting of oxide envelopes containing entrapped metal.

In aluminium melts containing magnesium this process is considerably enhanced. The rate of oxide production increases with the magnesium content of the alloy. The nodules have a similar morphology to those produced on pure aluminium, although more rapidly form a continuous layer of mixed oxides and entrapped metal. It is the formation of these nodular oxide morphologies containing entrapped metal that is responsible for the high metal losses reported when recycling thin section material and accounts for the major increase in metal loss when magnesium additions are present.

INTRODUCTION

Dross, a mixture of metal and oxide, can form on molten aluminium and its alloys during high production melting. The metal entrapped within the oxide is subsequently lost when the melt surface is skimmed. As a consequence metal yields are reduced by up to 15% due to dross formation [1]. This metal loss is a particular problem in the secondary aluminium industry especially when recycling high surface area, low volume material such as beverage cans.

To overcome these problems one must understand the underlying mechanisms by which dross formation occurs. This paper therefore examines the mechanism of dross formation on liquid aluminium-magnesium alloys during secondary melting.

EXPERIMENTAL PROCEDURE

Oxidation kinetics have been studied thermogravimetrically in dry (0.05 - 0.07KPa pH_2O) air and argon environments at 750°C. The observed oxidation rates are related to changes in oxide scale morphology which occur as a function of exposure time. The materials used for this study were pure aluminium and beverage can alloys of low and high magnesium content: Al-1%Mg (5005 alloy) and Al-5%Mg (5182 alloy) respectively. Typical examples are seen in Figs. 2 and 3 for pure aluminium and on Al-5%Mg alloy respectively.

A parallel series of exposure tests was performed in which oxide samples were removed from melts after times of 5, 60, 300 and 900 minutes using a copper loop technique [2]. These surface films, formed during the early stages of the oxidation process, were examined by transmission electron microscopy to determine their structure and composition of the scale formed as a function of exposure time. To examine the early stages of oxide growth on pure aluminium excess metal was removed with a 3% bromine-methanol solution. However, the duplex structure of the oxide developed on the Al-Mg alloys restricts the use of this method hence for these samples excess metal was removed by jet electropolishing in a 25% nitric acid, 75% methanol electrolyte.

The development of 'dross' during the oxidation process was studied by monitoring changes in surface morphology using optical and scanning electron microscopy. To study oxide cross sections nickel coat was necessary to help support the oxide layer. The structure of the oxide-melt interface was examined in the TEM using microtomed sections of solidified melts. In addition, the types of oxides formed were characterised using electron and X-ray diffraction.

RESULTS AND DISCUSSION

The Development of Oxide on Molten Aluminium

The first formed film on aluminium is an 'amorphous' form of alumina believed to be γ [3]. Nucleation and growth of crystalline transition aluminas (γ- or η-alumina) occurs rapidly within the original 'amorphous' alumina film. The crystals grow in size as both temperature and time increase until surface coverage is complete (Fig.1a). The crystalline monolayer then thickens with time.

The transition of γ- to α-alumina crystals occurs at the oxide-melt interface. This transformation is associated with a volume change and results in a local build up of stress. This stress is relieved by oxide rupture in weak areas thereby exposing the melt. Molten metal exudes via the ruptured oxide path to the surface where the newly exposed metal instantly combines with oxygen. This protruding structure of metal enveloped in an oxide film is described as a nodular oxide growth.

During the early stages of oxidation few growths are seen (Fig.1b). However, with longer exposure times rapid oxidation rates ensue as a consequence of growth formation. These growths occur by repeated transport of molten metal through fissures in the newly formed oxide at active sites. Thus each growth is made up of a network of oxide clusters and metal pathways (Fig.1e). Figure 1c shows the incipient stage of development of one of these nodular oxide growths, with Fig.1d illustrating well developed oxide growth structures after prolonged exposure (15 hours for this example) at 750°C.

Clearly, the oxide growths increase in number and size with time (Fig.1f). Their appearance is believed to be associated with the onset of breakaway oxidation illustrated by the weight gain/curve of Fig.1 and indicates the initial stage of dross formation. It can be shown that the time to breakaway and thus the rate of α-alumina formation increases with temperature, with decreasing water vapour pressure (3) and in the presence of fluoride and chloride (4).

The Influence of Magnesium on Aluminium Alloy Oxide Development

Studies of oxidation kinetics demonstrate that the oxidation curves for pure aluminium and the Al-Mg alloys containing between 1-5%Mg all exhibit similar trends. All alloys exhibit a sigmoidal shaped curves, however, when magnesium is present the rate of oxidation is greatly enhanced as illustrated in Fig. 2. Comparison of Figs. 1 and 2 demonstrates that the onset of rapid oxidation (breakaway) occurs sooner for magnesium containing alloys and that the absolute weight gains are far higher when magnesium is present.

The initial stage of oxide development on both aluminium and aluminium-magnesium alloys involves the nucleation and growth of oxide crystals in an 'amorphous' film. However from electron micrographs (Fig.2a) taken from a molten Al-Mg melt after 5 minutes it can be seen that magnesia crystals develop in an 'amorphous' alumina film and not α-alumina. The magnesia aggregates in clusters, such that nodules are observed on the surface (Fig. 2b) even following very short exposure times. These nodules are a consequence of crystal growth within the 'amorphous' oxide film and at this stage do not contain free metal, as is the case for the large growths that form early on pure aluminium.

TEM studies show that these coarse primary magnesia crystals form at the oxide-metal interface. Smaller secondary oxide also exists within the amorphous film and originate from the secondary reduction reactions of the 'amorphous' state of γ-Al$_2$O$_3$ namely.

$$3Mg + \gamma\text{-Al}_2O_3 \longrightarrow 3MgO + 2Al$$

This secondary reduction reaction has been observed by many investigators [5,6]. The oxide therefore consists of a duplex film of primary and secondary magnesia crystals as illustrated in Figs. 2b and 3a, which provides some level of protection. However, when the magnesium concentration at the melt-oxide interface falls to below a critical level localised formation of magnesium aluminate crystals (MgAl$_2$O$_4$) is favoured. This was demonstrated from TEM studies (Fig. 3c - 3e). MgAl$_2$O$_4$ nucleates and grows into very large crystals at the melt-oxide interface, as illustrated in Figs. 4a - 4c for an Al-1%Mg alloy.

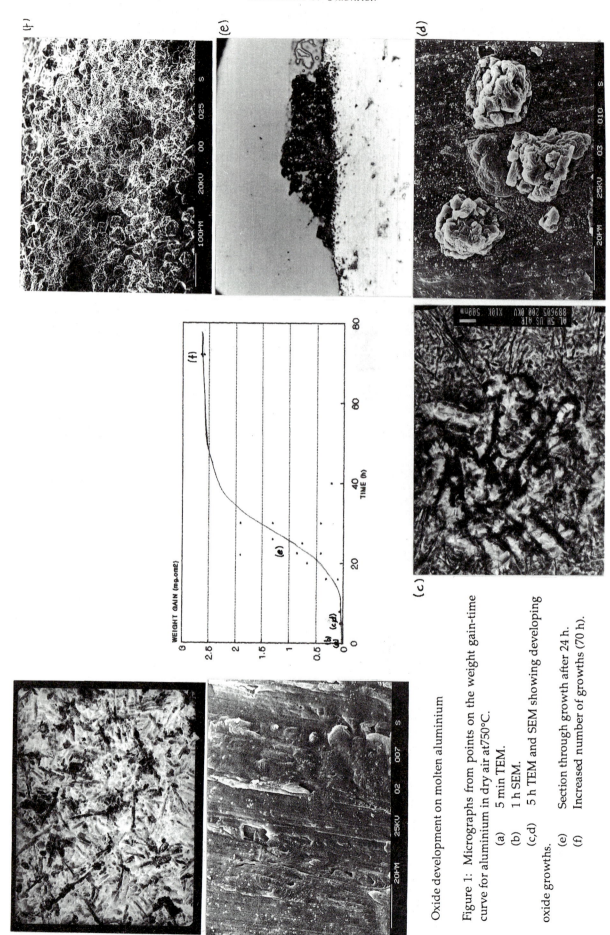

Oxide development on molten aluminium

Figure 1: Micrographs from points on the weight gain-time
curve for aluminium in dry air at 750°C.

    (a)   5 min TEM.
    (b)   1 h SEM.
    (c,d)   5 h TEM and SEM showing developing
oxide growths.
    (e)   Section through growth after 24 h.
    (f)   Increased number of growths (70 h).

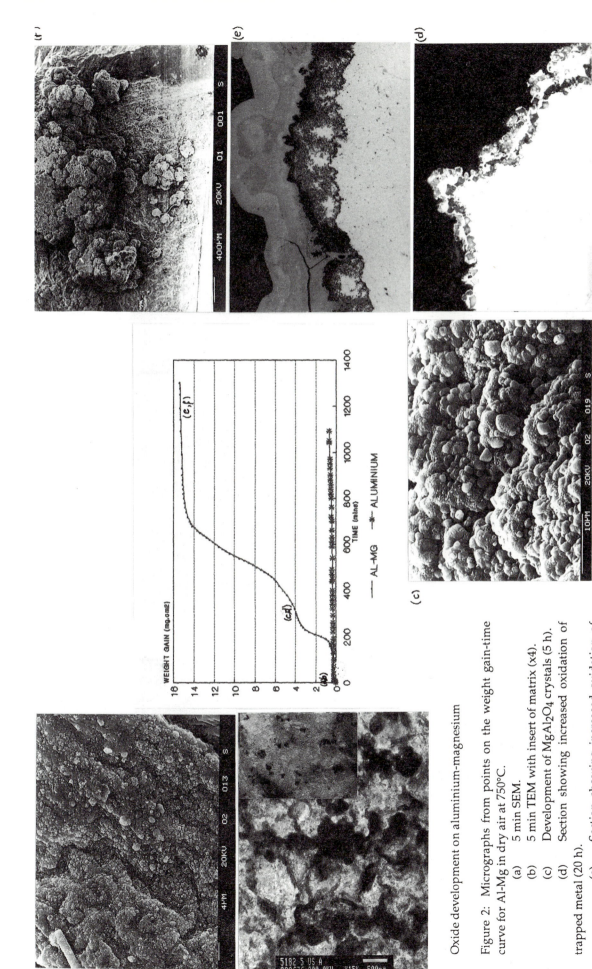

Oxide development on aluminium-magnesium

Figure 2: Micrographs from points on the weight gain-time curve for Al-Mg in dry air at 750°C.

(a) 5 min SEM.
(b) 5 min TEM with insert of matrix (x4).
(c) Development of $MgAl_2O_4$ crystals (5 h).
(d) Section showing increased oxidation of trapped metal (20 h).
(e) Section showing increased oxidation of trapped metal (20 h).
(f) Heavy oxide growth on surface (20 h).

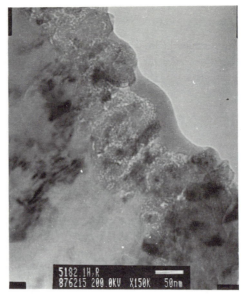

Primary Magnesia

Metal

Figure 3a: Microtome section through oxide film of secondary and primary magnesia.

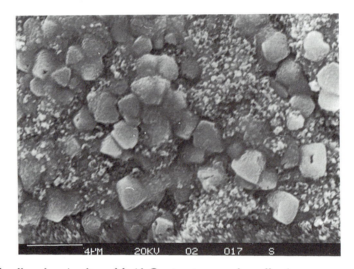

Figure 3b: Surface of Al-Mg alloy showing large $MgAl_2O_4$ structures and small primary magnesia crystals.

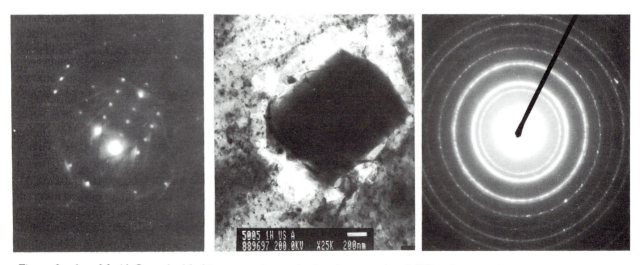

Figure 3c, d, e: $MgAl_2O_4$ embedded in magnesia matrix showing associated diffraction patterns.

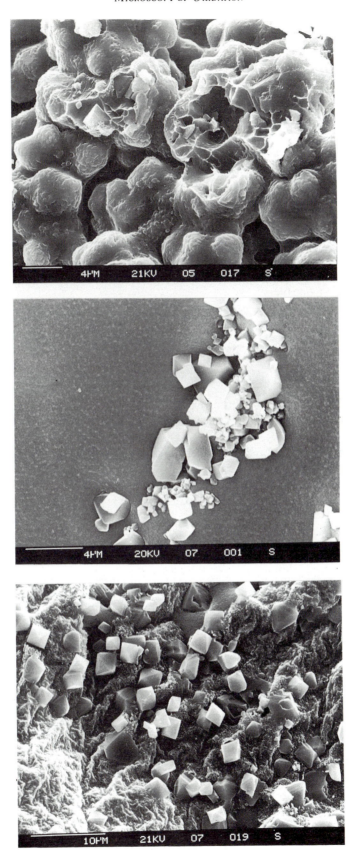

Figure 4: Magnesium aluminate crystals at:
      (a)   the oxide-air interface,
      (b)   the oxide-melt interface from metal side,
      (c)   the oxide-melt interface from oxide side.

The onset of breakaway oxidation in magnesium containing alloys closely follows that for pure aluminium, except that at the oxide-melt interface localised failure is initiated by the large $MgAl_2O_4$ crystals instead of α-alumina. The rapid growth of such crystals at the oxide-melt interface generates considerable stress within the oxide film causing scale failure. Molten metal can therefore exude through the ruptured scale and nodular growths develop. Examples are shown in Fig. 2c - 4f.

New crystals nucleate above the original oxide-melt interface (Fig. 2d). The metal trapped in the oxide tends to become severely depleted in magnesium and thus, the volume of trapped metal within each nodule decreases as the metal is consumed by spinel formation (Fig. 2f). Finally all the magnesium is removed from the melt by oxide development and, hence, during further exposure, the oxidation behaviour of magnesium containing alloys resembles pure aluminium.

SUMMARY AND CONCLUSIONS

1. The mechanism of dross development is similar on pure liquid alumimium and molten aluminium-magnesium alloys.

2. On pure aluminium γ-alumina or η-alumina nucleate at the oxide-metal interface in the first-formed 'amorphous' oxide film. Lateral growth occurs until these crystals touch, then the monolayer thickens. Ultimately, at longer exposure times, α-alumina develops within these oxide films.

3. On Al-Mg alloys, coarse primary magnesia crystals develop in the first formed 'amorphous' alumina film. As the magnesium content decreases, formation of $MgAl_2O_4$ is favoured. In low-magnesium-containing alloys or areas severely depleted in magnesium, rapid conversion through magnesia to spinel ($MgAl_2O_4$) predominates. If oxidation progresses still further, then α-alumina develops.

4. Magnesium additions greatly enhance the oxidation rate of molten aluminium. High oxidation rates are observed until the local magnesium concentration fails to a level below which only alumina scales form.

5. For both aluminium-magnesium alloys, formation of these oxide crystals beneath the protective 'amorphous' film is associated with considerable stress. This stress builds up locally, primarily due to the associated volume change. The 'amorphous' film must deform to accommodate the growing crystals or fracture. Therefore, as the degree of crystal growth increases, and hence the stress increases, the likelihood of oxide fracture is greater.

6. Molten metal exudation is associated with cracks of the oxide scales for both aluminium and Al-Mg alloy melts. At the surface, rapid oxidation ensues and the protruding metal is trapped in the oxide envelope which subsequently develops.

7. Progression of the mechanism leads to nodular oxide growth and ultimately dross formation. The early appearance of these growths is associated with the onset of breakaway oxidation and hence marks the initial stage of dross formation.

ACKNOWLEDGEMENTS

The authors wish to thank Alcan International Ltd for sponsorship of this project and one of the authors (SAI) is indebted to the Science and Engineering Research Council for the provision of a CASE studentship.

REFERENCES

1. S. Rao and P. R. Dawson, Report from Warren Spring Laboratory Report LR 359 (ME), 1980.

2. G. D. Preston and L. L. Bircumshaw, Phil. Mag. 22, 1936, p.654.

3. S. A. Impey, D. J. Stephenson and J. R. Nicholls, Mat. Sci. & Tech. 4, 1988, p.1126.

4. S. A. Impey, D. J. Stephenson and J. R. Nicholls. To be published.

5. D. J. Field, G. M. Scamans and E. P. Butler, Al-Li Alloys 19-21 May 1980, TMS/AIME, Eds. T.H. Saunders and E.A. Starke, p.325.

6. R. Grauer and P. Schmoker, Werkst. u. Korros. 27, 1976, p.1769.

28 THE EFFECTS OF SiC PARTICLES ON THE OXIDATION BEHAVIOUR OF COMPOSITES WITH Al, Al-Mg, AND Al-Li MATRICES

P. B. Prangnell, S. B. Newcomb and W. M. Stobbs

Department of Materials Science and Metallurgy,
Cambridge University, Pembroke Street, Cambridge CB2 3QZ, UK.

ABSTRACT

The effects of oxidising a range of composites with different Al alloy matrices are described as examined using "edge-on" TEM. The specimens studied were made by the spray deposition route and contained SiC particles. The primary aim of the work carried out was to ascertain whether the presence of SiC particles can have significant effects on the oxidation behaviour of the matrix alloy for the systems examined. It was found that the inclusions had strong effects only on the oxidation of the Al-Li matrix, and both the reasons for this and their significance are discussed.

INTRODUCTION

The incorporation of particulate SiC in Al alloys by spray deposition techniques provides an increasingly popular approach to the fabrication of metal matrix composites (MMCs) of high specific modulus. For the MMC to maintain an enhanced modulus, to reasonable stresses, it is important that the matrix should have a high yield strength so that the required load transfer [e.g. 1] does not cause premature failure at or near to the inclusion/matrix interface. This has led to the study of the effects which modifying the interface can have on the mechanical properties [e.g. 2] as well as to the clarification of the effects the reinforcement has on the precipitation behaviours of the typical age hardening alloys which are required to provide matrices of reasonable strength [e.g. 3]. In both these contexts there can be a requirement for the use of high temperature solution treatments and there is also increasing interest in the use of MMCs in relatively high temperature applications. Accordingly, although the oxidation behaviours of the standard Al alloys which are now being used as matrices for composites have been thoroughly characterised, there is a need for data on the effects which the reinforcing phase might have on the oxidation behaviours of the matrix alloys in question.

Our primary aim in this preliminary study has thus been to determine the specific rôles which SiC particles can have in modifying the oxidation behaviour of a set of model matrix alloys. The specific questions we have considered include:

i. whether or not the SiC/matrix interface can act as an effective diffusional pathway for the alloy constituents to the outer surface when the SiC intersects the oxidised surface;
ii. whether or not any such behaviour leads to enhanced oxidation within the interface regions;
iii. whether or not there can be active interactions between the alloy constituents and the SiC at the oxidised surface; and
iv. whether or not SiC particles beneath the oxidised surface can affect the process as a whole.

The approach we have taken to answering these problems has been to examine previously oxidised specimens of each of the alloys "edge-on" in the TEM, looking in particular at regions where SiC particles protrude to the oxidised surface.

EXPERIMENTAL DETAILS

The sprayed composites examined were all made by Alcan International Ltd. of Banbury using model alloy matrices, for which it was expected that there could well be differences in the behaviours as specified above. The specimen materials were provided as extruded bars and had the compositions:

Specimen A:	Al, "commercial purity (CP)"	+ ~15wt% SiC
Specimen B:	Al-1.9wt%Mg	+ ~15wt% SiC
Specimen C:	Al-3.5wt%Li	+ ~15wt% SiC

The specimen surfaces to be oxidised were ground and polished down to a 0.04μm final slurry polish taking care to keep the hard SiC particles as flush as possible with the surface. Each of the above specimens was then oxidised at 530°C in air and examined periodically in the SEM to see whether or not the SiC particles had any gross effects on the alloy oxidation behaviour in their immediate vicinity. In the SEM the specimens looked very similar prior to oxidation and only the preoxidised surface of specimen A is shown in Fig.1d. SEM images of the surfaces of specimens A and B are shown in Figs.1a and 1b respectively after 24 hours exposure while the image in Fig.1c shows the Al-Li alloy specimen (C) after only 25 minutes oxidation. No change can be seen for the Al(CP) matrix upon which a thin but highly protective oxide must have formed which is apparently unaffected by protruding SiC particles.

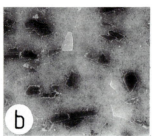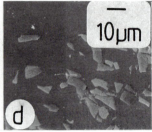

Figure 1: SEM images of the MMCs surfaces oxidised at 530°C: (a) Al(CP) 24h; (b) Al-Mg 24h; (c) Al-Li 25min; and in (d) Al(CP) prior to oxidation.

However the much thicker and more uneven oxide formed on the Al-Mg alloy appears to be more developed for distances of up to ~1μm from some of the SiC interfaces. The comparison of these two oxide scales with that formed after a much shorter period on the Al-Li matrix MMC is startling: thick reaction products completely cover the composite and the SiC particles can no longer be seen. Specimen C was thus examined in the TEM after only 25 minutes whereas the other two specimens were characterised after 24h oxidation. Prior to the preparation of TEM foils from the oxidised Al-Li matrix composite specimen, which had been "quenched" on a copper chill block from the oxidation temperature, an ageing treatment of 190°C for 5h was used to precipitate out the Li into δ' particles. This facilitated, through the absence of δ', the characterisation of the regions of the matrix which had been denuded of Li during the oxidation process. The edge-on TEM specimens were made by standard ion beam thinning methods, as have been described elsewhere [4], though it should be noted that particular care was taken to ensure that the oxidised surfaces were neither contaminated nor (in particular for the Al-Li alloy) came into contact with water, prior to their protection through being araldited together face to face in the first stage of the specimen preparation technique.

TEM RESULTS FOR THE EDGE-ON SPECIMENS

Specimen A (Al matrix) 530°C/24h

Unsurprisingly, a thin (~20nm) amorphous oxide was found on the surface of specimen A, the occasional sub-oxide pore as at A in Fig.2a probably being due to surface debris left during preoxidation polishing given that negligible cusping was observed where low angle boundaries (as at A in Fig.2b) met the surface. Surface steps were sometimes seen at high angle boundaries, but these appeared to have had their origin in deformation during the polish prior to oxidation and, even at such positions, no enhanced oxidation was seen. Figure 2c shows a small SiC particle protruding from the surface (perhaps indented into the matrix surface during the polish) and the alumina on the surface of the matrix may be seen to end abruptly where it meets the particle. Thus it would appear that the particle/matrix interfaces have no more effect than grain boundaries in modifying the surface oxidation behaviour of the Al(CP) matrix.

Specimen B (Al-Mg matrix) 530°C/24h

The MgO scale (Fig.3a) on this specimen varied in thickness between about 250nm and 350nm and consisted of equiaxed (~40nm) grains interspersed with retained unoxidised Al particles (Fig.3b). No evidence was seen of an amorphous alumina film at the alloy/oxide interface. Despite the extensive segregation of Mg to the upper surface there was surprisingly little extra oxidation at grain boundaries as may be seen from the image in Fig.3c obtained using MgO reflections, where there is a SiC particle interconnected with the surface by a grain boundary. It thus follows that in this temperature regime bulk diffusion of Mg must be dominant. The oxidation where SiC particles protrude from the surface may be understood from Fig.4. In Fig.4a the SiO_2 on the surface of the SiC is imaged as a thin (~4nm) layer on the outer surface and a still thinner (~1nm) layer where the SiC is adjacent to the MgO, extending into the material no further than does the MgO. It should be noted that the composite as sprayed has been shown, by methods described elsewhere [5],

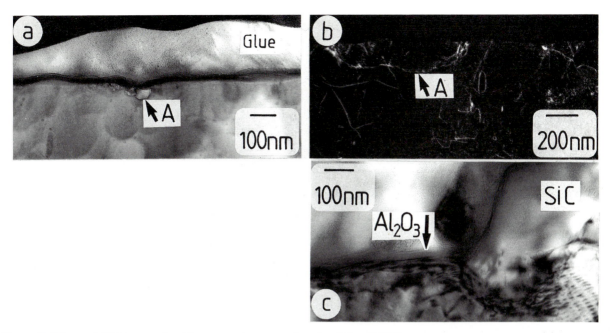

Figure 2: Edge on TEM images for $Al_{(CP)}$ matrix composite after 24h at 530°C:

  (a) Bright field showing Al_2O_3 protective oxide.
  (b) Weak beam image showing oxide not changed by sub grain boundaries.
  (c) Al_2O_3 stops at the SiC which has a thinner layer of SiO_2.

Figure 3: Edge on TEM images for Al-Mg matrix composite after 24h at 530°C:

  (a) Bright field showing uneven oxide of equiaxed MgO grains;
  (b) matrix dark field; note cusping at lower interface and Al retained in the oxide;
  (c) MgO dark field at matrix g.b.: sub-surface SiC has not affected the oxidation.

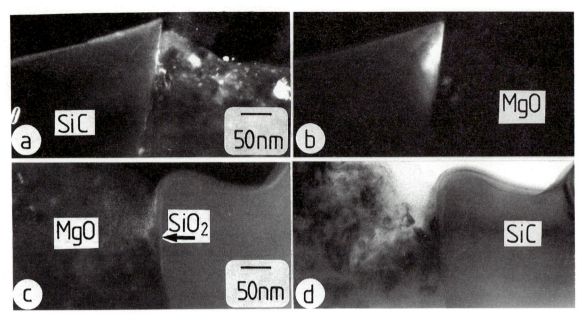

Figure 4: Two examples of protruding SiC on Al-Mg surface. For each, in (a) and (c) the SiO₂ is imaged. In (b) SiC is imaged demonstrating oxidation by Si diffusing out. (d) b.f.of (c) for comparison.

to have no oxide on the matrix/SiC interfaces so that where these interfaces meet the surfaces we are simply seeing a competition between the oxidation of SiC, through the movement of oxygen in along the interface boundary, and the oxidation of Mg as this element segregates to the surface. Other interfaces showed comparable behaviours and another example of the way the SiO_2 thickness can decrease to zero at the base of the MgO scale is shown in Figs.4c and 4d. It should be noted that in the latter example the MgO thickness decreases towards the interface of the SiC with the matrix, while the reverse is true for the region previously described.

Specimen C (AL-Li matrix) 530°C/25m

The oxidation behaviour of this composite contrasted strongly to that of the other two matrix alloys examined and it should be remembered that the extensive reaction observed in specimen C was after only about 1/50 of the oxidation exposure time given to the other specimens. Nevertheless, using super lattice reflections to image the δ' particles formed by the post-oxidation ageing treatment, it was apparent that there had been total denudation of Li to a depth of about 60μm and partial Li loss, as exhibited by there being progressively narrower precipitate free zones (moving inwards) at both SiC/matrix interface and grain boundaries, to depths of ~120μm. The way the grain boundaries and SiC/matrix interfaces are now important in the process may be seen from Fig.5. In Fig.5a and 5b we see the same region adjacent to a SiC particle some 120μm beneath the outer surface imaged in bright field and using a δ' reflection. Not only is there Li loss, but we also see substantial voiding through this loss of Li towards the outer surface. The effect becomes increasingly apparent closer to the surface as may be seen from the similar pair of images in Figs.5c and 5d taken still in the region of only local Li denudation. However from Figs.6a and 6b we see that only a little closer to the surface the SiC/matrix interfaces provide positions at which gross internal oxidation can occur, with further voiding and the retention of unoxidised remnants of Al within the internal oxides which appear to be partially amorphous and partially microcrystalline. The way in which this form of internal oxidation occurred at SiC/matrix interfaces, rather than at for example triple points in the matrix, is indicative of the way the SiC effectively "pegs" the local shape of the specimen to provide nucleation points for voids formed by the vacancy flux from the out diffusing Li. The voids thus formed at the SiC interfaces provide free surfaces for oxidation reactions to take place inside the specimen. This behaviour thus provides an interesting example of the way in which the (mechanical) presence of the substantially unchanged SiC particles might promote a greater degree of internal oxidation than would occur in the matrix alloy alone.

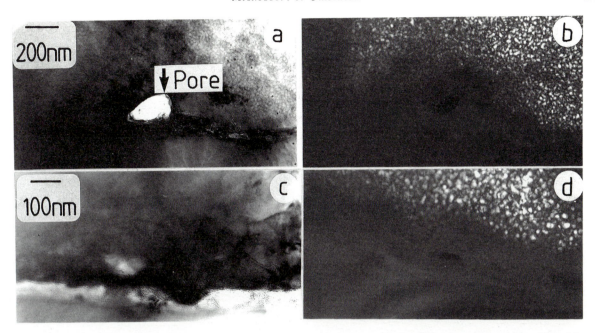

Figure 5: SiC/Al-Li alloy interface regions in (a) and (b) at a depth of about 120μm below the surface oxide and in (c) and (d) at about 90μm. (a) and (c) are bright field images and (b) and (d), obtained using δ' reflections, show the regions of Li depletion.

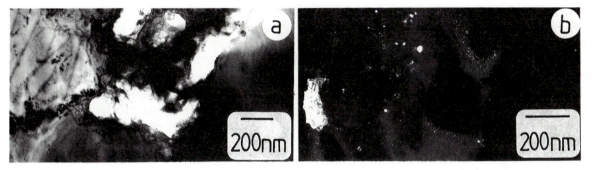

Figure 6: Images showing that at a depth of about 30μm, where the matrix is fully depleted of Li, there are both voids and reaction products, at the SiC interfaces, with retained Al in the oxides.

The outer scales which formed on this specimen consisted of several different phases and have not yet been fully characterised. The variations in their morphology are demonstrated by the micrograph in Fig.7a, from which it can be seen that the outermost oxide layers are extremely porous and that the local shapes of the fibrous like formations get coarser at the outer surface. While this points to the majority of the oxidation occurring (at the stage of the reaction which has been observed) within the outer layers and progressively more slowly at the composite surface it is clear, from the oxide microstructure found to depths of 35μm in the matrix, that there is also substantial internal oxidation. The temporal balance of the different processes will need further study. It might further be noted that the facets on these outer oxides belie the fact that they are in the main amorphous or microcrystalline.

Our more interesting observations on the outer oxides were made where they overlapped a SiC particle that had apparently once been on the outer surface. One such region is shown in a dark field image (Fig.7b) taken using part of the microcrystalline and amorphous halo of the oxides, and using a SiC reflection (as well as other oxide reflections) in Fig.7c. What we believe to be the position of the original outer surface of the carbide is marked at A and it is clear that there have been several competitive reactions involving the desorption of carbon from the SiC in association with a reaction with Li diffusing out of the matrix. The overall structure is bilayered, the inner layer being microcrystalline and amorphous and showing no particular morphology, unlike the outer layer which has a different composition (as well as having a columnar form which is reminiscent of anodisation) and contains at least a few grains of increased size despite again being in the main amorphous. Both micrographs are indicative of substantial SiC attack though the reason for the "worm hole"

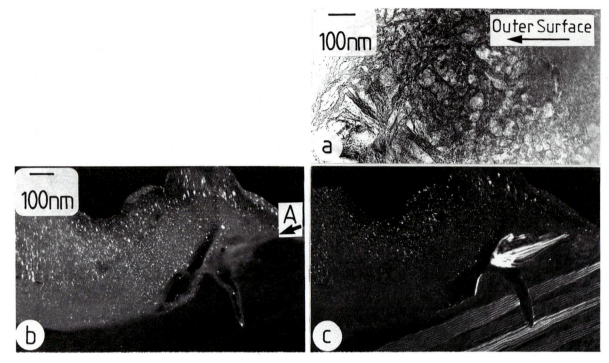

Figure 7: Surface regions on the Al-Li alloy after 25 mins at 530°C; (a) shows the outer oxides; (b) and (c) show the reaction products on a SiC particle protruding to the outer surface imaged in (b) using an oxide reflection and in (c) in a SiC reflection.

like pore running into the SiC, with its substantially increased local porosity, is not clear and surprisingly there does not seem to be any associated preferential attack of the SiC faults intersecting this pore. The overall microstructure suggests the formation, at least in the inner layer, of lithium carbonate early in the interaction as Li diffuses across the surface. Speculatively it is likely that later the oxidation will occur higher and require carbon and lithium to diffuse up into the fine "columnar" oxidation zone while the process will presumably be slowed by the sublayer formation of silica. The progress of the oxidation beyond this point leads to the observed increased porosity near to the SiC surface.

A substantial amount of further work will be needed to determine the temporal development of the various chemistries present, using Parallel EELS, while comparing the oxidation process above and to the sides of the protruding SiC particles. The effects of oxidation on the Al-Li composite are perhaps best visualised from the summary diagram in Fig.8.

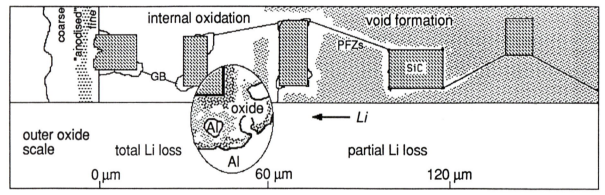

Figure 8: Summary diagram of the various microstructural changes seen on and below the surface of the Al-Li composite as oxidised for 25 mins at 530°C.

CONCLUSIONS

We are only too well aware that we have only scratched the surface of a set of research problems of considerable interest, both technologically and scientifically. Nevertheless we can answer the questions which we set ourselves for this exploratory project on the synergistic effects of the SiC particles on the oxidation behaviour of characteristically different types of matrices.

If we first consider the gross effects seen for the Al-Li system it is clear that the SiC affects the oxidation process in two distinct fashions. In one rôle, the SiC simply by its sub-surface presence promotes interface voiding as the volume of the surrounding matrix is decreased through primarily (at least lower in the material) grain boundary diffusion. This in turn promotes internal oxidation in a way which is not seen at grain boundary triple points, where it must be presumed that there is less tendency to form cavities, and thus nucleate oxide, mainly because vacancy annihilation there is more easily accommodated through climb and grain boundary sliding. It is also possible that at the internal silicon carbides there is a form of the reaction seen at the external SiC surfaces but, given the way there is always gross voiding it would appear that the oxygen partial pressure is too low for the reaction to take place, so that the main tendency is just for the Li to use these interfaces in its diffusion to the outer surface. The rôle of the SiC particles at the outer surfaces is clearly different, in that where the oxygen partial pressure is high, we now have clear evidence of the way the normally passivating oxidation of the SiC, by silica formation, is changed radically to the formation of oxides incorporating carbon because of the the flux of lithium to the surface. That the behaviour is seen at all, given both the way the lithium reaches the external surfaces at best at the edges of the SiC particles, and the fact that the competitive reactions the lithium can undergo to form both lithium oxide and a range of other Al containing oxides are all strongly exothermic, indicates that the processes involved at the SiC surface are highly favoured.

At first sight the behaviours of the commercially pure Al and AL-Mg matrices are considerably simpler than those seen for the Al-Li matrix composite. It should however be noted that it was surprising that the Mg diffused to the external surface in the main by bulk rather than by grain boundary diffusion, as can be deduced from the minimal modification of the outer oxides at the grain boundaries. It would be expected that for lower oxidation temperatures grain boundary diffusion would become more favoured, and under these circumstances one would again expect the SiC to affect the process more strongly (if less so than in the Al-Li alloys) at least in the passive rôle discussed above in which internal voiding is promoted.

We should conclude by noting that the oxidation behaviours we have described, even for the Al-Li system, will not cause problems in the solution treatment of MMCs since scalping is standard practice.

ACKNOWLEDGEMENTS

We are grateful to Prof. D. Hull for the provision of laboratory facilities and thank Alcan International Ltd both for the supply of the test pieces and for financial support (P.B.P).

REFERENCES

1. P.J. Withers, W.M. Stobbs and O.B. Pederson, Acta Met., 37, 1989, p.3061.

2. T.J.Warner and W.M.Stobbs, Proc.VIIth ICCM, Beijing, V.1, 503, 1989, p.508.

3. P.B.Prangnell and W.M.Stobbs, Proc.VIIth ICCM, Beijing, V.1, 573, 1989, p.578.

4. S.B.Newcomb, C.B.Boothroyd and W.M.Stobbs, J.Microsc., 140, 1985, p.195.

5. P.J.Withers, W.M.Stobbs and A.J.Bourdillon, J.Microsc., 151, 1988, p.159.

29 THE USE OF IMAGING SIMS IN A STUDY OF AN OXIDISED CoNiCrAlY OVERLAY COATING

D. G.Lees and D. Johnson*

Manchester Materials Science Centre, University of Manchester and UMIST, Grosvenor Street, Manchester M1 7HS, UK.

** Centre for Surface and Materials Analysis,
UMIST, Manchester, M60 1QD, UK.*

ABSTRACT

An oxidised CoNiCrAlY overlay coating has been studied in cross-section by means of imaging SIMS. The scale consisted mainly of aluminium oxide, with chromium in the outer part. The yttrium was very unevenly distributed in the scale and in the underlying metal; the possible implications of this for the adhesion of the scale are discussed. The results show the power of imaging SIMS for studying the distribution of elements which are present in very low concentrations. A preliminary experiment using oxygen-18 as a tracer showed that the scale grew primarily by means of oxygen difusion inwards.

INTRODUCTION

The beneficial effect of reactive elements such as cerium and yttrium on the adhesion of chromia and alumina scales to the underlying metal has been known for many years. The subject has been reviewed in detail by Whittle and Stringer [1]. In this programme we wanted to investigate the distribution of yttrium in the oxide scale formed on an overlay coating. SIMS is needed for this purpose in order to detect the low levels of yttrium that are present. We also wished to study the growth-mechanisms of the scale. For this purpose we used oxygen-18 as a tracer and determined its position in the scale by means of imaging SIMS. The use of imaging SIMS in this way has previously been shown to be a powerful method for investigating oxide growth mechanisms [2,3].

EXPERIMENTAL PROCEDURE

The substrate was in the form of a rod; the coating was sprayed on to it in an argon atmosphere. The nominal composition of the coating was 37%Co. 32%Ni, 21%Cr, 8%Al, 0.3-0.5%Y and that of the substrate was 60.25%Ni, 10%Co 10%W, 9%Cr, 5.5%Al. 2.5%Ta, 1.25%Hf, 1.5%Ti. All compositions are in weight per cent. The coated rod was heat-treated in an argon atmosphere for 1h at 1100C and for a further 16h at 870C. Two pieces were cut from the rod for oxidation. Both pieces were oxidised in air for 116h at 1050C and cooled to room temperature. One of them (Specimen 1) was not oxidised further. The other (Specimen 2) was oxidised for a further 27h in natural oxygen and then, without cooling to room temperature, for a further 70h in gas enriched in oxygen-18. The specimen was oxidised in natural oxygen before being oxidised in oxygen-18 because the oxide may have been cracked by cooling to room temperature after oxidation in air and this would have complicated the interpretation of the oxygen-18 distribution in the scale. The oxidation in the natural oxygen would have healed any cracks before the specimen was exposed to the oxygen-18. The specimens were vacuum-coated with gold and then plated with nickel; discs were cut from them for examination, then ground and polished.

The distributions of the various elements and of the oxygen-18 and oxygen-16 isotopes in the scale and the coating were determined by means of imaging SIMS in the Centre for Surface and Materials Analysis, UMIST. The equipment was manufactured by VG Ionex Ltd. A 10keV, 0.5nA beam of gallium ions was used; this was coupled to a quadrupole mass spectrometer and a digital scan unit. A framestore imaging system allowed rapid acquisition and processing of the data. The minimum diameter of the beam was less than 0.5 microns. Positive ions were used to locate the metallic elements and negative ions to locate the oxygen isotopes.

RESULTS

Figure 1 shows the distributions of Cr+ (mass 52), Al+ (mass 27), Y+ (mass 89) and YO+ (mass 105) in the scale and underlying coating on Specimen 1. The yttrium was very unevenly distributed in the scale and in the underlying coating.

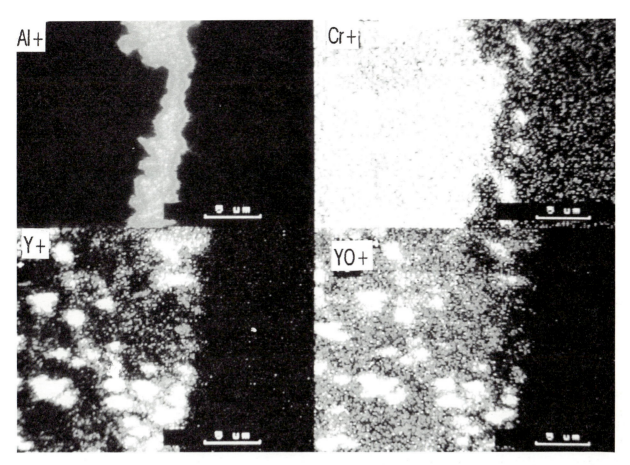

Figure 1: Specimen 1. The aluminium, chromium, yttrium and YO+ distributions in the scale and underlying coating.

Figure 2 shows the results of a real-time linescan across the scale in the upper part of the field, just below the thick oxide. The chromium and the yttrium were both located in the outer part of the scale.

Figure 3 shows oxygen-16 and oxygen-18 maps for a thin part of the scale on Specimen 2 and Fig. 4 shows one of the linescans across the scale on another part of this specimen. The oxygen-18 was located primarily in the inner part of the scale but it was also present at a lower level as far as the scale-gas interface.

DISCUSSION

A notable feature of these results is the highly uneven distribution of yttrium in the scale and in the underlying coating. It is interesting to note that this could not be detected by EDAX because the yttrium levels were too low. Preliminary studies suggest that the uneven distribution existed in the coating before the specimen was oxidised. Therefore it could have arisen during the deposition of the coating. We do not know whether this uneven distribution has a significant effect upon the adhesion of the scale. According to the "Sulphur Effect" theory proposed by Lees [4-6], Ikeda et al.[7] and by Smeggil et al. [8-10] the yttrium exerts its beneficial effect upon scale adhesion by preventing sulphur from segregating to the scale/metal interface and thus weakening the bond between scale and metal. From the point of view of sulphur removal it will not matter if the yttrium is unevenly distributed, provided that there are no areas which are so far from the yttrium that it cannot react with any sulphur that is there. However, Kvernes [11] has shown that for the case of an Ni-9%Cr-6%Al alloy, increasing the yttrium content above 0.1% increased the isothermal oxidation rate. Also, Barnes et al. [12] have shown that in the case of an alloy which formed a chromia scale, there is an optimum value for the cerium content as far as scale adhesion is concerned; if this value is exceeded, the adhesion deteriorates. If this is the case with the alloy in the present study then the uneven distribution of the yttrium could have a deleterious effect upon scale adhesion.

The oxygen-18 tracer results are consistent with growth of oxide at the scale-metal interface by short-circuit transport of oxygen, in agreement with the results of Reddy et al. for a Ni-14Cr-24Al-0.27Y (at%) alloy [13]. Young and de Wit found

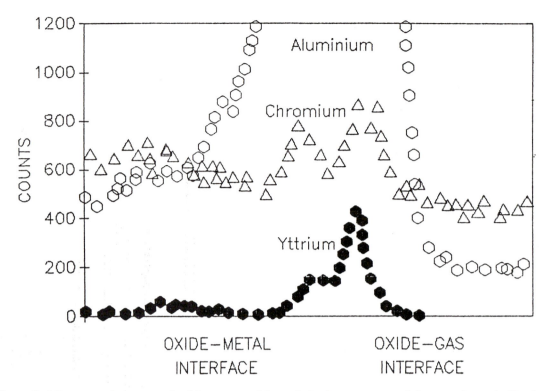

Figure 2: A linescan taken across the thinner part of the scale in the upper part of the area shown in Fig. 1.

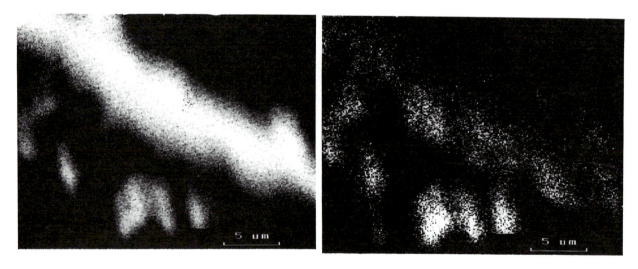

Figure 3: Specimen 2. The oxygen-16 and oxygen-18 distributions in the scale.

that the scale on pure NiAl grew by aluminium transport, with some oxygen diffusion; that when 0.07%Y was added, the scale grew almost entirely by aluminium transport, and that when 0.5%Y was added, the scale grew by substantial transport of both species [14]. Further work will be needed to see if the uneven yttrium distribution in the coating in the present study gives rise to diferent growth-mechanisms.

CONCLUSIONS

This work has shown that imaging SIMS is a powerful technique for the examination of cross-sectioned scales. It can be used to determine the distribution of elements which are present at levels which are too low to be detected by EDAX, and

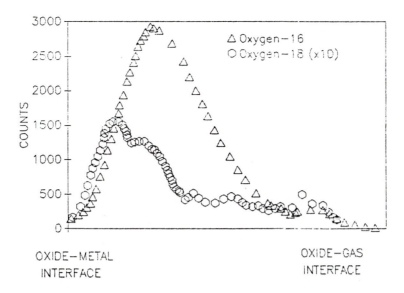

Figure 4: A linescan taken across the scale on Specimen 2, showing the oxygen-16 and oxygen-18 distributions.

it can used to distinguish between isotopes. The distribution of yttrium in the scale on this CoNiCrAlY overlay coating and in the coating immediately beneath the scale has been found to be very non-uniform. By using oxygen-18 as a tracer the scale has been shown to grow primarily by oxygen transport.

REFERENCES

1. D. P. Whittle and J. Stringer, Phil. Trans. R. Soc. Lond A295, 1980, p. 309.

2. D. G. Lees, P. C. Rowlands and A. Brown, Inst.Phys.Conf. Ser. No.90, 1987, p57.

3. D. G. Lees, P.C. Rowlands and P. Humphrey, J.Mat.Sci.Tech. 4, 1988, p. 1117.

4. D.G.Lees, I.S.Grant and G.W.Lorimer,"An Investigation 0f The Mechanism By Which Reactive Elements Improve Scale-Metal Adhesion", Science and Engineering Research Council Proposal, September 1982.

5. D.G.Lees, Oxid.Met.27,1987, p. 75.

6. I.Melas and D.G.Lees, J.Mat.Sci.Tech.4, 1988, p. 455.

7. Y.Ikeda, K.Nii and K.Yoshihara, Proceedings JIMIS-3(1983): High Temperature Corrosion. Transactions of the Japan Insititute of Metals, Supplement, p.207.

8. A.W.Funkenbusch, J.G. Smeggil and N.S. Bornstein, "Proposal for the Study of Adherent Oxide Scales", Technical Proposal (P83-19), Office of Naval Research Contract N00014-82-C-0618, January 1983.

9. A.W. Funkenbusch, J.W. Smeggil and N.S. Bornstein, Met.Trans,16A, 1985, p. 1164.

10. J.G. Smeggil, Mat.Sci.Eng. 87, 1987, p. 261.

11. I.A. Kvernes, Oxid. Met. 6, 1973, p.45.

12. D.G. Barnes, J.M. Calvert, D.G. Lees and P.J. Parry, Scripta Met. 8, 1974, p. 1209.

13. K.P.R. Reddy, J.L. Smialek and A.R.Cooper, Oxid.Met.17, 1982, p. 429.

14. E.W.A.Young and J.H.W.de Wit, Oxid.Met.26, 1986, p. 351.

SECTION FIVE
Reactive element effects

30 THE EFFECT OF SULFUR CONTENT ON Al₂O₃ SCALE ADHESION

J. L. Smialek

NASA Lewis Research Center,
Cleveland OH 44135, USA.

ABSTRACT

The results are summarized from a number of investigators who focussed on the effects of sulfur content on scale adhesion. Beneficial effects were observed by using higher purity alloys. Sulfur removal was accomplished by vacuum annealing, hydrogen annealing, and by repeated segregation and polishing. Some degree of improved adhesion resulted in all cases. Hydrogen annealing essentially eliminated spallation for NiCrAl foil and bulk PWA 1480. An adherence map, constructed on the basis the amount of sulfur available for segregation, indicated between 1/4 to 4 monolayers as the boundary between spalling and adherence. Sulfur additions also destroyed adhesion for NiCrAl doped with reactive elements. The preponderance of the data is in concert with a sulfur dominated adhesion mechanism.

INTRODUCTION

This paper addresses the widely studied topic of the reactive element effect on the adhesion of Al₂O₃ scales to MCrAl alloys. A number of mechanisms have been proposed over the last 20 years; the sulfur effect has been studied in the last 5 years. The assumptions of this model are that low levels (10-50 ppmw) of unintentional sulfur impurities selectively segregate at the oxide-metal interface, weaken the bond, and produce interfacial spalling upon thermal cycling. Reactive element additions bind to these sulfur impurities, as predicted from the high thermodynamic stabilities of their sulfides, which then prevents segregation and alters sulfur's ability to poison the chemical bond at the interface. This model is supported by an increasing number of investigators. The purpose of this report is to review some of these results with special emphasis toward critical experiments.

First, some comment on the need for critical experiments is in order. Reactive elements have been typically associated with one or more effects upon oxidation: production of oxide intrusions, decrease in grain size, change in growth mechanism, elimination of buckling and interfacial voids, and a strengthening of the chemical bond at the interface. Because a number of these may occur simultaneously, it is difficult, if not impossible, to conclude which mechanism controls adhesion just by examining some of the changes that occur by reactive element doping.

The first order effect should be both necessary and sufficient for adhesion. That is, adhesion should be achieved if this mechanism is operative alone (sufficient), and only if it is operative (necessary). Ironically, these conditions can best be tested by studying the reactive element effects separate from the reactive elements themselves (e.g. independently vary stress on the scale, grain size, sulfur content, etc.). The following discussion attempts to show that major changes in adhesion do occur by independently varying the sulfur content of the alloy.

DISCUSSION

Effect of Initial Sulfur Content

We begin with one of the earliest published studies (1983) which proposed and successfully demonstrated the sulfur effect [1]. Although this is a study of Cr₂O₃ adhesion to stainless steels, the basic factors apply to the case of Al₂O₃ adhesion. Ikeda et al. measured segregation at the free surface of four stainless steels by hot stage Auger spectroscopy at 727°C [1]. They also measured weight change in cyclic oxidation at 827 and 1000°C. The amount of spalling varied with the sulfur saturation level found by Auger, Fig. 1. The best adhesion and lowest sulfur segregation was obtained for a type 304 stainless containing boron and nitrogen, which were found to segregate themselves and displace sulfur. These results provide convincing evidence for the control of scale spallation by sulfur segregation, assuming that spallation occurred at the oxide-metal interface and that no major changes in the scale or interface morphology occurred.

It is also assumed that interfacial segregation during oxidation occurs to a degree commensurate with that observed for free surfaces in the Auger experiments. This observation was nicely verified by Ikeda et al. [2]. Here they sputter deposited a 1μm Al_2O_3 film onto a type 310 stainless steel. After cycling to 827°C in the Auger microscope, the film spalled to the metal surface and revealed a large new sulfur peak. Thus interfacial sulfur segregation was shown directly.

Also, the amount of film spallation after a few oxidation cycles was correlated with the free surface sulfur segregation levels for three heats of 304 stainless, Fig. 2. These levels in turn were related to the bulk sulfur contents, as indicated on the figure. Other experiments on doped alloys showed the expected reduction of sulfur segregation and film spallation.

Control of scale adhesion by sulfur content alone was also achieved rather dramatically by Smeggil for NiCrAl alloys [3]. By using high purity charge materials, an alloy with about 1-2 ppmw sulfur was produced. The oxidation behavior of this alloy clearly resembled more the adherent behavior of a NiCrAlY than the non-adherent behavior of a normal purity (30 ppmw S) NiCrAl, Fig. 3. The proposed sulfur mechanism offers the most consistent explanation.

Segregation/oxidation data has been reported by Luthra and Briant that contradicts the above findings [4,5]. A very high purity NiCrAl alloy showed little or no sulfur segregation (0 to 1.1 at. %) in 1100°C Auger, but spalled quite excessively in 1150°C, 168 h, cyclic oxidation. The bulk sulfur content was reported as 1-8 ppmw. An attempt to resolve this apparent contradiction was unsuccessful [6].

Effect of Sulfur Removal

Adhesion has also been produced by removing sulfur from conventionally prepared alloys. Annealing NiCrAl in a 5% H_2/Ar gas mixture at 1200°C for 100 h resulted in partially adherent scales during 1100°C, 200 h cyclic oxidation, Fig. 4. The adhesion behavior (-2 mg/cm^2) more closely resembled that of a Zr-doped NiCrAl (+1 mg/cm^2) than that of unannealed NiCrAl (-22 mg/cm^2). This family of curves bears a remarkable similarity to those in Fig. 3 for NiCrAl's of different purity. As originally suggested by Lees for a similar effect on pure Cr [7,8], this result is believed to arise from a reduction in the sulfur content. (NiCrAl annealed in air and Cr annealed in vacuum, then repolished and cyclicly oxidized, did not show any beneficial effects in these two studies.)

A similar effect was produced on the same alloy composition produced as a melt spun ribbon [6]. Annealing in 1 atm. H_2 for 100 h at 1000, 1100, or 1200°C reduced the sulfur content from 2.5 ppmw, as-spun, to 0.62, 0.19, and 0.12 ppmw, respectively. The subsequent 1100°C cyclic oxidation behavior was excellent for 500 h, Fig. 5. The superior behavior of the foil samples compared to that of bulk samples is believed to be due to the better sulfur removal for thinner samples [6]. Very few, minute regions of spalling to bare metal were observed, primarily on the 1000°C sample.

The as-spun sample spalled the most, but again not as great as did the bulk samples. This may be due to the lower initial sulfur content. Nevertheless, after 60 h this sample was sufficiently degraded to change from a light grey scale to a dark greenish oxide (presumably from Al_2O_3 to $NiCr_2O_4$). This resulted in an increase in oxidation weight after 80 cycles.

Similar discolorations were observed in the latter part of the test for some regions of the H_2 annealed samples. This degradation occurred because 1/3 of the total aluminum content was depleted by oxidation alone, without spalling. Thus further testing could not be used to demonstrate longer term Al_2O_3 adhesion because the scale phases were changing.

One atm. H_2 was also used to anneal a nickel base superalloy, PWA 1480, in both 0.8 mm thick single crystal and 1.6 mm polycrystal form [11]. The degree of adhesion in 1100°C, 200 h cyclic oxidation tests of the single crystals was found to be a direct function of the annealing temperature, Fig. 6. The 1200°C and 1300°C weight change curves showed very slight gains (0.3 mg/cm^2) and a positive slope at the end of the test. This is consistent with the little or no spalling to bare metal observed visually. Similar results were obtained for the polycrystal material, except the degree of adhesion was somewhat less.

Annealing reduced the sulfur content of the polycrystal samples from 11.0 to as low as 0.035 ppmw. These values were in basic agreement with those calculated from a simple diffusion model for loss of solute in a thin slab. The single crystal values were not measured, but were calculated to be considerably less because of a more limited reserve. The weight change after 200 h was found to be strongly correlated with the sulfur content of the single crystal (calculated) and polycrystal (measured) samples, Fig. 7. These curves strongly suggest that scale adhesion is remarkable for uncoated PWA 1480 if the sulfur content is reduced below about 0.1 ppmw.

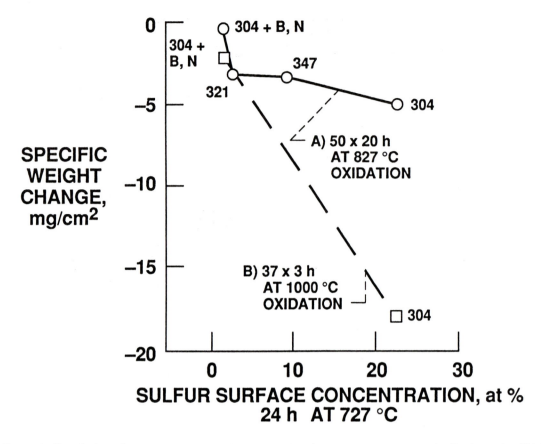

Figure 1: Correlation of weight loss with sulfur segregation for various stainless steels (Ikeda et al., 1983 [1]).

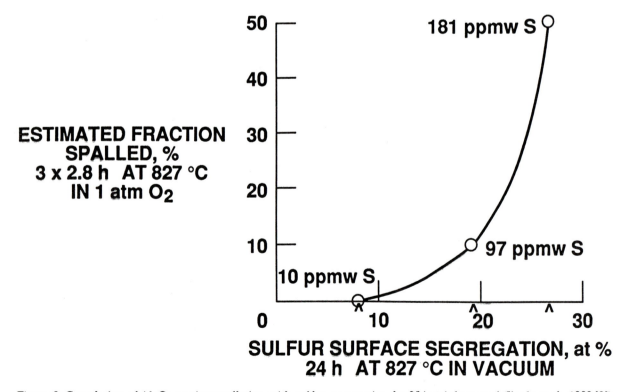

Figure 2: Correlation of Al_2O_3 coating spallation with sulfur segregation for 304 stainless steel (Ikeda et al., 1989 [2]).

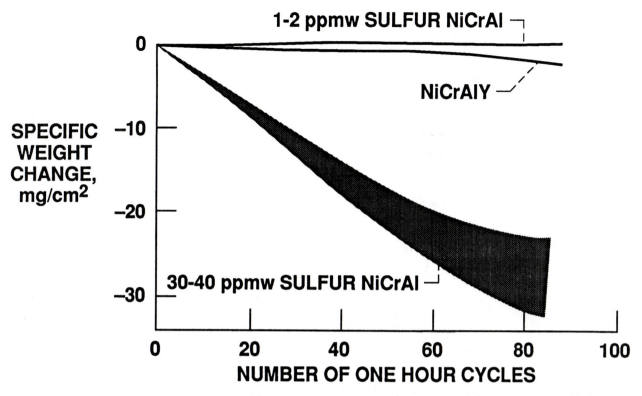

Figure 3: Improvement in scale adhesion produced by high purity, low sulfur Ni, Cr, and Al charge materials; 1180°C cyclic oxidation (Smeggil, 1987 [3]).

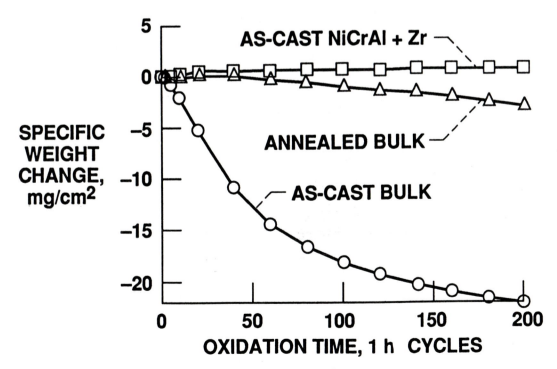

Figure 4: Effect of annealing undoped NiCrAl in 5% H_2 (1200°C, 100 h) on 1100°C cyclic oxidation.

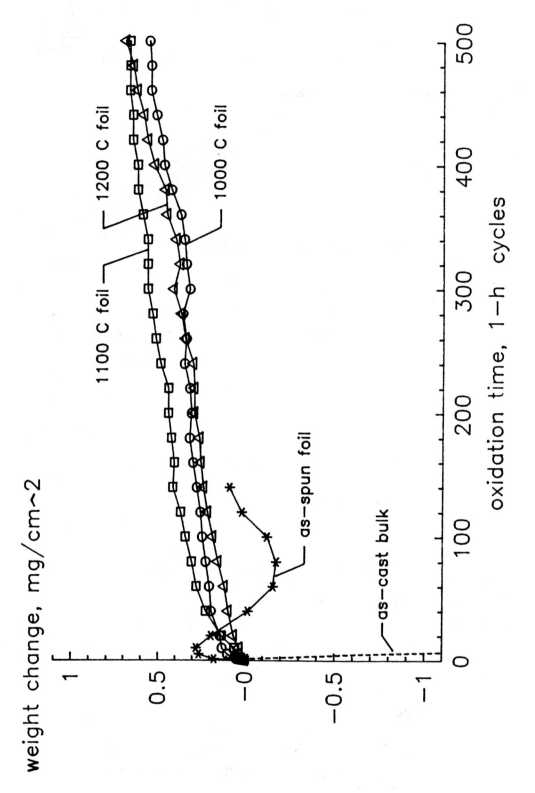

Figure 5: Effect of annealing in 1 atm. H₂ for 100 h on the 1100°C cyclic oxidation of undoped NiCrAl foil [6].

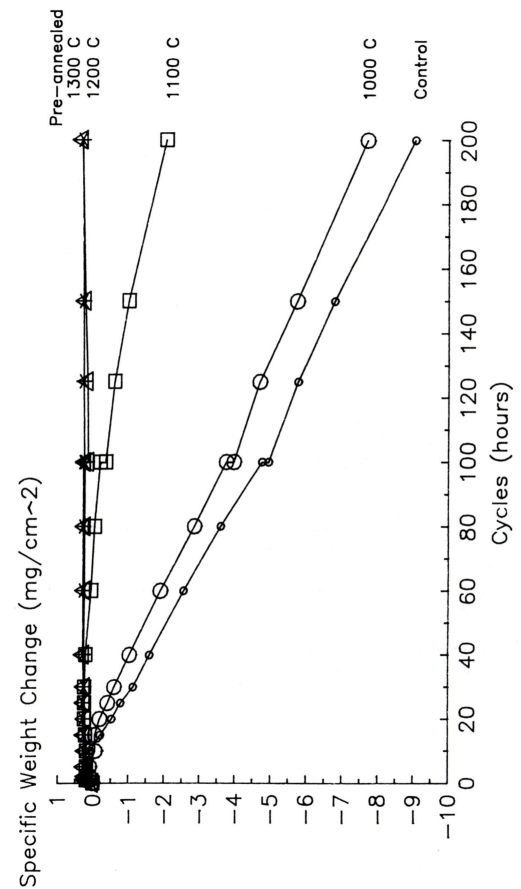

Figure 6: Effect of annealing in 1 atm. H_2 for 100 h on the 1100°C cyclic oxidation resistance of single crystal PWA 1480 [11]).

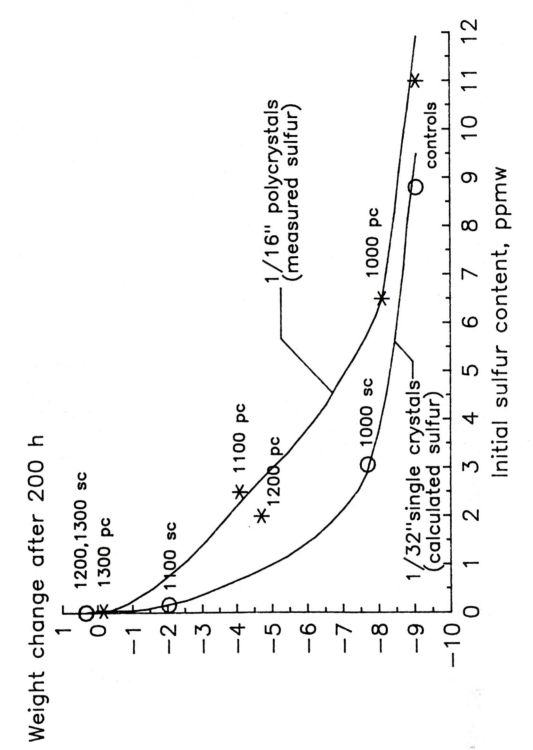

Figure 7: Correlation of adherent behavior in 1100°C cyclic oxidation with low sulfur content for H$_2$-annealed PWA 1480 [11].

Similar conclusions were reached by plotting the cumulative area fraction of spalling to bare metal as a function of sulfur content [11]. Extrapolation to 0 ppmw sulfur indicated a spall fraction of nearly zero. The major difficulty with these conclusions is that in addition to Al_2O_3, $CrTaO_4$ and $NiAl_2O_4$ were also formed on the adherent samples. Al_2O_3 was the oxide in contact with the metal. The non-adherent samples contained Cr_2O_3 and $NiCr_2O_4$ in addition to the other oxides.

Another method of sulfur removal, although less effective, was oxidation/polishing. This technique allowed sulfur to segregate in a 1 h oxidation treatment at 1120°C, then the scale and segregation layer were removed by light polishing. Visual observation, weight change, XRD, and SEM/EDS revealed that complete spallation changed to complete adhesion from 8 to 13 such cycles without any apparent changes in the scale and alloy morphologies, phases, or composition, Fig. 8 [6,9,10]. After adhesion did occur, the rate of weight gain was equal to that for adherent scales formed on NiCrAl+Y,Zr. The only change that was observed was the reduction in sulfur content from about 10 to 2 ppmw after 25 such cycles, with adherent behavior beginning at about 5 ppmw. This correlation, and the absence of any other, leads to the conclusion that sulfur removal produced adhesion without reactive element dopants.

This experiment was repeated using a 0.2 mm thick sample rather than 0.9 mm [6]. Adhesion was again produced, but after only 4 cycles (cf. 13 cycles) because of a more limited supply of sulfur. This behavior was modeled on the basis of a critical flux of sulfur required to saturate the interface each oxidation cycle. Adhesion was predicted for both samples in general agreement with the observed times. Again, no other explanation of a thickness effect seemed plausible.

The samples from Fig. 8 were repolished and cyclically oxidized at 1120°C with no more polishing. The resultant weight losses due to spalling were found to correlate well with the sulfur content before this exposure, Fig. 9 [6]. Extrapolation of the curve to 0 ppmw sulfur indicated that a weight <u>gain</u> could be achieved, close to that observed for Y or Zr-doped NiCrAl, if high enough purity is attained.

The control of adhesion by sulfur content suggests that a critical sulfur content may exist for NiCrAl alloys, below which good adhesion may be expected. The data from the above experiments was examined in the context of the total number of sulfur monolayers that are possible for a given sample size and sulfur content [6]. The relationship between the bulk sulfur content, (F_s, in ppmw), the number of monolayers, (N_m), the sample area, (A, in cm^2), and sample weight, (W, in gm), was found to be:

$$F_s = 8.27 \times 10^{-8} gm/cm^2 \cdot N_m \cdot A/W$$

For a fixed value of N_m, and a NiCrAl density of 7.1 gm/cm^3, the inverse relationship between sulfur content and sample thickness is shown as a family of curves in Fig. 10. Thus, for example, a sample thickness of 1 mm and a maximum sulfur segregation level of a single 1/2 monolayer requires that the bulk sulfur content be 0.14 ppmw or less, point A. A value of 1/2 monolayer is an important baseline level because this is about the equilibrium saturation level for these NiCrAl alloys [12,13].

The data points are for actual experiments in which the sulfur content (reduced by oxidation/polishing, H_2 annealing, or high purity charge materials) was measured and the cyclic oxidation weight change determined. An approximate measure of the degree of adhesion is given by the solid filling of each data point. This was determined as the difference between the final weight of the sample and a totally non-adherent NiCrAl, relative to the difference between a fully adherent NiCrAlY or NiCrAlZr and the non-adherent NiCrAl. The code for the individual data points is given as:

circles	: oxidation/polishing,	1120°C/25 h [6]
squares	: H_2 annealed foil,	1100°C/500 h [6]
diamond	: high purity elements,	1180°C/90 h [3]
hexagons	: H_2 annealed PWA 1480,	1100°C/200 h [11]
triangles	: same, polycrystals,	1100°C/200 h [11]

The general trend is that, as expected, total adhesion was observed for low sulfur, thin samples and total spallation for high sulfur, thick samples. The data are too sparse to accurately define a specific locus of adherent behavior. Furthermore, at these low sulfur levels, adhesion is a matter of degree and not a yes or no proposition. Different cyclic exposures will obviously influence the measured degree of adhesion. A realistic expectation might be that some adhesion will be observed below 4 monolayers and some spallation will be observed above 1/4 monolayer. The important finding here is

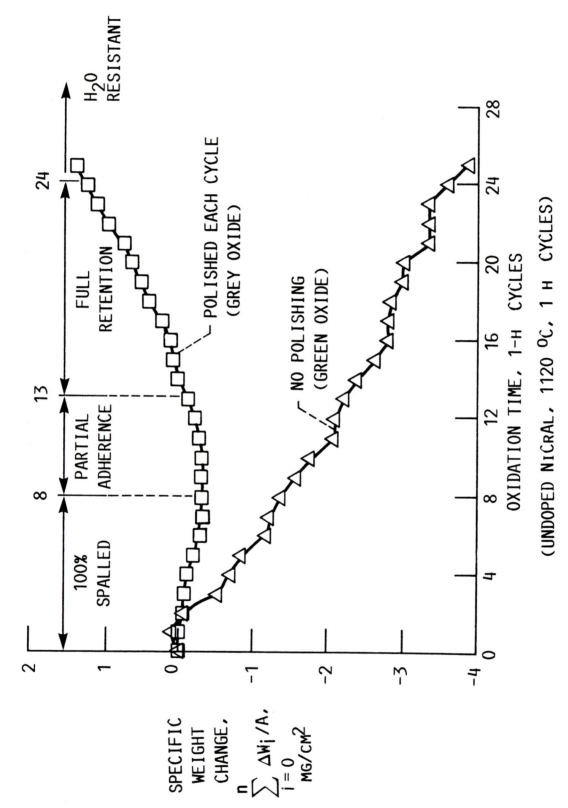

Figure 8: Effect of repeated 1120°C oxidation/polishing on Al_2O_3 scale adhesion for undoped NiCrAl [10].

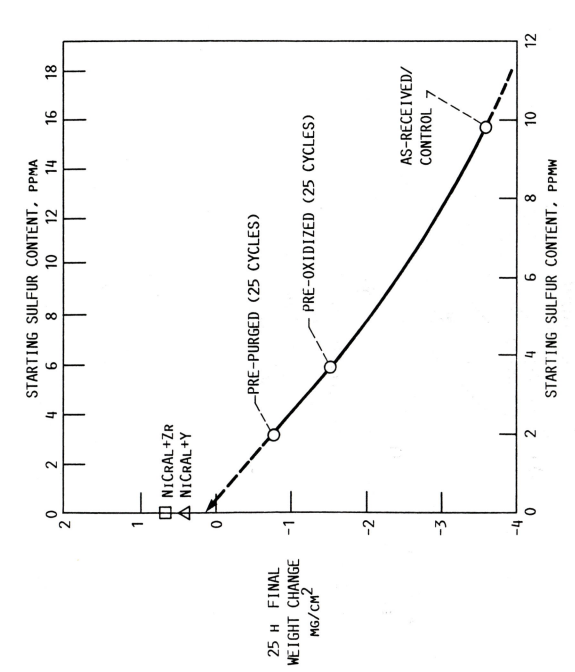

Figure 9: Correlation of adherent behavior in 1120°C cyclic oxidation with low sulfur contents for pretreated undoped NiCrAl specimens [6].

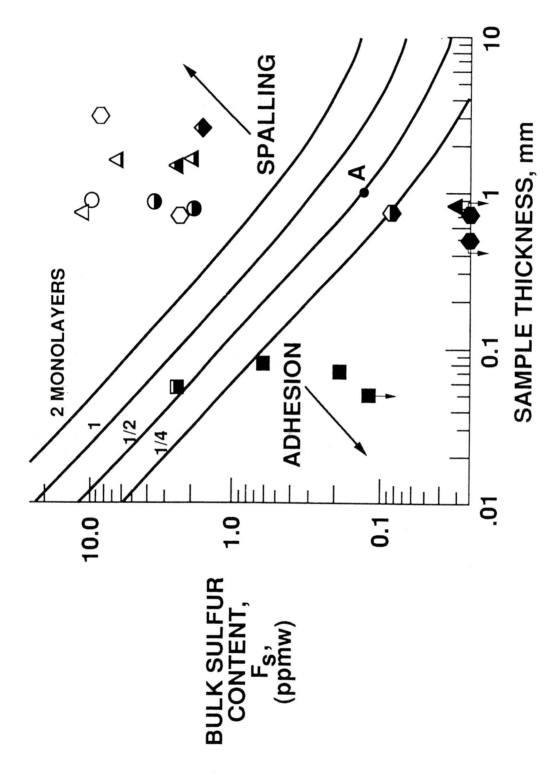

Figure 10: Oxide adhesion map based on critical amounts of sulfur segregation: curves predict total possible number of segregated monolayers; actual degree of adhesion shown by the degree of solid filling of data points [6]. See text for code.

that only a small amount of sulfur segregation (in terms of monolayers) is necessary to produce a clear manifestation of decreased adhesion.

The Effect of Sulfur Additions to Doped Alloys

Finally, this section describes some of the adverse effects that sulfur has when added to doped alloys that already demonstrate excellent adhesion. Because of space limitations, no additional figures can be presented. The key experimental findings are summarized here. Smeggil et al. dealt with the effects of ^+S ion implantation on the isothermal exposure of Ni-19Cr-13Al-0.07Y at 1050°C for 24 h [14]. These scales were cracked and showed some regions of spallation in contrast to the intact scales formed on an aluminum-implanted control sample. It was concluded that some spalling could be achieved just by adding sulfur to the surface of a NiCrAlY alloy.

In related studies by Smeggil et al. [15,16], the co-addition of sulfur and yttrium as 0.15 or 0.45 wt.% Y_2S_3 to Ni-20Cr-12Al produced weight losses more severe than with no additions at all in 1050°C and 1180°C cyclic oxidation tests. Thus sulfur is sufficient to render the yttrium ineffective for adhesion, assuming that the yttrium was dispersed uniformly. A similar study by Luthra and Briant, however, showed that 300 ppmw sulfur additions to Ni-20Cr-12Al-0.3Y did not adversely affect adhesion in 1150°C cyclic tests, whereas 2000 ppmw most certainly did [14].

In contrast, our own studies of successive sulfur additions to Ni-15Cr-13Al-0.18Zr found that scale adhesion in 1100°C cyclic tests was substantially reduced, even at the low level of 60 ppmw. And finally, Quadakkers et al. found that 80 ppmw additions destroyed adhesion of a Ni-10Cr-9Al-0.03Y alloy in 1000°C tests, as did 240 ppmw additions to Ni-10Cr-9Al-0.4Y at 1100°C [17,18]. Spalling in the latter study occurred only after an increased growth rate was sustained because of the greater amounts of Ni and Cr oxides formed with the sulfur addition. Increased Ni and Cr oxides have been noticed in the other studies of non-adherent and higher sulfur materials, however this was not a precurser or prerequisite to spalling [6,9,10,11,17].

SUMMARY AND CONCLUSIONS

This commentary has summarized a number of experiments designed to isolate just one of the reactive element effects, namely prevention of sulfur segregation. It is seen that, for the most part, a direct correlation exists between the sulfur content of a NiCrAl alloy and the propensity for non-adherent Al_2O_3 scale formation. This finding was corroborated by a number of independent studies and by a number of ways: alloys of different initial purities, sulfur removal by annealing in H_2, and sulfur removal by repeated oxidation/polishing.

Furthermore, the degree of adhesion varied with the degree of sulfur removal, as controlled by annealing temperature, sample thickness, or polishing cycle. Fully adherent behavior, i.e. no apparent spalling to bare metal and weight changes equivalent to NiCrAlY(Zr) was indeed approached for specimens having less than 0.1 ppmw sulfur. An adhesion map suggests a transition between adherent and non-adherent behavior for a sulfur content equivalent to 1/4 - 4 monolayers of segregation.

Conversely, sulfur additions on the order of 100 ppmw destroyed the otherwise adherent behavior of NiCrAlY,Zr alloys. The following conclusions are therefore claimed:

1. Decreasing the sulfur content and segregation is sufficient to impart a high degree of Al_2O_3 scale adhesion in normal cyclic oxidation tests of NiCrAl, without reactive element dopants.

2. Preventing sulfur segregation effects, either in doped or undoped NiCrAl, is necessary to maintain Al_2O_3 scale adhesion.

3. Preventing sulfur effects is a first order effect of reactive element dopants on Al_2O_3 scale adhesion to NiCrAl alloys.

4. These concepts allow a better understanding of oxide-metal adhesion, in general, and of the rationale for improved alloy design of current and future materials.

REFERENCES

1. Y. Ikeda, K. Nii and K. Yoshihara, Trans. Japan Inst. Met. Supplement, 1983, p.207.

2. Y. Ikeda, M. Tosa, K. Yoshihara and K. Nii, ISIJ International 3, 1989, p.966.

3. J.G. Smeggil, Mat. Sci. Eng. 87, 1987, p.261.

4. K. Luthra and C.L. Briant, Oxid. Met. 26, 1988, p.397.

5. C.L. Briant and K.L. Luthra, Metall. Trans. 19A, 1988, p.2099.

6. J.L. Smialek, Corrosion and Particle Erosion at High Temp., Eds.V. Srinivasan and K. Vedula, TMS-AIME, 1989, p.425.

7. D.G. Lees, Oxid. Met. 27, 1987, p.75.

8. D.G. Lees, Mat. Sci. Tech. 4, 1988, p.455.

9. J.L. Smialek, Metall. Trans. 18A, 1987, p.164.

10. J.L. Smialek, Proc. N.L. Peterson Mem. Symp. Oxid. Met. and Mass Trans., Eds. M.A. Dayananda, S.J. Rothman and W.E. King, TMS-AIME, 1986, p.297.

11. B.K. Tubbs and J.L. Smialek, Corrosion and Particle Erosion at High Temp., Eds.V. Srinivasan and K. Vedula, TMS-AIME, 1989, p.459.

12. J.L. Smialek and R. Browning, Proc. High Temp. Mat. Chem. III, Eds. Z.A. Munir and D. Cubicciotti, Electrochem. Soc., 1986, p.258.

13. C.G.H. Walker and M.M. el Gomati, Appl. Surf. Sci. 35, 1988/1989, p.164.

14. J.G. Smeggil, E.L. Paradis, A.J. Shuskus and N.S. Bornstein, J. Vac. Sci. Tech. A3, 1985, p.2569.

15. A.W. Funkenbusch, J.G. Smeggil and N.S. Bornstein, Metall. Trans. 16A, 1985, p.1164.

16. J.G. Smeggil, A.W. Funkenbusch and N.S. Bornstein, Metall. Trans. 17A, 1986, p.923.

17. J.L. Smialek, Proc. High Temp. Mat. Chem. IV, Eds. Z.A. Munir and D. Cubicciotti, Electrochem. Soc., 1988, p.241.

18. W.J. Quadakkers, C. Wasserfuhr, A.S. Khanna and H. Nickel, Mat. Sci. Tech. 4, 1988, p.1119.

19. A.S. Khanna, W. J. Quadakkers, C. Wasserfuhr and H. Nickel, Role of Active Elem. on Oxidation, Ed. E. Lang, Elsevier, London, 1988, p.287.

31 EFFECTS OF REACTIVE ELEMENTS ON OXIDE SCALE DEFORMATION AND CRACKING BASED ON SUBMICRON INDENTATION TESTING

P.F. Tortorelli, J.R. Keiser, K.R. Willson* and W.C. Oliver

Oak Ridge National Laboratory, Oak Ridge, TN 37831, USA.
*Geneva College, Beaver Falls, PA 15010, USA.

ABSTRACT

A method for characterizing the mechanical response of oxide scales by a depth-sensing submicron indentation technique is described for application to chromia-rich scales grown on a stainless steel. Results on room-temperature micromechanical properties (elastic modulus, hardness, etc.) are presented to provide data for determining scale fracture criteria and to evaluate models in which reactive element additions to alloys modify the mechanical behavior of oxide scales and thereby improve scale adherence. Average values of elastic modulus for thin chromia-rich scales formed on 20-25-Nb steels was in good agreement with that for the bulk chromia. The addition of 0.13% Ce to the steel, which significantly improves scale adherence, did not affect the elastic modulus hardness, strength, or plasticity of the oxide scale. These findings therefore did not support the subject models of scale adherence for cerium. However, the technique of depth-sensing submicron indentation testing appears to offer promise as a method of characterizing oxide scales.

INTRODUCTION

The ability of a metal or alloy to withstand high-temperature oxidation is directly related to the presence of a protective surface oxide scale that is either inert to, or slowly reactive with, the corrosive environment. Accordingly, the mechanical integrity and properties of the oxide are of considerable importance in understanding and controlling oxidation, scale adhesion, and the interaction between chemical and deformation effects [1-7]. To this end, measurements of certain mechanical properties of oxide scales by a submicron indentation technique are presented for stainless steels. Such results can be used in calculations of fracture energies and toughness of oxide scales and in the determination of the limits to scale adherence [3-8].

As discussed in recent overviews [9-11], the presence of relatively small amounts of rare earth (and other) oxide dispersoids or reactive elements in certain alloys can promote the formation of more adherent oxide scales. While there is no doubt that these reactive element/dispersoid additions (hereafter referred to as "REs") do generally improve scale adherence in many circumstances, there is considerable debate regarding the mechanism(s) by which they do so. Some of the proposed mechanisms suggest that REs decrease the propensity for scale cracking and spallation by modifying growth stresses, scale plasticity, and fracture toughness [4, 11-13]. Indeed, a recent review cited the need for new or improved techniques for measuring the mechanical properties of oxide scales to differentiate between various proposed RE models [11]. This paper represents an initial attempt to evaluate possible effects of RE additions on the micromechanical characteristics of oxide scales using the unique approach of submicron load-displacement measurements.

EXPERIMENTAL PROCEDURES

The elastic and plastic properties of chromia-rich scales on a stainless steel were measured with a Mechanical Properties Microprobe (MPM), which utilizes a depth-sensing submicron indentation technique [14] that has recently been successfully applied to mechanical characterization of thin films [15]. The MPM is a modified version of a commercial instrument known as a Nanoindenter[16]. It is a load-controlled device in which the force on the indenter is applied via magnetic coupling and the resulting displacement is measured by a capacitance gage. The smallest load increment is approximately 0.3 μN (with a maximum load of 120 mN) and the displacement resolution is about 0.2 nm. A light microscope is used to position the indenter on the specimen surface, but subsequent movements in all three dimensions are controlled by a computer, which also is used for data acquisition and analysis. A Berkovitch type of diamond indenter was used. It produces an indent with the same relationship between area and penetration depth as a Vickers indentation.

A typical MPM indentation load-displacement curve is shown in Fig. 1. Normally, both elastic and plastic deformation occur during indentation [14,15]. Hardness is determined by the indentation load and the contact area of the indenter and thus can be calculated as a function of displacement depth. The elastic modulus of the material can be obtained from the slope of the unloading curve when the indenter is pushed back by the elastic restoring forces. Alternatively, as is done with the present MPM system, a small superimposed ac signal causes a brief, limited load reduction at a prescribed frequency during the entire loading phase of the indentation event (Fig. 1). This procedure allows the modulus to also be measured as a function of displacement. Extrapolation of the initial slope of the final unloading curve to zero load yields the plastic depth, h_{plas} as shown in Fig. 1. (This value also includes any friction contribution.) While the presence of a substrate should affect the hardness and modulus measurements of thin layers, it has been shown that such an influence is minor if the total depth of the indenter penetration is a small fraction of the thickness of the film [15]. In the present case of thin oxide scales on a steel, all indentations were made on normal or tapered polished cross sections so that substrate influence was minimal.

Disks of Fe-20Cr-25Ni-0.7Mn-0.7Si (wt %) niobium-stabilized steels were obtained from Harwell Laboratory. One of these steels (designated 20-25-Nb) was of the composition indicated, while the second (20-25-Nb+Ce) differed only in that it also contained 0.13 wt % Ce. The oxidation behavior of these steels and the positive influence of cerium on scale spallation have been well documented [2,3,17-19]. Specimens were oxidized in air at 930-950°C to grow scales (composed principally of Cr_2O_3) for subsequent indentation testing. Because the addition of cerium results in slower scale growth, oxidation times were varied (24-96 h) to obtain scales of approximately the same thickness (1-2 μm). After oxidation, the specimens were sectioned, metallographically polished, and examined. Later specimens were polished with a tapered cross section in order to have a thicker effective scale for indentation testing. After an appropriate section of adherent scale was located, a matrix of indent positions covering both metal and oxide was selected and a MPM run was initiated. Under computer control, the series of steps shown in Table 1 was automatically performed at each indent position.

Table 1. Typical indent sequence

Step	Description
1	Indenter approaches surface at >10 nm/s
2	Loading at 2 nm/s to depth of 200 nm
3	Hold at load, record 30 displacements at 1/s
4	Hold at 90% maximum load, record 30 displacements at 1/s
5	Unloading at 100 μN/s to 10% maximum load
6	Hold at 10% load, record 50 displacements at 0.5/s
7	Unloading at 10 μN/s to zero load

RESULTS

The application of the MPM to the characterization of the micromechanical properties of oxide scales was straightforward in principle, but somewhat difficult in practice. The relative thinness of the adhering scales (particularly those formed on the 20-25-Nb+Ce steel) and the thermal drift of the specimen and instrument hindered the placement of the indenter squarely in the scale. Furthermore, if the indenter location coincided with that of a pre-existing scale defect, an atypical load-displacement curve was obtained and was subsequently neglected in the data compilation. (It is actually an advantage of this indentation technique that spurious results can often be detected and discarded, which is an important concern regarding the accuracy in which the mechanical properties of scales can be measured [7].) Therefore, only those indentations that yielded curves like that of Fig. 1 and were (usually) verified by scanning electron microscopy to be in the surface oxide and not distorted by defects were used in the determination of the scale properties. The micrograph in Fig. 2 is an example of an indentation in the oxide scale that was acceptable for analysis. The difference between load-displacement curves for the steel substrate (Fig.1) and the chromia scale (Fig. 3) was readily apparent. The load at a given displacement (indenter contact area) was substantially greater for indentations in the scale, thereby indicating a higher hardness for the oxide relative to the steel. The scale also showed a much greater elastic recovery, as evident from the respective shapes of the load-displacement curves in Fig. 3 and its correspondingly lower value of plastic depth (h_{plas}). Scanning electron microscope observations were consistent with these differences in hardness and plastic depth: indentations in the steel were much larger than those in the scales (see Figs. 1 and 3).

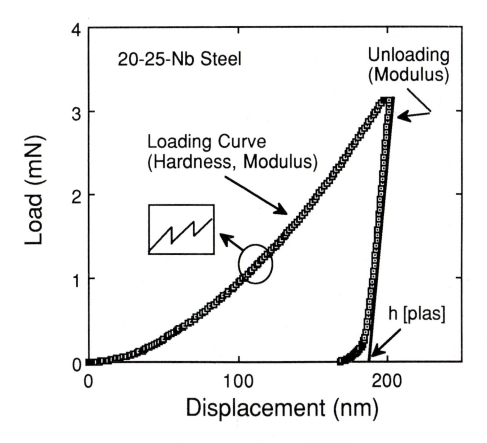

Figure 1: Typical load-displacement curve from indentation test on 20-25-Nb steel.

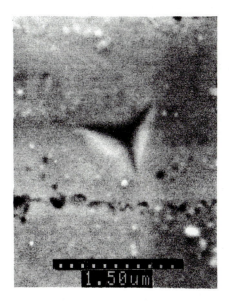

Figure 2: Indentation in cross-section of scale formed on 20-25-Nb steel.

As described in the Experimental Procedures section, the MPM can yield hardness and modulus data as a function of indenter penetration depth. (In the present case, this depth is the distance below the polished cross section and is normal to the scale growth direction.) Therefore, the hardness and modulus from an individual indent experiment were taken as averages of values determined from data at four depths between 140 and 170 nm, inclusively. In general, the hardness

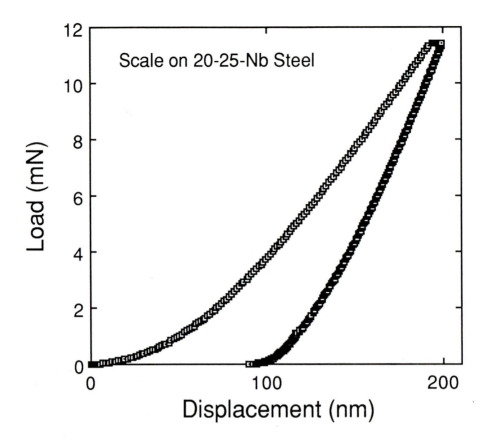

Figure 3: Typical load-displacement curve from indentation test on scale grown on 20-25-Nb steel.

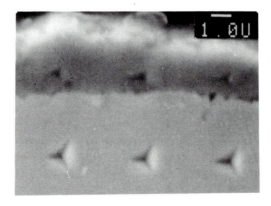

Figure 4: Array of indentations made in scale and substrate (20-25-Nb+Ce steel).

and modulus approached asymptotic values for indenter depths greater than about 80 nm. (Higher hardness values at small displacements from an arbitrary surface are observed for many materials [14].) Table 2 presents the averages and standard deviations of the hardness, elastic modulus, and plastic depth (h_{plas}) as determined from multiple indents in chromia scales grown on both 20-25-Nb and 20-25-Nb+Ce steels. Similar averages and standard deviations for the respective metallic substrates are also shown. No significant effect of cerium in 20-25-Nb steel on the hardness, modulus, or h_{plas} of the scale (and steel) was found.

DISCUSSION

Knowledge of the mechanical characteristics of oxide scales is important for calculations and predictions of scale cracking and such properties have been measured and applied to oxidation phenomena for many years [1-8]. Therefore, the present results are not "new"; many of these scale properties have been measured previously. Rather, it is the method of measurement and the application of the resulting mechanical properties data to evaluation of the ways in which REs affect scale adherence that are unique to this investigation.

Elastic moduli of certain oxide scales measured by vibrational techniques have been previously reported (see, for example, refs. 1, 7, 20, and 21). However, a recent study of the limiting conditions for scale adherence [7] used elastic moduli determined for bulk oxides in the appropriate calculations because of concern with the accuracy of at least some of the reported measurements on scales. It is therefore significant that the present measurements of the elastic modulus for chromia scales resulted in values (273 ± 18 and 296 ± 26 GPa, see Table 2) that agree quite well with that used for bulk Cr_2O_3 (283 GPa) in the aforementioned work [7]. This agreement supports the argument that elastic moduli for scales should be similar to those for bulk oxides of the same type [7]. (In actuality, the elastic moduli measured in this study are within a factor of about 1.3 of many of those measured for scales grown on nickel- and iron-based alloys [20,21].) The reasonably good accuracy and precision of the MPM measurements is helped by the ability, mentioned above, to discriminate between normal and atypical indents based on the load-displacement behavior of each event and associated scanning electron microscopy.

The finding that scale moduli determined from MPM measurements are well within the expected range of values (suggested by measurements on scales and bulk oxides) is most important. It contributes to validation of the use of submicron indentation testing for obtaining mechanical characteristics of thin oxide scales grown on metallic materials and increases the confidence in the technique as a way to evaluate relative changes in selected mechanical properties as a function of compositional changes (such as RE additions). In this regard, while the results in Table 2 may appear to suggest that the scale on the 20-25-Nb+Ce steel has a higher modulus than the scale on the 20-25-Nb steel, the standard deviations indicate that the difference between the two averages cannot be supported on the basis of these data. A variation in elastic modulus should not be expected since small changes in compositions or microstructure do not usually affect this property [22]. On the other hand, such small compositional modifications often affect properties such as hardness

Table 2. Mechanical properties microprobe results from analyses of polished cross-sections of oxidized 20-25-Nb steels

Steel	Material	Hardness (GPa)	Modulus (GPa)	Max. Load (mN)	h_{plas} (nm)
20-25-Nb	Oxide	23.4 ± 2.0	273 ± 18	11.0 ± 0.8	132 ± 7
	Alloy	2.9 ± 0.2	229 ± 5	2.8 ± 0.1	194 ± 4
20-25-Nb+Ce	Oxide	25.3 ± 1.3	296 ± 26	11.7 ± 0.8	134 ± 6
	Alloy	3.1 ± 0.1	236 ± 12	3.0 ± 0.2	194 ± 2

and strength. However, the results shown in Table 2 indicated that, as with the modulus, the difference in hardness between the scales formed on the two steels was within the overlap of the standard deviations of the respective measurements. Furthermore, the amount of room-temperature plasticity (as measured by h_{plas}. see Fig. 1) was the same for both scales (see Table 2). It thus appears from these initial results that the presence of cerium, while significantly improving scale adherence, does not affect room-temperature micromechanical properties in a way that influences spallation. Therefore, the present results do not support models in which RE additions modify the mechanical response of oxide scales to the extent that such models depend on changes in room-temperature elastic modulus, hardness, strength, or plasticity.

These initial MPM results are encouraging with respect to the potential of depth-sensing submicron indentation testing for characterizing thin oxide scales on metallic substrates and for evaluating certain models of improved scale adherence. However, there are two important caveats with respect to these measurements. First, only parts of scales that remained attached to the substrate after an oxidation treatment and metallographic preparation were used for the measurements. Scale material that spalled was not characterized and it may be just such scales that are most interesting for comparison and for understanding failure mechanisms. Second, the measurements were made at room temperature and they may not be characteristic of the scales at service temperatures, particularly with respect to plastic deformation. The present technique is useful for examining RE effects only to the extent that any relative changes in room-temperature mechanical properties of the scales are also operative at elevated temperatures in an oxidizing environment. A new type of MPM, designed for measurements at elevated temperature, is currently under development at Oak Ridge National Laboratory. When available and applied to the characterization of oxide scales, it will be possible to overcome these two limitations to the present approach by allowing measurements of the micromechanical properties of oxide scales at temperatures of interest and at which most protective scales are completely adherent.

SUMMARY AND CONCLUSIONS

Depth-sensing submicron indentation testing was used to measure room-temperature elastic and plastic properties of thin chromia scales formed on 20-25-Nb steels. The values of elastic modulus determined in this way were in good agreement with that for bulk Cr_2O_3.

The addition of 0.13% Ce to the steel, which significantly improves scale adherence, did not affect the room-temperature elastic modulus, hardness, strength, or plasticity of the oxide scale. Therefore, on the assumption that relative changes in room-temperature mechanical properties induced by compositional modifications are maintained at elevated temperatures in an oxygen environment, these findings do not support, for this alloy system, certain models of scale adherence in which RE additions modify the mechanical response of oxide scales.

The Mechanical Properties Microprobe appears to offer promise as a method to measure the micromechanical properties of oxide scales and to test models of scale growth and adherence based on their mechanical response.

ACKNOWLEDGEMENT

Research sponsored by the Fossil Energy AR&TD Materials Program, U.S. Department of Energy, under contract DE-AC05-84OR21400 with Martin Marietta Energy Systems, Inc.

REFERENCES

1. P. Hancock, Werkst. und Korros. 21, 1970, p.1002.

2. H.E. Evans and R.C. Lobb, Corros. Sci. 24, 1984, p.209.

3. H.E. Evans, Mater. Sci. Eng. A120, 1989, p.139.

4. F.H. Stott, Mater. Sci. Technol. 4, 1988, p.431.

5. M. Schutze, Mater. Sci. Technol. 4, 1988, p.407.

6. M. Schutze, Mater. Sci. Eng. A121, 1989, p.563.

7. J. Robertson and M. I. Manning, "Limits to the Adherence of Oxide Scales," to be published in Mater. Sci. Technol., 1990.

8. P. Hancock and J.R. Nicholls, Mater. Sci. Technol. 4, 1988, p.398.

9. H.S. Hsu, Oxid. Met. 28, 1987, p.213.

10. J. Stringer and P.Y. Hou, in Corrosion and Particle Erosion at High Temperatures, Eds.V. Srinivasan and K. Vedula TMS, 1989.
11. D.P. Moon, Mater. Sci. Technol. 5 , 1989, p.754.

12. G.C. Wood and F.H. Stott, High Temperature Corrosion, Ed. R.A. Rapp, National Association of Corrosion Engineers, Houston, 227.

13. T.A. Ramanarayanan, R. Ayer, R. Petkovic-Luton and D.P. Leta, Oxid. Met. 29, 1988, p.445.

14. J.B. Pethica, R. Hutchings and W.C. Oliver, Phil. Mag. A 48, 1983, p.593.

15. W.D. Nix, Metall. Trans. A 20A, 1989, p.2217.

16. Nano Instruments, Inc., P.O. Box 14211, Knoxville, TN 37914, U.S.A.

17. M.J. Bennett, G.W. Horsley and M.R. Houlton, Fundamental Aspects of Corrosion Protection by Surface Modification, Eds. E. McCafferty, C.R. Clayton and J. Oudar, Proc. Electrochem. Soc., 1984, 84-3, 1984, 282.

18. M.J. Bennett, B.A. Bellamy, C.F. Knights, N. Meadows and N.J. Eyre, Mater. Sci. Eng. 69, 1985, p.359.

19. M.J. Bennett, D.J. Buttle, P.D. Colledge, J.B. Price, C.B. Scruby and K.A. Stacey, Mater. Sci. Eng. A120, 1989, p.199.

20. D. Bruce and P. Hancock, J. Inst. Met. 97, 1969, p.148.

21. R.C. Hurst and P. Hancock, Werkst. und Korros. 23 , 1972, p.773.

22. A.G. Guy, Introduction to Materials Science, McGraw-Hill, New York, 1972.

32 THE EFFECT OF REACTIVE ELEMENTS ON THE OXIDATION BEHAVIOUR OF Fe-Cr-Al ALLOYS

J. Jedlinski* and G.Borchardt

Institut für Allgemeine Metallurgie, Technische Universitut Clausthal, Robert-Koch-Str.42, D-3392 Clausthal-Zellorfeld, FRG
* On leave from the Academy of Mining and Metallurgy, Kracków, Poland

and S.Mrowec

Institute of Materials Science, Academy of Mining and Metallurgy, al. Mickiewicza 30, 30-059 Kraków, Poland

ABSTRACT

The modification of the composition of a ferritic Fe-23Cr-5Al alloy by addition of reactive elements in different forms was used in order to study their effect on the commencement and extent of scale spalling. The reactive elements have been added as their natural mixture ('mischmetal'), as elemental hafnium and yttrium and as implanted yttrium. In addition, zirconium was used as an alloying element.

Oxidation was carried out in air in the temperature range 1473 to 1623 K. Short and long term experiments were performed.

It was found that the reactive elements improve the resistance of the scale to spalling. Their efficiency depended on the form of their presence in the substrate and on the reaction conditions.

INTRODUCTION

The beneficial effect of small amounts of certain reactive elements on the oxidation behaviour of FeCrAl alumina-forming alloys is well documented [1-8]. It manifests itself mainly in an increased resistance of the scale to spalling.

The former studies of the oxidation behaviour of the FeCrAl alloys were carried out at relatively low temperatures. In most caseg the temperatures did not exceed 1470 K, which is well below the liquidus temperature of these materials. The results of an investigation carried out at higher temperatures are presented in this paper.

MATERIALS AND EXPERIMENTAL PROCEDURE

The wrought alloys with the compositions given in Table 1 were obtained in the form of rods with a diameter of 15 mm. The details of the sample characteristics are given elsewhere [9].

The oxidation experiments were carried out in the temperature range 1473 to 1623 K. The scale composition and morphology were studied by means of SEM and EDX.

Thermogravimetry was used to follow the extents of reaction.

RESULTS AND DISCUSSION

The thermal cycling experiments have shown that all materials gain but do not lose weight at temperatures up to 1600 K. The only temperature at which some differences in the scale resistance to spalling were observed is 1623 K (Fig.1). The base alloy and the zirconium containing alloy degraded by spalling of the scale while breakaway oxidation was observed in the case of the alloy containing 'mischmetal'. No indications of accelerated degradation were found for the alloy which contained hafnium and yttrium. Implanted yttrium was found to improve the scale resistance to spalling in the case of alloys 0 and 1, its effect being, however, not of long duration.

Table 1. The composition of the studied alloys (in wt. %)

Element	Alloy			
	O	1	2	3
Fe	bal.	bal.	bal.	bal.
Cr	22.70	23.40	22.94	23.18
Al	4.80	4.90	4.92	4.47
Mn	0.68	0.55	0.61	0.60
Si	1.20	0.87	0.84	0.81
Ti	0.025	0.009	0.009	0.009
Ni	0.19	0.10	0.11	0.11
Cu	0.07	0.02	0.04	0.02
C	0.06	0.04	0.05	0.05
Zr	—	0.23	0.25	0.20
Mm*	—	—	0.035	0.05
Y	—	—	—	0.04
Hf	—	—	—	0.01
S	0.011	0.010	0.005	0.007
P	0.017	0.007	0.008	0.008

* - Mm denotes 'mischmetal'

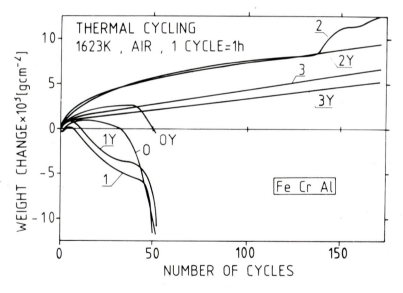

Figure 1: Weight changes observed during thermal cycling at 1623 K. (The notation corresponds to that in Table 1, while 'Y'denotes yttrium-implanted (2 x10 ions/cm) alloys).

As violent spalling of the scales occured after the first oxidation cycle short term oxidations were carried out in order to determine the beginning and the extent of this process. Oxidation periods of 1, 5 and 40 minutes were chosen.

The scale was found to spall away from the base alloy during cooling already after 1 minute oxidation time, as illustrated in Fig.2 (a). Resistance of the scale to spalling is lower after prolonged oxidation. The case of 5 minute oxidation time is shown in Fig. 2(b).

In contrast, a well adherent scale was formed on the yttrium implanted base alloy oxidised for 1 and 5 minutes at 1623 K (Fig.3a). Small areas of the substrate surface were observed after 40 minutes oxidation time (Fig.3b).

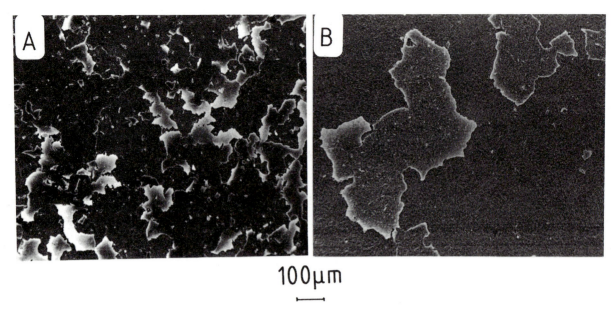

Figure 2: Surface of the samples of the base alloy oxidised at 1623 X for : A: 1 min, B: 5 min.

More intensive spalling of the scale was observed for the Zr-containing alloy (number 1 in Table 1). The only larger remnant of oxide on unimplanted material oxidised for 40 minutes at 1623 K is shown in Fig.4(a). Implanted yttrium reduced the extent of spalling (Fig.4b), this process being, however, more intensive than that observed for the yttrium implanted base alloy (number 0 in Table 1).

From the these results it follows that yttrium implanted into the base alloy as well as into the Zr doped alloy retards the commencement of the scale spalling and reduces its extent. Its efficiency in global improvement of the scale resistance to spalling is, however, not persistent.

It should be also noted that the scale spalled away from yttrium implanted alloys 0 and 1 although a gain of weight was observed in both cases. Thus, global weight changes do not give precise informations concerning the real resistance of the scale to spalling which locally could be low.

Adherent scales developed on mischmetal containing material (number 2 in Table 1) as well as on the material containing

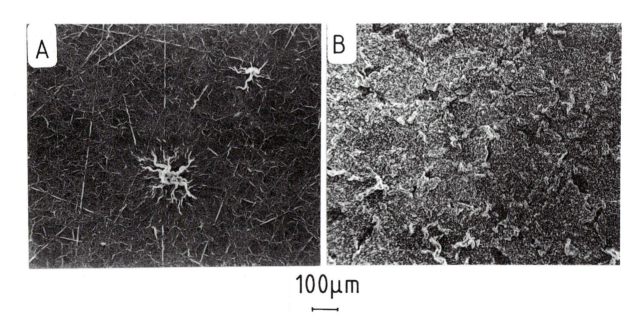

Figure 3: Surface of the samples of yttrium implanted (2 x10 ions/ cm) base alloy oxidised at 1623 K for : A: 1 min, B: 40 min.

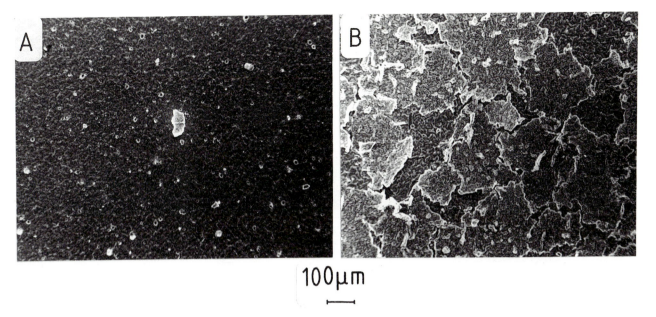

100µm

Figure 4: Surface of the zirconium containing alloy oxidised at 1623 X for 40 min : A: unimplanted, B: yttrium implanted (2 x10^{17} Y/cm^2).

metallic hafnium and yttrium (number 3 in Table 1). This finding is consistent with the results of thermal cycling experiments (Fig.1).

Implanted yttrium has not affected the resistance of the scale to spalling but it significantly modified its morphology. Porous scales were, namely, observed on unimplanted materials, despite their resistance to spalling which is illustrated for Zr and Hf+Y containing alloys in Fig. 5. As can be seen in both cases the integrity of the scale is far away from ideal.

In contrast, fairly compact scales were produced during oxidation of yttrium implanted alloys. At lower temperatures they were, however, convoluted (Fig.6a and b). The size of convolutions was smaller on the alloy implanted with a higher dose of yttrium (Fig.6b). At 1623 K they did not appear for oxidation periods up to 40 minutes (Fig.6c). Both these results indicate that the extent of convolutions can be reduced either by applying higher yttriun doses or by increasing the temperature.

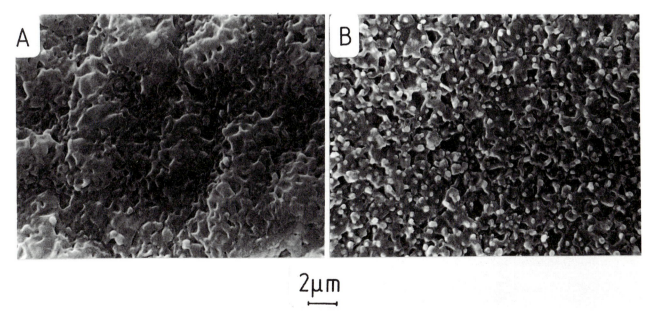

2µm

Figure 5: Surface of the scales formed at 1623 X : A: zirconium containing alloy (oxidation time 5 min), B: hafnium and yttrium containing alloy (oxidation time 40 min).

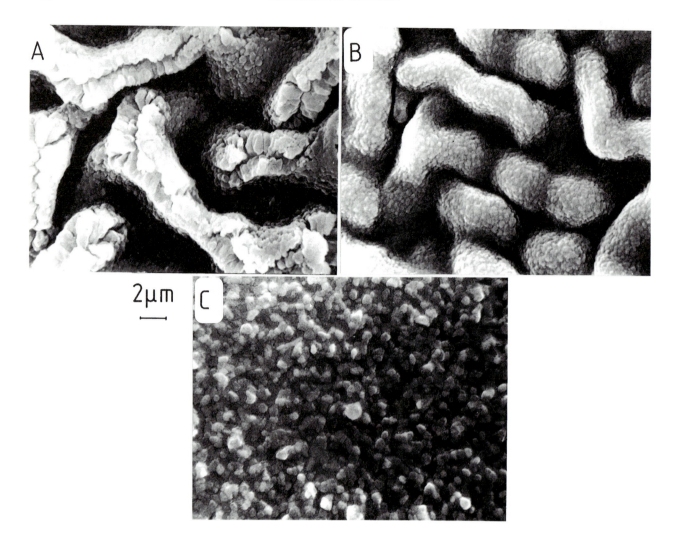

Figure 6: Surface of the scale formed on the yttrium implanted alloy containing 'mischmetal' : A (2 x10 Y/cm) and B (2 x10 Y/cm²) : 1473 X, 60 min, C (2 x10^{17} Y/cm²) :1623 X, 40 min.

CONCLUSIONS

The short and long term oxidation experiments carried out at 1473-1623 K have shown that :

1. The scales formed on 'pure' and on zirconium containing Fe-23Cr-5Al alloys are poorly resistant to spalling while 'mischmetal' and elemental hafnium and yttrium additions provide very good resistance.

2. Spalling occurs more extensively on the Zr containing alloy than from the base material.

3. Implanted yttrium retards the commencement of the spalling and reduces its extent, its effect being, however, not of long duration.

4. No simple correlation between the scale morphology and its resistance to spalling can be inferred.

ACKNOWLEDGEMENTS

The authors thank Mr.E.Ebeling, Mr.T.Weiss and Ms.T.Krasovec (TU Clausthal) for their assistance with the experiments. One of us (J.J.) is grateful to the Alexander von Humboldt-Foundation for their financial support.

REFERENCES

1. J.K.Tien and F.S. Pettit, Metal.Trans. 3, 1972, p.1587.

2. F.A.Golightly, F.H. Stott and G. C. Wood, Oxid. Met. 10, 1976, p.163.

3. F.H. Stott,F.A. Golightly and G.C. Wood, Corros. Sci. 19, 1979, p.889.

4. J.G. Smeggil, and A.G. Shuskus, J.Vac. Sci.Technol.A 4, 1986, p.2577.

5. T.A.Ramanarayanan, R. Ayer, R.Petkovic-Luton and D.P. Leta, Oxid. Met. 29, 1988, p.445.

6. W. J. Quadakkers, H. Holzbrecher, K.G. Briefs and H. Beske, ibid. 32, 1989, p.67.

7. A.M. Huntz, G.Ben Abderrazik, G. Moulin, E.W. A.Young and J.H.W. de Wit, Applied Surface Science 28, 1987, p.345.

8. D.R. Sigler, Oxid.Met. 29, 1988, p.23.

9. J. Jedlinski and G. Borchardt, submitted to Werkstoffe und Korrosion.

SECTION SIX
Growth and fracture of scales on Ti, Zr, Si, Ni and C-based materials

33 HOT SALT STRESS CORROSION CRACKING IN TITANIUM ALLOYS

K. Seebaruth, S.B. Newcomb and W.M. Stobbs

Department of Materials Science and Metallurgy,
Cambridge University, Pembroke Street, Cambridge CB2 3QZ, UK.

ABSTRACT

We describe the results of a preliminary TEM study aimed at characterising the microstructural changes which take place when titanium alloys are corroded under stress in the presence of salt. The formation of voids within the sub-scale alloy is shown to be important in promoting inward scaling which can involve chloride formation ahead of regions of the Ti which are degraded through the development of a fibrous structure.

INTRODUCTION

The high strength to weight ratio of titanium alloys as well as their general resistance to oxidation has led to the widespread use of these materials in the aircraft industry. While it is well known that commercial titanium alloys are subject to hot salt stress corrosion cracking (HSSCC)[1], the incidence of in-service cracking worldwide, however, is very low and is not what would be predicted from laboratory test data. This discrepancy in the performance of titanium alloys highlights one of the ways in which HSSCC is not fully understood. We have been studying HSSCC in titanium alloys and here present some preliminary TEM results on the scale microstructures which are formed when Ti is corroded in the presence of salt. Our primary aim has been to characterise how either low cycle fatiguing or the application of a constant load during corrosion can determine what microstructures are seen and thus how HSSCC is both initiated and controlled. The rôle played by various alloy constituents has also been investigated by comparing the scales formed on two alloys of differing Sn, Al and Zr content.

EXPERIMENTAL

The nominal composition of the two near a Ti-based alloys (IMI 679 and 829) investigated are shown below:

IMI 679	Bal. Ti	11Sn	5Zr	2.5Al	1Mo	0.2Si	
IMI 829	Bal. Ti	3.5Sn	3Zr	5.5Al	0.25Mo	0.3Si	1Nb

Both alloys exhibited a coarse grained lath microstructure prior to oxidation as well as interlath hydride formation though the latter is likely to have occurred during thin foil electropolishing [2] and is not significant in the HSSCC process. Ground specimens were coated with salt (1.5 mg cm^{-2}) and oxidised at 500°C either under constant stress or with low cycle fatiguing. We have also compared the corrosion scales formed in this way in the presence of salt with the oxides formed in the absence of salt under the same load conditions by the simple expedient of coating only one surface of a given specimen. In this way we hoped to be in a good position to compare the various synergistic effects in the alloy surface region due to either constant loads or low cycle fatiguing both with and without the different corrosion products that then form in the absence and the presence of the salt. Six specimens have been examined, the details of which are tabulated below, A,B,C being oxidised in the absence of salt while D,E and F were salt coated:

Specimen	Alloy	Treatment
A	IMI 679	Constant Stress 250MPa: 1h
B	IMI 679	Constant Stress 250MPa: 10h
C	IMI 679	LCF: 250MPa: 50h
D	IMI 679	Constant Stress 250MPa: 10h + Salt
E	IMI 679	LCF 250MPa: 50h + Salt
F	IMI 829	LCF 250MPa: 50h + Salt

While a range of techniques have been used to characterise the different oxides formed, here we describe the results obtained only from both cross-sectional and through film TEM specimens prepared using standard techniques [3].

RESULTS AND DISCUSSION

First we describe the oxides formed on IMI 679 at 500°C in the absence of salt and highlight the ways in which the microstructures of the sub-scale alloy and oxides can change both as a function of time and applied stress condition. We then describe the salt corrosion scales formed on both IMI 679 and 829. As an aid to our descriptions a summary diagram of the different microstructures is shown in Figs. 1a to 1f for specimens A-F respectively.

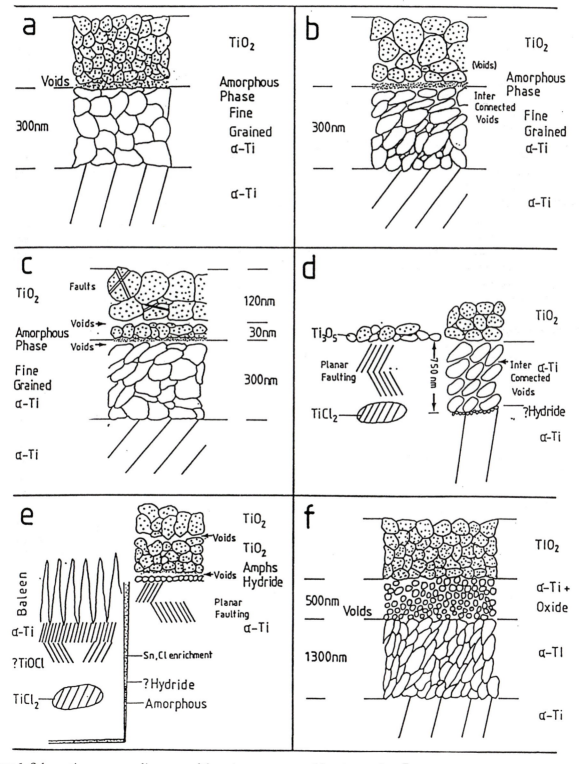

Figure 1: Schematic summary diagrams of the microstructures of Specimens A to F.

Oxidation Scales

The oxide formed under constant load after 1 hour's exposure to air at 500°C (Specimen A) was found to be rutile (TiO_2). The morphology of the ~20nm grain size oxide layer, which was ~100nm in thickness, is shown in Fig. 2a where a distribution of fine pores may also be seen at the metal/oxide interface (as arrowed) as well as an amorphous phase (arrowed in Fig. 2b). While the composition of the amorphous phase remains to be determined by PEELS, the TiO_2 was found to contain ~7wt% Sn but only very minor concentrations of Al and Zr. Beneath the oxide and to a depth of ~300nm we see fine grained α-Ti (as in all the oxidised specimens, Fig.2c) which exhibited the bulk alloy composition except for minor Al depletion (to ~1.5wt%). That the fine grained α-Ti was found not to develop either with time or the applied stress condition suggests that it formed during the pre-oxidation grinding treatment.

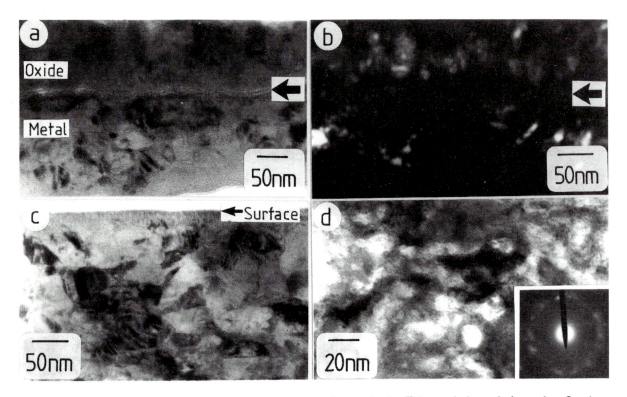

Figure 2: (a,b) Metal/oxide interface region for Specimen A, (c) fine grained α-Ti beneath the scale formed on Specimen B, and (d) interconnected porosity found in the α-Ti, as seen in plan-view.

Specimen B, as oxidised under constant load for 10 hours, was again found to have formed TiO_2 as well as an amorphous phase at the metal/oxide interface. The oxide was found to contain approximately 6wt% Sn though the sub-scale alloy was not found to be significantly depleted or enriched in Sn or Al. The fine grained α-Ti immediately beneath the scale was now found to contain an interconnected distribution of voids. These voids are shown in Fig. 2d, as seen in plan view. The appearance of voids within the Ti is indicative of a clear way in which a constant load can cause the coarse disruption of the alloy microstructure so that oxidant can presumably then diffuse into the metal with ease. The outer TiO_2 layer would normally be regarded as highly protective but the formation of both the amorphous phase and the voids at the metal/oxide interface means that it can spall readily which, in the presence of salt, would lead to an increased rate of scaling. The effects of low cycle fatiguing on the microstructural development of the oxide formed in the absence of salt (Specimen C) were also examined. In some ways the oxide scale was rather similar to those described above. For example, we again see a layer of TiO_2 but it is now rather more developed and appears to be bi-layered. The metal/oxide interface region, as viewed in cross-section, is shown in Fig. 3 where a double layer of voids may be seen (as arrowed), suggesting that we are seeing both inward and outward growth of the scale though further work is required to determine whether this is due to low cycle fatiguing or to the relatively long (50 hours) oxidation treatment used here. It does however further indicate that, despite the general protective properties of the TiO_2 scale, the diffusion of oxidant can quite readily occur through it, the diffusional properties of the oxide possibly being modified by the incorporation of divalent Sn.

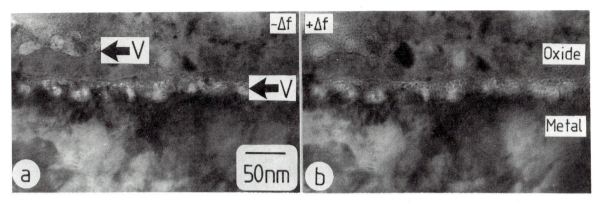

Figure 3: The metal/oxide interface in Specimen C. Voids may be seen at V.

Corrosion Scales

Turning now to the scales formed in the presence of salt we first consider the oxide formed after 10 hours under constant load (Specimen D). Two distinct oxide and sub-scale microstructures were found though the morphological similarity of one of them to that on Specimen B suggests that the scaling is not necessarily modified by the salt. For example, we find that TiO_2 has been formed which has an approximate 40nm grain size at its outer surface and a porous zone at its base. The oxide was found to contain very low concentrations of Al, Zr and Mo but approximately 12wt% Sn, a much higher concentration than was seen either in the scale on Specimen B or in any of the other oxide scales. The alloy immediately beneath the scale again consisted of fine grained α-Ti which was not surprisingly slightly depleted in Sn (to ~9wt%), but was now rather more developed than for Specimen B (see Fig. 1). The fine grained α-Ti zone, as shown in Fig. 4a in plan view, is very porous, and was also found to contain some amorphous or microcrystalline oxide, which further indicates how the incorporation of Sn into the TiO_2 scale may be important in promoting inward diffusion of oxidant and thus internal oxidation. Cl was not, however, found within the voided layer. Beneath this zone, where the α-Ti had a coarse lath structure, no oxides were seen. The second type of oxide seen in Specimen D was comparatively fine grained (~25nm) and was found to be Ti_3O_5 (Fig. 4c). Beneath this part of the scale the α-Ti exhibited planar faulting (Fig. 4b) though the observation of similar microstructures in other specimens (see Fig.1) means that this type of faulting cannot be attributed solely to either the loading conditions used here or to Ti_3O_5 formation. While diffraction patterns from this faulted zone were necessarily diffuse other structurally distinct regions were seen (Fig. 4d) which, though not unambiguously identified, were possibly either $TiCl_2$ or TiOCl.

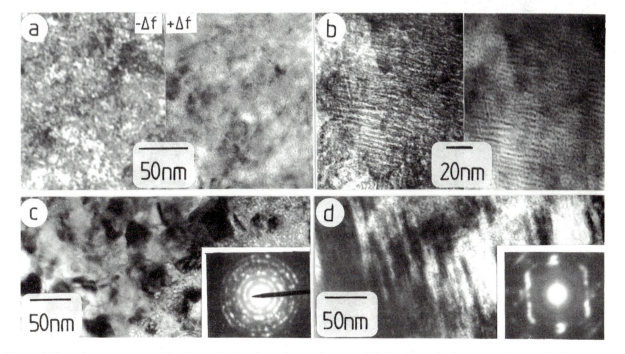

Figure 4: The microstructure of Specimen D showing (a) porosity in the Ti, (b) planar faulting as well as (c) Ti_3O_5 and (d) $TiCl_2$ formation.

Turning to Specimen E, which was subject to low cycle fatigue, we now find that the IMI 679 alloy has been grossly disrupted both in bulk and at alloy grain boundaries. The scale itself, when retained in thin foil TEM specimens, was found to be generally stratified, as was the oxide formed in the absence of salt, although as many as three different layers were now seen, as shown in the dark field micrograph in Fig.5a, as well as local nodule formation. The oxide was generally fine grained (~25nm) and contained both pores and amorphous material (as at A, Fig.5a). Diffraction patterns taken from the base of the scale where the α-Ti reflections were weak showed an fcc phase (a_o=0.456nm). While this lattice parameter is ~3% larger than for TiH_2 it should be noted that the transformation chemistry for the hydride can involve a variable Ti:H ratio. The possibility that hydrides have been formed or that hydrogen has been incorporated into the Ti alloy is important because of the tendency of hydrogen to cause embrittlement. Other regions of the alloy examined, where we do not see an outer scale, were found to be both chemically and structurally different. Here we find that the alloy has a fibrous morphology. A micrograph, where an alloy grain boundary may also be seen, is shown in Fig. 5b which demonstrates that this attack has taken place to a depth many microns beneath the pre-corroded surface of the alloy. Planar degradation of the Ti may also be seen where some of the boundaries between the layers may be associated with prior α-Ti grain boundaries. The way in which enhanced corrosion has taken place around the intermetallic particle (as at A, Fig. 5b) further suggests that the attack is diffusion controlled. Diffraction patterns taken from the fibrous zone (see Fig. 5b, inset) indicated that it was essentially α-Ti but the patterns also exhibited weak diffraction both from a larger lattice parameter material (possibly TiOCl) in microcrystalline form as well as from an amorphous phase. The low magnification dark field micrograph shown in Fig. 5c indicates that the alloy grain boundary where corrosive attack has taken place is both grossly porous and contains microcrystalline as well as amorphous phases, where the concentration of Sn could be as high as 30wt% and where Cl was also found. Well beneath (>1μm) the fibrous zone described we again find α-Ti, but it now has a sub-lath morphology (Fig. 5d) rather like the sub-scale region seen in Specimen D. Diffraction patterns (see

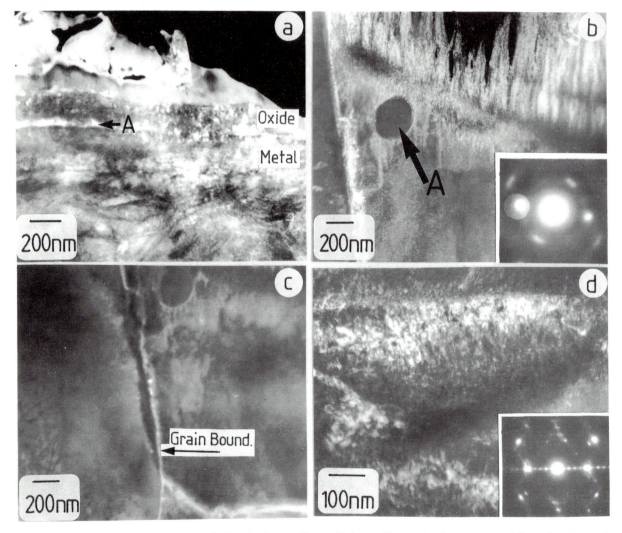

Figure 5: Specimen E showing (a) a stratified scale, (b) the fibrous Ti formed beneath other regions of the scale, (c) a grain boundary zone, and (d) $TiCl_2$ formation.

inset, Fig. 5d) were consistent with $TiCl_2$ though this compound has not been unequivocally identified. The result does however demonstrate that Cl can diffuse into the Ti alloy ahead of the fibrous structure described and possibly provide precursory attack through the formation of a chloride.

The region of Specimen F examined in the TEM showed many similarities in both its morphology and chemistry to the 679 alloys described above despite there being significant compositional differences between IMI 679 and 829. The sub-scale alloy was again particularly interesting in exhibiting a clear boundary ~1.8μm beneath the scale between the coarse grained matrix and finer lath α-Ti above. This region is shown in Fig. 6a where both the boundary (arrowed) and the fine 1-2μm laths may be seen. Here the α-Ti was marginally depleted in both Sn (to ~2wt%) and Al (to ~4wt%) by comparison with the alloy matrix, whereas the α-Ti above it and immediately beneath the scale contained 6.5wt% Sn and only low Al concentrations. Here the α-Ti was extremely porous (Fig. 6b) and, although we found no direct evidence from the diffraction patterns for oxide formation within it, the locally increased Sn concentration here relative to that in the laths beneath it and the oxide above it suggests a degree of internal oxidation like in Specimen D which was similarly voided. The α-Ti grains here are, however, generally finer (~50nm) than those in Specimen D while the TiO_2 formed above is similarly fine grained (20-100nm) as well as being porous (Fig. 6c) containing low concentrations (<2.5wt%) of Sn, Al, Zr and Nb.

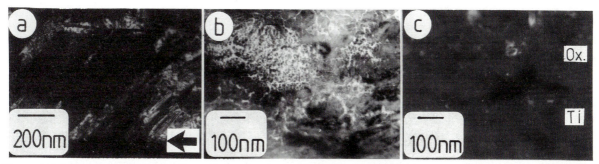

Figure 6: Sub-scale changes in Specimen F showing (a) fine laths in sub-scale Ti, (b) porous Ti, and (c) the 'open' metal/oxide interface.

CONCLUSIONS

While the results which we have presented are largely preliminary in nature, they do point to a number of possible ways in which HSSCC is both initiated and controlled. We have shown, for example, that even in the absence of salt the sub-scale metal can become voided when oxidation takes place under constant load and that inward oxidation can be promoted in such areas. Equally it would appear that Cl is able to diffuse into these alloys with apparent ease where the formation of $TiCl_2$ may facilitate the formation either of the fibrous structure seen or of planar faulting within the α-Ti. We are less clear, however, about the possible synergistic relationship between the incorporation of Sn either into the outer scales or into internal oxides or oxychlorides and the way in which Cl diffuses into the alloy. That Sn does play some rôle here is strongly supported by the observation of ~7wt% in the TiO_2 formed on the lower Sn content alloy (IMI 829). The part other elements such as Zr and, particularly, Al play is even less clear but our data indicates that Al is denuded from the substrate and yet not obviously segregated into the general scale. This would suggest that Al might be incorporated into inhomogeneously distributed nodules but it is equally possible that there is localised layer segregation at a spatial resolution beneath that analysable by conventional TEM approaches. Work is currently in progress on the further characterisation of these fascinatingly different scales which should further clarify the changes in mechanism.

ACKNOWLEDGEMENTS

We thank Professor D. Hull FRS for the provision of laboratory facilities as well as the SERC and Rolls Royce plc (Leavesden) for financial support. We are also grateful to R.Wing for useful discussions.

REFERENCES

1. Proc. of Fundamental Aspects of Stress Corrosion Cracking, 1967, Ed. R.W. Staehle, NACE, Texas 1969.

2. C.Shelton, Ph.D. Thesis, Cambridge University,1984.

3. S.B.Newcomb, C.B.Boothroyd and W.M.Stobbs, J.Microsc. 140, 1985, p.195.

34 THE MICROSTRUCTURAL AND COMPOSITIONAL ANALYSIS OF THE OXIDES ON Zr-2.5wt% Nb IN RELATION TO THEIR RÔLE AS PERMEATION BARRIERS TO DEUTERIUM UPTAKE IN REACTOR APPLICATIONS

B.D. Warr, E.M. Rasile, A.M. Brennenstuhl, M.B. Elmoselhi,
N.S. McIntyre*, S.B. Newcomb** and W.M. Stobbs**

Ontario Hydro Research Div., (KR182) 800 Kipling Ave., Toronto, Ontario, Canada M8Z 5S4.
*University of Western Ontario, London, Ontario, Canada N6A 5B7.
**Department of Materials Science and Metallurgy, Cambridge University, Pembroke Street, Cambridge CB2 3QZ, UK.

ABSTRACT

Zirconium alloys are widely used to make core components such as fuel cladding and pressure tubes in nuclear reactors where the rate of hydrogen uptake is largely controlled by the effectiveness of the oxide as a permeation barrier. The structure of the oxides formed on the CANDU reactor pressure tube material, Zr-2.5 wt%Nb, was investigated using electron microscopy and, following exposures both within and outside the reactor, SIMS was used to characterise the deuterium uptake profile in the oxide which it was found could be different from point to point. At regions of high bulk deuterium content, deuterium concentrations were high throughout the oxide on the outside surface of the pressure tube; conversely, where bulk alloy deuterium concentrations were low, much steeper diffusional profiles were observed in the oxide. The data obtained allowed a diffusion coefficient to be inferred for the low uptake regions and the implications of the results are discussed in relation to the different oxide structures seen in the scale.

INTRODUCTION

Zr-2.5 wt%Nb and other zirconium alloys such as the Zircaloys have a low neutron cross-section and high strength and are thus often used to make nuclear reactor core components such as fuel cladding and pressure tubes. The material's major disadvantage in this application is, however, that it is embrittled by hydrogen absorption during operation. Accordingly components are used with an oxide coating which can act as an hydrogen permeation barrier, though the effectiveness of such coatings may be reduced as a result of structural and compositional changes occurring during inreactor exposures. The work described here forms part of a study both of the structure of these oxides and of their rôle as permeation barriers.

The pressure tubes which form the primary pressure boundary in CANDU reactors are made of Zr-2.5 wt%Nb as cold worked with a microstructure consisting of elongated grains of α-Zr (~1wt% Nb) which are approximately 0.3 μm thick, 3 μm wide and ~10 μm long, surrounded by a thin continuous filament of β-Zr (~20wt% Nb). In the as-received condition, CANDU pressure tubes have a black adherent oxide film of ~1 to 2 μm in thickness, as a result of autoclaving in steam at 400°C. Average hydrogen uptake rates in Ontario Hydro's pressure tubes are generally low (< 2ppm by weight/annum), although a few tubes have shown local regions of anomalously high hydrogen uptake[1-3]. On pressure tube inside surfaces, which are in contact with lithiated heavy water (D_2O) coolant at up to 300°C, between 5 and 10 % of the hydrogen produced by the corrosion reaction may be absorbed during normal operation. The oxides on the outer surfaces of the pressure tubes are in contact with a dry (as low as -50°C dew point) annulus gas consisting of N_2 or CO_2 with <25ppmO_2 and up to 1wt% deuterium, at approximately atmospheric pressure. Increased localized rates of deuterium ingress could well result from the oxide degradation which is likely in these environments in the reactor. Our aims have thus been to develop an improved understanding of the relationship between oxide structure and composition, and oxidation and hydrogen uptake in Zr-2.5 wt% Nb, in order to predict and improve pressure tube performance. The structure of the oxide near its interface on zirconium alloys has been investigated in the past, using electron microscopy, for pure zirconium and Zircaloys (4-8), and limited data have also been reported for Zr-2.5wt% Nb [4,9,10]. Near-surface oxide chemical concentration profiles have also recently been obtained using SIMS [10-12]. Here our approach has been to attempt to correlate the structural data we have obtained by "edge-on" TEM and other EM methods on the oxide structure with the oxide's diffusional characteristics as inferred from SIMS data.

THE EXPERIMENTAL METHODS USED

Thin rectangular samples were cut from inside and outside surfaces of irradiated pressure tubes at selected axial locations corresponding to differing hydrogen uptake rates and local reactor environments. Samples were also taken from coupons of Zr-2.5 wt% Nb and pure zirconium following exposures in elevated temperature aqueous/steam environments outside the reactor. Selected samples were also exposed to vacuum (~10^{-8} Torr) at elevated temperature (400°C-500°C) in order to investigate possible changes occurring in the oxide in a simulated reducing reactor annulus gas environment.

A Cameca IMS-3f ion microscope was used for the SIMS study described below of the deuterium diffusion behaviour in the scale. A 60 µm diameter caesium ion beam with a current of 0.8 - 2.0 µA and an acceleration voltage of 10kV was used as rastered over an area of 250 x 250 µm. Secondary ions were collected through a 150 x 150 µm area at the centre of the sputtered region and high sputtering rates gave high sensitivity and good spatial resolution. The secondary ion yields were quantified using hydrogen ion implants into thin ZrO_2 films and clean Zr-2.5wt%Nb metal surfaces. Further details of the SIMS procedure have been reported elsewhere [11]. Oxide undersides were exposed and examined with a JEOL 35C SEM, using a slightly modified approach to that developed by Cox [8] as described elsewhere [9,10]. For the TEM examination of the microstructure near to and across the metal/oxide interface, "edge-on" foils, made using methods described by Newcomb et al. [13], were examined in a JEOL2000FX.

THROUGH-THICKNESS DEUTERIUM CONCENTRATION PROFILES

All samples, prefilmed in steam at 400°C, and subsequently exposed in aqueous D_2O or gaseous environments containing deuterium showed up to ~0.5 wt% deuterium at the outer oxide surface. Also in most, but not all of the samples examined, very low levels of deuterium were found at the metal-oxide interface. The differences in the diffusivity of the films which may be inferred may relate to differences in the scale and metal/oxide interface microstructure and it was the investigation of possible causes for these observations which was the primary goal of the EM work described below. Here we will first describe the variations we have found, using SIMS, in through-thickness depth concentration profiles for deuterium, and the way these variations can be related to deuterium uptake in the bulk metal.

Figure 1 shows typical deuterium concentration profiles in the oxide and oxide/metal interface region in samples taken from the outside surfaces of irradiated pressure tubes removed from high temperature outlet regions (at ~300°C) in the reactor. Where deuterium concentrations in the bulk alloy are low (<20 ppm), the deuterium content in the oxide drops sharply by up to between 2 and 3 orders of magnitude before the oxide/metal interface is reached (Fig.1a). In samples taken from regions with high bulk alloy deuterium concentrations, the deuterium contents in the oxide are also higher (~0.5 wt%), and remain high throughout the oxide, dropping significantly only at the metal/oxide interface (Fig.1b). A possible explanation for the latter behaviour, in which the diffusion of deuterium into the specimen is apparently not limited by the normally low diffusivity of the oxide, is that increased and interconnected porosity has developed within it, at such locations, during longterm exposures in the reactor to the relatively reducing environment there.

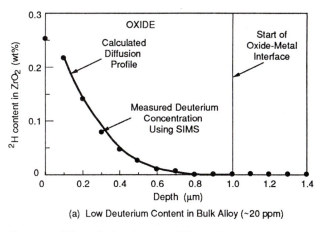

(a) Low Deuterium Content in Bulk Alloy (~20 ppm)

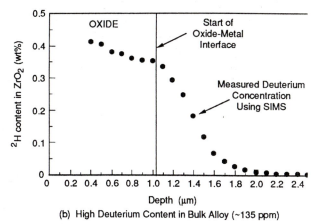

(b) High Deuterium Content in Bulk Alloy (~135 ppm)

Figure 1: The relation between Deuterium concentration profiles through the outside surface oxide and the bulk Deuterium content in irradiated pressure tubes.

For the diffusion characteristics shown in Fig.1a it is possible to calculate the concentration profile which would be expected on certain limiting assumptions and on the basis that the deuterium ingress is indeed diffusion controlled. This allows the best fitting diffusion coefficient for the process to be determined. The calculated concentration profile shown in Fig.1a was obtained by solving the diffusion equation assuming cylindrically symmetric diffusion under the boundary conditions that the concentration at the outer surface remained constant while the metal/oxide interface was taken to be an absolute barrier to hydrogen. The diffusion coefficient that best fits the observed deuterium concentration profile in the oxide in Fig.1a was found, under these assumptions, to be approximately 1×10^{-22} m^2s^{-1}. This value is considerably lower than that obtained by the characterisation of hydrogen ion implants in this oxide for a temperature of 300°C but not inconsistent with values measured in the same way for oxides on pure Zr [14]. Given that it may be seen that the deuterium concentration falls to a very low value somewhat before the oxide/metal interface (as well as that the over all fit is good) our low diffusivity would not appear to be explained by there being an impermeable barrier at only the metal interface (as would be the case if the oxide in this area is amorphous and continuous). On the other hand our data (Fig.1b) also indicates unequivocally that, in other areas, the oxide has pathways allowing relatively easier deuterium ingress, at least to the metal/oxide interface, in a manner more consistent with the ion implant spreading results quoted above [14] for the oxide on Zr-2.5wt%Nb. In our structural examination of the material we must thus seek explanations for both these behaviours, paying particular attention to any microstructural features of the oxide scale which might inhibit grain boundary diffusion as well as to others, such as interconnected porosity, which might aid it.

OXIDE MICROSTRUCTURE

The Topography of the Metal/Oxide Interface

This was examined, as described above, by the SEM of the underside of oxides after the removal of the alloy. Figure 2 shows typical oxide undersides, representing the interfacial topography, on pure polycrystalline Zr and Zr-2.5 wt% Nb following steam exposures (24 h, 400°C). For both the pure material and the Nb-containing alloy, a series of growth fronts is indicated by the morphology. The interface topography above the α-Zr grains in Zr-2.5 wt% Nb appears to indicate finer hemispherical growth fronts, generally >100nm but down to ~50nm in diameter, than are found above the pure Zr where they are between ~500 and 2500 nm in diameter. As we will see below the sizes of the finest growth fronts for Zr-2.5wt%Nb are similar to the diameter of the columnar grains identified by TEM with which they may thus reasonably be associated. Preferential oxidation of the grain boundaries is evident, through the topography of the ridges which correspond to the grain boundaries, and in the case of Zr-2.5 wt% Nb, in the regions of the intergranular Nb-rich, β-Zr phase.

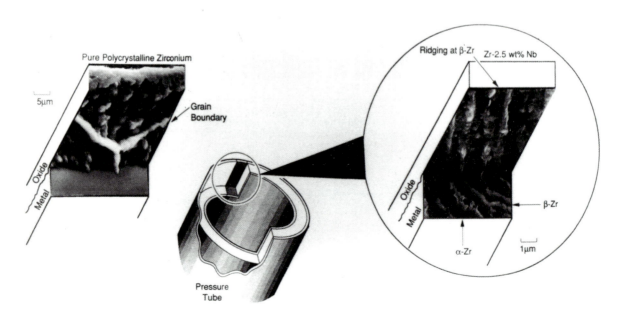

Figure 2: Oxide/metal interfacial topography of pure zirconium (left) and Zr-2.5wt%Nb (right) following exposure in steam (400°C, 24 h).

Figure 3 shows the differences in the interface structures which were observed for samples of Zr-2.5wt%Nb in the as-received steam prefilmed condition (Fig.3a) and following both exposure to vacuum (~10-8 Torr, 500°C, 48 h) and exposures in the reactor at ~300°C, for 12 years (Figs.3b and 3c respectively). The irradiated sample was taken from the outside surface of a pressure tube removed at a location of relatively high bulk deuterium concentrations. Oxide undersides on this sample show a different structure by comparison with that for unirradiated samples of pressure tubes, with a geometrical array of 'holes'. These holes may result from oxide dissolution into the base material at these locations, arising during the prolonged exposure at the reactor operating temperatures to the weakly oxidising atmosphere of the annulus gas. This inference is strengthened by the way similar holes have been observed in some samples exposed to vacuum at high temperature, as shown in Fig.3b. More work is required to clarify the mechanisms of oxide degradation.

(a) 1μm   (b) 1μm   (c) 1μm

Figure 3: Interface structures in Zr-2.5wt%Nb; (a) before and (b) after exposure to vacuum while (c) shows the interface structure following in-reactor annulus gas exposure.

The Bulk Oxide as Examined by TEM "Edge-On"

A flat sample machined from the wall of a Zr-2.5 wt% Nb pressure tube was characterised "edge-on" using TEM following oxidation in steam at 400°C for 24h. with the primary aim of clarifying those features of the scale which might be associated with the disparate diffusion characteristics it can develop given extended reactor exposure. A schematic diagram of the main features found in the scale is shown in Fig.4a. The particular region of the scale described, which was about 1.5μm thick, was above a region where there were α-Zr grains which were more equiaxed than in the bulk microstructure. The images in Figs.4b,c and d show a region of the metal/oxide interface under different conditions. A layer of amorphous material was seen at the marked interface and although in some regions such zones could be up to ~10 nm thick the amorphous layers were not everywhere continuous and separated the metal from colonies of columnar oxide grains (typically ~40 nm in diameter and >200 nm long). These tended to show a textured orientation relation with the metal substrate perhaps indicating that the amorphous oxide developed only after the initiation of the scaling process. In relation to the low diffusion coefficient found for some parts of this type of oxide it is probably significant that amorphous regions were found not only at the metal/oxide interface but also fairly generally at grain boundaries as well as quite high in the scale at a boundary separating an upper scale (which was more equiaxed) from the lower scale (see Fig.4e). On the other hand the oxide was also relatively highly porous with the larger pores tending to lie relatively parallel to the interface. Examples of such pores are marked in Figs.5a and b which are images of the same area of the scale taken at different defoci so that the pore visibility be enhanced by their Fresnel contrast. Some of the pores may be associated with heterogeneities caused by the distribution of the Nb-rich β-Zr phase. It would appear that we have evidence here for the possible causes of both the different diffusion behaviours which have been observed. However, the heterogeneity of the scales would have to have developed in other regions than those examined still further (as is not uncommon for the type of in-situ exposures met) to exhibit the disparate diffusional behaviours seen. The porosity would have to have become interconnected, for example, in the regions exhibiting high diffusivity.

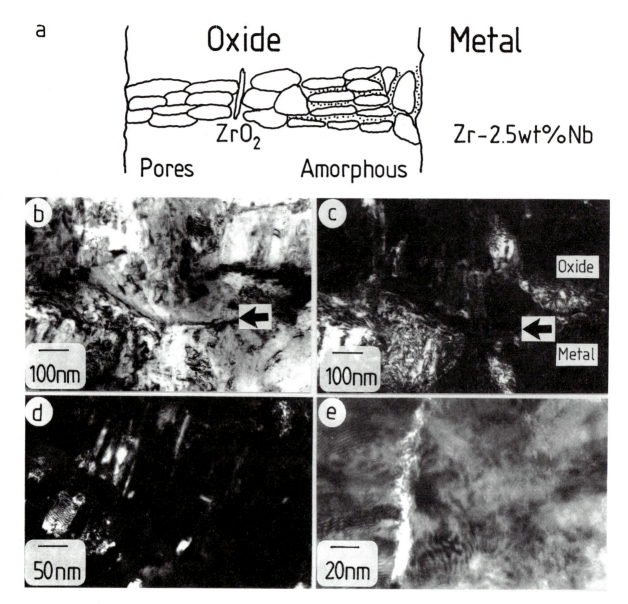

Figure 4: (a) Schematic diagram of scale features. (b) b.f. and (c) d.f.images of metal/oxide interface region. (d) d.f. image for amorhous contrast above interface. (e) b.f. image from near to the top of the scale showing amorhous region disrupting columnar oxide structure.

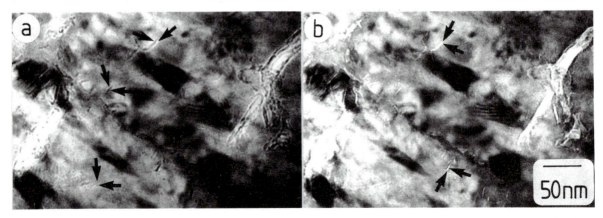

Figure 5: Above and below focus b.f. images of the same area about 1μm up from the metal oxide interface. The Fresnel contrast due to some of the smaller voids is arrowed.

SUMMARY

The work we have done on the diffusivity of Zr-2.5 wt%Nb pressure tube material oxide scales to deuterium may be summarised:

1. A relationship has been found between the deuterium concentrations in the bulk alloy and the deuterium concentration profiles in the surface oxides on the outside of CANDU pressure tubes removed from the reactor. At regions of relatively high bulk deuterium content, deuterium concentrations were high throughout the oxide. Where, conversely, bulk alloy deuterium concentrations were low much steeper diffusional profiles were observed in the oxide, from which a diffusion coefficient of $1 \times 10^{-22} \mathrm{m}^2\mathrm{s}^{-1}$ was inferred. These data suggest that under some circumstances the oxide has anomalously high diffusivity and under others has a lower diffusion coefficient than would be expected from other measurements [14].

2. Reasons for the above disparate behaviours of the scale were sought in its substructure as examined both by the SEM of the underside of the oxide and in the microstructure of the scales as originally formed and characterised by their "edge-on" examination in the TEM. The data obtained indicated that the scales were both porous and contained amorphous regions at various depths as well as on grain boundaries generally.

It is clear that much remains to be done in order to clarify how the scales can have developed inhomogeneously to exhibit either, on the one hand, the properties required of them in inhibiting the ingress of deuterium or alternatively relatively high diffusivity. The characterisation of the heterogeneity of the oxide microstructure, which may develop in the in-situ environmental exposures to which the scale is subjected or could, alternatively, be present within it as formed will be as interesting academically as it is technologically worthwhile.

ACKNOWLEDGEMENTS

We would like to thank Ontario Hydro and the Candu Owners Group (COG) for funding the above work. Thanks are also due to Ontario Hydro Research Division Staff, Dr P. Lichtenberger and Messrs J. Mummenhoff, L. Grant, J. De Luca, F. Turone and C. Ashby. Dr B. Cox (University of Toronto) provided useful advice and encouragement.

REFERENCES

1. P.C. Lichtenberger, An Overview of Hydrogen Ingress Mechanisms in Zr-2.5 wt%Nb Pressure Tubes. Fuel Channel Technology Seminar, Alliston, Ontario, Canada, November 12-15, 1989.

2. V.F. Urbanic, B.D. Warr, A. Manolescu, P. Chow and M. Shanahan, Oxidation and Hydrogen Uptake of Zr-2.5wt%Nb Pressure Tubes in CANDU-PHW Reactors, Eighth International Symposium on Zirconium in the Nuclear Industry, American Society for Testing and Materials, San Diego, CA, US, 19-23 June 1988.

3. B.D. Warr, M.W. Hardie, V.F. Urbanic and D.O. Northwood, Fifth Asian-Pacific Corrosion Control Conference, Melbourne, Australia, November 22-28, 1987.

4. G.P. Sabol, S.G. MacDonald and G.P. Airey, ASTM,STP 551, American Society for Testing and Materials, 1974, p.435.

5. A.W. Urquhart and D.A. Vemmilyea, ASTM STP-551, 1974, p.463.

6. H. Stehle, F. Garzorolli, A.M. Garde and P.G. Smerd, ASTM STP 824, American Society for Testing and Materials, 1984, p.483.

7. R.A. Ploc and M.A. Miller, J. Nuc. Mat. 64, 1977, p.71.

8. B. Cox, J. Aust. Inst. Met. 3, August 1969, p.14.

9. B.D. Warr, E.M. Rasile and A.M. Brennenstuhl, Electron Microscopical analyses of oxides in Zr-2.5wt%Nb, Technical Committee Meeting on Fundamental Aspects of Corrosion of Zirconium-Base Alloys in Water Reactor Environments, Portland, Oregon, U.S., September 11-15, 1989.

10. B.D. Warr, M.B. Elmoselhi, E.M. Rasile and A.M. Brennenstuhl, "Electron Microscopical analyses of oxides in Zr-2.5wt%Nb", Fuel Channel Technology Seminar, Alliston, Ontario, Canada, November 12-15, 1989.

11. A.M. Brennenstuhl, B.D. Warr, N.S. McIntyre, C.G. Weisener and R.D. Davidson, SIMS Seventh Int. Conf. on Secondary Mass Spectrometry, Monterey, CA, US, September 3-8, 1989.

12. N.S. McIntyre, A.M. Brennenstuhl and B.D. Warr, Deuterium in depth analyses of oxidised Zr-2.5wt%Nb Pressure Tube Samples removed from an Operating Nuclear Reactor. Fuel Channel Technology Seminar, Alliston, Ontario, Canada, November 12-15, 1989.

13. S.B. Newcomb, C.S. Baxter and E.G. Bithell, IOPCS 93, 1988, p. 43.

14. D. Khatamian and D. Manchester, J. Nuc. Mat. 166, 1989, p.300.

35 DEFECTS AND INTERFACIAL STRUCTURE IN Ge/Si LAYERS

D. Fathy and M. Sayah

Nuclear Materials Group, A.E.O.I.
& Physics Department Amir-Kabir University,
P.O.Box 14155-1339,
Tehran, Iran

ABSTRACT

During thermal oxidation of Ge - ion implanted Si, Ge is completely rejected from the growing oxide and forms an intermediate crystal layer between the Si and SiO_2. The segregated Ge rich layer is epitaxial with the Si substrate and leads to enhanced oxidation rates. The increased oxidation rates are shown to arise because the segregated Ge modifies the interfacial growth kinetics and lowers strain in the oxide layer. Kinetic and interfacial structure information are provided over a range of temperatures and oxidation ambients. Enhanced oxidation rates studied were found to be consistent with a modified reaction at the oxide/Si interface. An anomalous behaviour is exhibited for steam oxidation at 800 °C, due to the viscous flow of the intermediate Ge layer. This behaviour is characterised and the mechanisms responsible are discussed. Also a comparison using HRTEM is made between Ge/SiO_2 and Si/SiO_2 interface. This shows that the undulations at the interface are different indicating a reduced binding energy due to the presence of substitutional Ge atoms.

INTRODUCTION

It has recently been shown [1,2] that an epitaxial film of Ge forms on Si during thermal oxidation of Ge-ion implanted Si. During oxidation Ge is completely rejected at the growing oxide interface, where it accumulates forming a distinct, Ge-rich layer. The presence of this intermediary layer between the oxide and Si substrate leads to substantially enhanced oxidation rates.

Deal and Grove [3] modelled oxidation of Si using two rate-limiting process: (a) a diffusive process which transports oxidant from the surface to the growing oxide boundary and (b) reaction at the oxide/Si interface which forms the oxide phase. In this case it was shown that a greatly increased interfacial reaction due to Ge-implantation accounts for the enhanced oxidation [4]. Successful modelling of this effect was achieved by assuming a step-function dependence of the interfacial reaction rate on the amount of segregated Ge- Si of lower binding energy than that of Si - Si. No dependence was observed until approximately a single mono-layer of Ge is segregated after which additional accumulation of Ge at the interface produces no effect.

The use of Ge-implantation to modify thermal oxidation rates raises interesting possibilities, since Ge is isoelectric with Si. Thermal oxides are extensively used in the fabrication of integrated circuits in Si for isolation and masking purposes. It is anticipated that both processes could benefit directly from the effects of Ge-implantation. Also, the differential rate of oxide growth between implanted and non-implanted Si is used to selectively mask against subsequent process steps.

In this paper, the influence of Ge-implantation on thermal oxidation is investigated over a range of temperature and oxidation ambient. Kinetic data is generated over this range and is related to the mechanisms for oxidation enhancement by Ge-implantation. Oxidation is shown to be consistent with the mechanism described [4]. except for steam oxidation at low temperature (800 °C), where a lower compressive strain in the oxide layer and lower binding energy at the interface together increases the oxidation rate. A high resolution comparison is made in order to qualitatively determine the influence of the Ge intermediate layer on the roughness of the SiO_2/Ge_xSi_{1-x} compared to that of Si/SiO_2 interface.

EXPERIMENTAL PROCEDURE

Rutherford backscattering spectroscopy (RBS) was used to determine oxide thicknesses and characterize the behavior of Ge during oxidation. The detailed microstructure in the oxidized samples was determined by cross-sectional transmission-electron microscopy (TEM). Oxidations were done in a standard tube furnace in both steam and wet ambient. The wet ambient was produced by bubbling O_2 through 95 °C water. Single crystal, n-type Si < 100 > was used in this investigation. A 0.20 μm screen oxide was grown on the Si before implantation to passivate the surface and also reduce the effective range of the implanted Ge-ions in Si. Samples were implanted with 30 keV Ge-ions to a dose of 10^{16} atoms/cm^2. Implantation through the screen oxide at this energy placed the peak of the implant profile near the Si surface and therefore resulted in a high surface concentration of Ge.

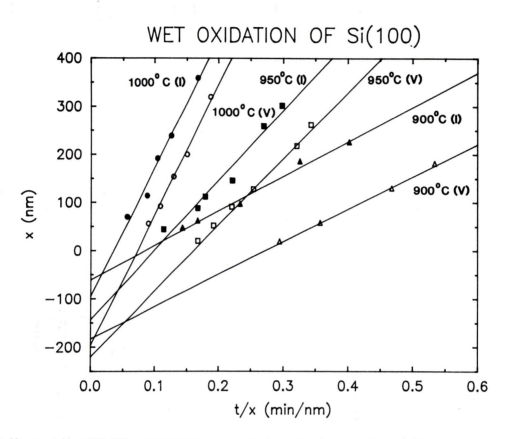

Figure 1: Kinetic data at 900, 950 and 1000 °C for wet oxidation of implanted and virgin Si.

In the Deal and Grove model [3], oxide thickness is related to oxidation time by

$$ t = \frac{x}{B/A} \quad x \quad \frac{x^2}{B} \qquad (1) $$

where x is the oxide thickness, t the time, and A and B are the constants specific to a given set of oxidation conditions.

Two different regimes of growth are generally identified, a linear regime when $t << \frac{A^2}{4B}$ which leads to the relation $x = t\frac{B}{A}$, and a parabolic regime when $t >> \frac{A^2}{4B}$ which gives $x^2 = Bt$. The constant B, referred to as the diffusion rate constant, and $\frac{B}{A}$, the linear rate constant, express the rates of the two limiting processes which control the thermal oxidation. Plots of x versus t/x are shown in Fig.1 for wet oxidation of both implanted and virgin Si.

This manner of plotting allows the kinetic parameters to be determined. The diffusion rate constant B is given by the slope of the curves and the y-intercept gives the value of (-A). A consistent behavior is observed over the temperature range investigated. The kinetic data from each sample lies along a straight line which is expected from Eqn.(1) so long as the rate constants do not vary with time (i.e. oxide thicknesss). The absence of non-linearity in the implant data shows that the oxidations were "well-behaved" at each temperature with constant reaction rates. Implanted and non-implanted data at each temperature have the same slopes but different intercepts which demonstrates that the interfacial rate constants are substantially affected by implantation but the diffusion rates are not. The rate parameters determined from the data in Fig.1 are tabulated in Table 1.

Table 1. Rate constants for wet oxidation of Si

Oxidation Temp.(°C)	A(μm)	Parabolic rate constant B(μm²/h)	Linear rate constant B/A(μm/h)
1000 (I)*	0.094	0.160	1.702
1000 (V)	0.196	0.162	0.826
950 (I)	0.061	0.081	1.328
950 (V)	0.183	0 080	0.437
900 (I)	0.120	0.043	0.358
900 (V)	0.206	0.040	0.194

* "I" denotes implanted and "V", Virgin.

An Arrhenius plot for the interfacial reaction in both implanted and virgin samples generally indicates a simple activated process; exp(-Ea/kT), characterized different slopes for various temperatures (Fig.2). But plots of the data for implanted and virgin Si at low temperature indicate different slopes at 800 °C. The slope of 800 °C implanted sample is the same as 950°C virgin sample. Therefore not only the activation energy but also other factors such as the tension free energy relieved by the viscous flow should be considered.

Other studies have shown an increase in oxidation rates on annealing thick oxides [9]. Relieving the compressive stress would increase interface kinetics by satisfying the volume requirements for conversion of Si to SiO_2. Also, a crossover is observed in oxidation rates at the viscous flow point of the virgin oxide [10]. This accounts for the difference in the activation energies measured at low (780-930 °C . 2.2 eV) and high (> 950°C ,1.3 eV) temperatures.

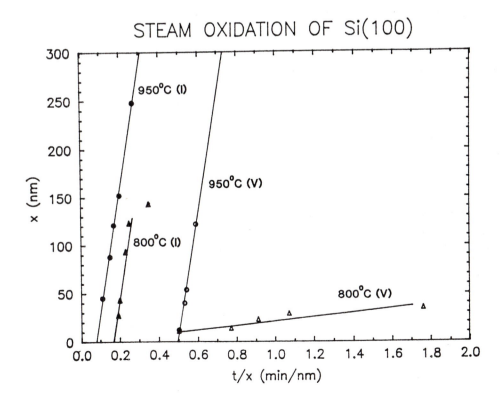

Figure 2: Kinetic data at 800 and 950 °C for steam oxidation of implanted and virgin Si.

The change in the binding energy of Si-Si(78.1 kCal/mole) to that of Ge-Si (72 kCal/mole) accounts for implanted samples stress relaxation at lower temperatures. It is reasonable at this stage to assume that the associative effect with the viscous flow can lead to the total 40% enhancement rates observed. Maxwell viscoelastic flow properties [5,6] explain the occurrence of increasing stress at reduced oxidation temperatures(< 950° C) for virgin Si. A 120% volume increase in Si conversion to SiO_2 leads to a limiting factor specially at lower temperature which is also in agreement with the observation of lower oxidation rate of virgin Si in our experiments. Similarly we have observed that at temperatures above 1000 °C no enhancement takes place. Therefore the major role of Ge is in strain relief.

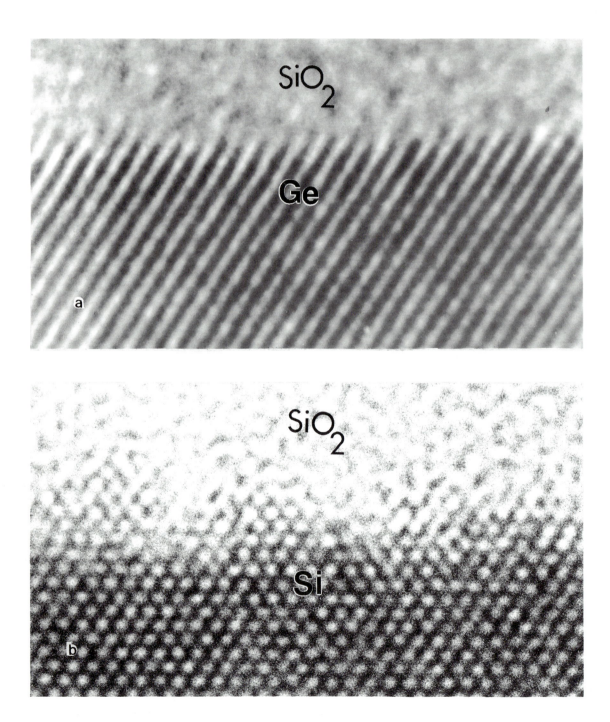

Figure 3: High resolution image of oxidized samples under the same condition but (a) Ge-implanted (b) virgin.

Kinetic data for steam oxidation is given in Fig.2 for two different temperatures. The figure shows that at the higher temperature the implanted data parallels that from virgin Si. This indicates that the diffusion rate constant is the same in both samples but the interfacial reaction rate is different. However, a totally different behavior is observed in samples oxidized at 800 °C. The plot in Fig.2 shows that both the slopes and intercept of the two data sets (implanted and non-implanted) differ. The reason for this behavior as discussed is due to reduced strain in the oxide and at the interface. A similar low-temperature oxidation effect in As⁺-implanted Si was recently reported [7].

A comparison of the high resolution images of the interface is shown in Fig.3 indicating a smoother interface in the case of Ge implanted samples. An interesting morphology was observed in samples with a thick Ge layer which had undergone multiple oxidations at 800 °C. The oxidations were done by cycling the sample in and the out of the furnace.

A TEM micrograph of a sample cycled three times at 800 °C is shown in Fig 4(b). Crystal islands of Ge are observed within the oxide corresponding to the location of the oxide/Si interface after each cycle. Also, a continuous Ge layer (epitaxial on Si) is seen at the final oxide/Si boundary. The presence of Ge precipitates within the oxide can be a result of Ge precipitation with stable sizes in the Ge - Si layer due to supercooling effect.
These Ge precipitates form at the interface where the Ge concentration is highest. These islands are left inside the oxide by the preferential oxidation that takes place at the Ge - Si interface during the cooling cycle as shown in Fig.4. It must be mentioned at this stage that Ge trapping in the oxide only occurs in temperature cycling.

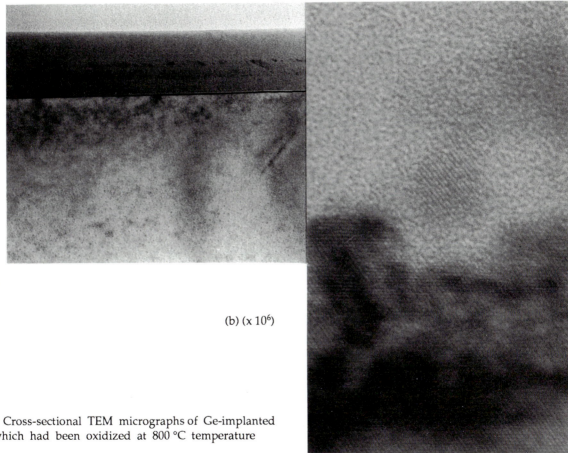

(a) (x 10⁴)

(b) (x 10⁶)

Figure 4: Cross-sectional TEM micrographs of Ge-implanted sample which had been oxidized at 800 °C temperature cycling.

CONCLUSION

It has been shown that, over most of the range of temperature and ambient studied, the effect of Ge-implantation on Si oxidation rates can be accounted for by a modified interfacial reaction rate, and oxide strain relief. An anomalous behavior was observed during 800 °C oxidation in steam. Rates were greatly enhanced over their intrinsic values and can be explained by interface strain relief. These conditions, which produce the largest differential growth rates between implanted and virgin Si, seem ideally suited for masking applications. It should be noted that the effect of Ge-implantation on the interfacial reaction was observed to be less for wet oxidation than for steam oxidation. Virtually no difference was observed for dry oxidation or temperatures > 1000°C where the oxide is already viscoelastic.

4. ACKNOWLEDGEMENT

The authors gratefully acknowledge Dr. A. A. Bidokhti for assistance in preparing this manuscript.

5 REFERENCES

1. D. Fathy, C. W. White and O. W. Holland, Proceedings of SPIE, Vol.797, Ed. Sayan D. Mukherjee, SPIE, Bellingham, Washington, 1987, p.83.

2. D. Fathy, O.W. Holland and G.W. White, Appl. Phys. Lett. 51, 1987, p. 1337.

3. B. E. Deal and A. S. Grove, J. Appl. Phys. 36, 1965, p. 3770.

4. O.W. Holland, C.W. White and D. Fathy, Appl. Phys. Lett. 51, 1987, p. 520.

5. E. Kobeda and E. A. Irene, J. Vac. Sci. Technol. B4, 1986, p. 720.

6. E. Kobeda and E. A. Irene, J. Vac. Sci. Technol. B6 (2), 1988, p. 574.

7. M. Seong.Choi, M.Z.Numan and W.K.Chu, Appl. Phys. Lett. 51, 1987, p. 1001.

8. D. Fathy and M. Sayah, A.E.O. I. Scientific Journal (to be published).

9. J. K. Srivastava and E.A.Irene, J.Electrochem. Soc.132, 1985, p. 2815.

10. A. Fargeix and G. Ghibaudo, J. Appl. Phys. 56, 1984, p. 589.

36 OXIDATION AND LIQUID-ALUMINIUM DEGRADATION OF ENGINEERING CERAMICS

M.W. Johnston and J. A. Little

Dept. of Materials Science and Metallurgy, University of Cambridge, Pembroke Street, Cambridge CB2 3QZ, UK.

ABSTRACT

The oxidation of a silicon nitride-bonded silicon carbide ceramic was studied using thermogravimetric analysis (TGA) and scanning electron microscopy (SEM). Reaction occurs mainly in the nitride phase and the silica scale formed was imaged. The effect of varying SiC and Si particle sizes in the material was investigated. Corrosion of the ceramic by molten pure aluminium after varying degrees of oxidation was investigated. The most severe degradation was observed in the nitride phase. The reaction areas imaged can be explained by considering the relevant thermodynamic relations.

INTRODUCTION

Ceramic materials are attractive candidates for use at high temperatures in environments where metallic alloys would suffer extensive degradation. One such application is gas-fired immersion tube heating, used especially in non-ferrous foundry situations.

Such a heating system allows major improvement in performance over conventional radiative heating units. Since heat transfer occurs by conduction, the tube operates at a temperature very similar to that of the molten metal, giving fuel savings of up to 70% [1]. In addition, for an immersion tube system thermal gradients in the bath are much smaller, allowing deeper, more compact structures to be constructed.

Because of the aggressive nature of the molten metal, the use of metal tubes is precluded, and ceramic materials must be substituted. Silicon carbide (SiC) ceramics have a high thermal conductivity and excellent resistance to thermal shock. When mixed with a suitable binding phase, they are also resistant to liquid-metal attack, relatively easy to fabricate and competitively priced.

One recent application of this technology uses a silicon nitride-bonded silicon carbide ceramic, manufactured by reaction-bonding. Silicon carbide particles are mixed with silicon powder and a clay-based binder. This green ceramic is formed to the desired shape (a closed-ended tube) and then fired in a nitrogen atmosphere at 1400°C so that a skeleton of silicon nitride forms. This process gives a material which has 15-20% porosity.

During service in an aluminium foundry, tube performance is found to be inconsistent, with some premature failures occurring [1]. The tubes operate by burning natural gas within them, and the combustion products can oxidize the ceramic to silica, SiO_2. Silica is wetted by, and reacts with, molten aluminium:

$$3SiO_2 + 4\underline{Al}_{melt} = 2Al_2O_3 + 3\underline{Si}_{melt} \qquad (1)$$

This work investigates the oxidation behaviour of the nitride-bonded silicon carbide and considers the effect of altering the silicon and silicon carbide particle sizes during manufacture. The reaction between molten aluminium and both oxidized and unoxidized ceramic is also studied, using scanning electron microscopy (SEM).

EXPERIMENTAL METHOD

Samples of four different ceramics were used in this work:

(i) S, the standard tube-making grade.
(ii) F, as for S but with finer silicon carbide particles.
(iii) C, as for S but with coarser silicon carbide particles.
(iv) M, as for C but using a finer initial silicon powder.

The manufacturing process was the same in each case, and is described above.

For each of the materials, specimens were cut as a cross-section of the tube wall thickness.

Coupons for oxidation and corrosion tests were cut to approximate dimensions of 17mm x 13mm x 6mm, and ground on all faces using 180 and 240 grit silicon carbide paper to allow reproducibility of surface finish throughout the tests, whilst retaining similarity to the as-used condition of the material. The coupons were cleaned with distilled water in an ultrasonic bath. An oil-based cutting fluid contaminant from the initial fabrication process persisted, but was eliminated by heating the coupons to 1000°C for one minute. Some coupons were notched to facilitate later fracture.

Oxidation Tests

Oxidation of the ceramics was studied using a thermogravimetric analysis (TGA) system, run in the isothermal mode. An Oertling NA264 electronic balance, measuring to a tolerance of 0. lmg, was linked to a BBC Master microcomputer to allow continuous data-logging. A controlled gas-flow from the bottom of the furnace tube was applied in each case, the gas used being compressed atmospheric mix (21% O_2, bal. N_2). Variation of gas flow-rate from 80 cm³/min to 400 cm³/min showed no effect on oxidation, so a standard rate of 150cm³/min was chosen. Tests were run for either 6 hours or 72 hours at 1000°C.

Also, coupons of S were oxidized at 1000°C in a tube furnace for periods of up to 500 hours. These were weighed before and after oxidation using the Oertling balance.

Liquid Aluminium Corrosion

Liquid aluminium corrosion of the S ceramic was investigated by dipping coupons oxidized for different times into molten aluminium. Pure Al was melted in an alumina capsule and held at 750°C in a box furnace. Coupons oxidized for 0 and 120 hours at 1000°C were held in this melt for 150 hours, using a stainless steel support rig. After removal from the melt, some excess adherent metal was removed by grinding and the coupons were then fractured. The reaction zones were imaged using SEM to investigate the extent of corrosive degradation.

RESULTS AND DISCUSSION

Unoxidized material

A scanning electron micrograph of unoxidized S ceramic is shown below (Fig. 1). The extensive porosity in the nitride phase (A) is evident. The silicon carbide particles (B) have a trimodal size distribution, measuring ~500, ~125 and ~25 microns. X-ray diffraction shows that the silicon carbide exists as β-SiC, and the nitride phase contains α- and β-Si_3N_4 and Si_2N_2O. The F and C ceramics are similar in appearance, with the appropriate distribution of SiC particle sizes. The nitride phase in the M (finer Si) ceramic has a different, more log-like structure (Fig. 2) which can occur with pure β-Si_3N_4. However, this material had a higher Ca content than the standard ceramic, as determined by energy-dispersive X-ray analysis (EDX) and it is possible that a Ca-rich phase exists.

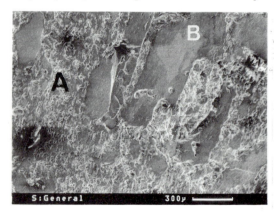

Figure 1 Unoxidized ceramic S (SEM)
A: nitride B:carbide

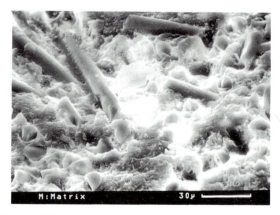

Figure 2 Log-like grains in mitride phase of ceramic M (SEM)

Oxidation

The weight gain ($\Delta m/m_{init}$) $\underline{v}$ time curves for oxidation at 1000°C are shown below (Figs. 3,4). The TGA curves show all ceramics, while the long-term test applies to S ceramic.

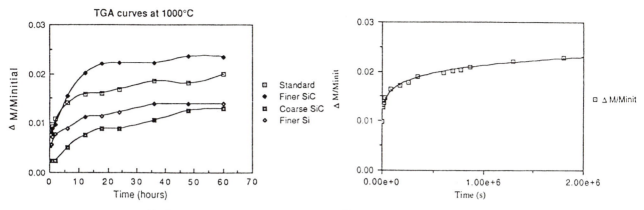

Figure 3 TGA traces for oxidation

Figure 4 Oxidation of ceramics in tube furnaces

Figures 5 and 6 are scanning electron micrographs of material oxidized for 6 hours. Patches of silica (**O**), identified from an X-ray diffraction trace, can be seen as a glassy layer upon the ceramic surface, and it is noted that reaction has occurred mainly in the nitride phase (**A**).

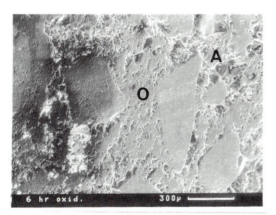

Figure 5 Ceramics after oxidation for 6 hours at 1000°C (SEM)

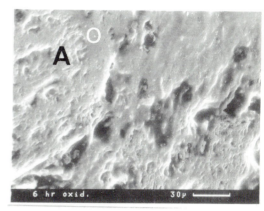

Figure 6 Ceramics after oxidation for 6 hours at 1000°C (SEM)

After 240 hours oxidation, the silica (**O**) is clearly visible (Fig. 7). It appears to have formed in the nitride phase and spread laterally across the silicon carbide grains. This silica layer is up to 15 microns thick (Fig. 8).

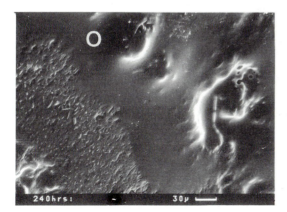

Figure 7 Silica spreading across SiC grain (SEM)

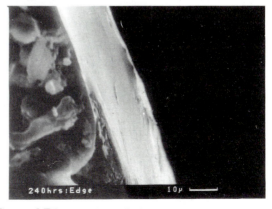

Figure 8 Edge-on view of silica layer on ceramic oxidized for 240 hours at 1000°C (SEM)

Both silicon carbide and silicon nitride are thermodynamically unstable in air at the temperature of interest [2]. The silicon carbide in this material is essentially fully dense and thus oxidizes in a protective manner [3,4] giving a thin, glassy film visible by interference tints. The silicon nitride is very porous, so oxidation occurs extensively within the pore network [5,6]. Porz and Thümmler [7] proposed a model for reaction-bonded Si_3N_4 oxidation with initial reaction occurring on all free surfaces and then the pores being progressively blocked by silica product, which gives asymptotic kinetics:

$$\Delta m = \Delta m\infty[\ 1- exp(-k_a t)] \qquad (2)$$

where Δm is the mass gain, $\Delta m\infty$ is the mass gain at $t = \infty$, t is the time in seconds and k_a is a time constant. The data in Figs. 3 and 4 can be described using this model, with the parameters shown in Table 1 below. Pore-filling occurs more rapidly in the flowing gas of the TGA system, presumably due to removal of gaseous reaction products.

Table 1

Material	$\Delta m\infty\ (m_{initial})$	$k_a\ (s^{-1})$
S	0.021	6×10^{-6}
F	0.025	8×10^{-6}
C	0.012	13×10^{-6}
M	0.015	5×10^{-6}
S, 500 h	0.024	1.3×10^{-6}

The ceramic containing fine SiC particles oxidized more than the standard material, and less reaction occurred for the ceramic containing coarse SiC particles. This is because of differences in actual surface area of the nitride phase. While finer SiC particles would be expected to pack more closely, the manufacturing process used means that this does not happen, and optical micrographs show a larger fraction of nitride phase (when compared to the standard material) within which reaction can occur. The use of coarser SiC particles reduces the fraction of nitride phase (as determined by optical microscopy) and hence the extent of reaction. The use of finer Si powder with coarse SiC caused an increase in oxidation. This could be attributed to the altered nitride phase structure giving a different pore network, and hence surface area, but it can be noted that the free energy change, and thus driving force, for oxidation of β-Si_3N_4 is greater than for α-Si_3N_4 [9].

3.3 Liquid - Aluminium Corrosion

For the unoxidized ceramic, limited reaction occurred at the material surface with a thin (~ 10μ) layer of aluminium bonding to the ceramic (Fig. 9).

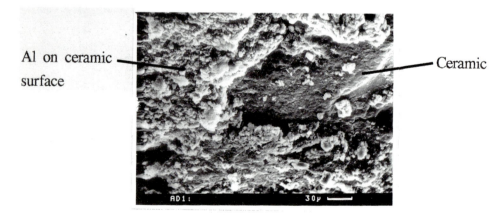

Figure 9 Aluminium adherent to unoxidized ceramic

There was a much greater degree of reaction for the oxidized ceramic, and these coupons fractured easily on removal from the corrosion rig. The degradation is greatest at the near-surface of the nitride phase, as shown in Figs. 10 and 11. The initial ceramic surface level is marked with a dashed line, and the reaction zone is indicated with R.

These results are fully consistent. With no oxidation, there is no silica film for the aluminium to wet, so adhesion requires a reaction with the ceramic. Thus, for SiC:

$$3SiC\ (s) + 4Al\ (l) == Al_4C_3\ (s) + 3Si\ (s) \qquad (3)$$

Figure 10 Reaction zone (R) in ceramic which had been oxidized and dipped in Al. Dashed line: original surface level (SEM)

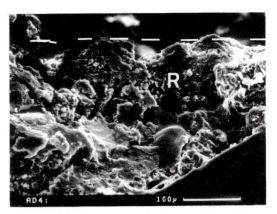

Figure 11 Reaction zone (R) in oxidized and dipped ceramic. Dashed line:original surface level (SEM)

The thermodynamics of this reaction have been studied by Iseki et al. [9] for Al-Si alloys, who found that for the reaction to occur at 750°C, the activity of silicon in liquid aluminium should be less than 0.096. Previous work by the present authors [10] has shown that reaction (3) does not occur with an Al-Si alloy having an activity of 0.105 (calculated using the data of [9] and [11]). A similar process is believed to occur for nitride ceramics [12].

For the oxidized ceramic, aluminium wets the silica and reaction (1) occurs. Since the nitride is more extensively oxidized, it is in this phase that corrosion is concentrated. Thus, the nitride is effectively 'etched' by the molten metal, leading to degradation of physical and mechanical properties.

CONCLUSIONS

The oxidation behaviour of a group of silicon nitride-bonded silicon carbide ceramics was investigated using TGA in a dry N_2/O_2 mixture, and SEM techniques. Reaction was found to occur predominantly in the nitride phase, with a silica scale of up to 15μ thickness growing out across the carbide. The oxidation kinetics are controlled by product-blocking of the pores in the nitride phase. A rate equation for this process was applied. Ceramics with fine SiC particles oxidized more than those with coarser SiC particles because of the former's increased inpore surface area. When finer Si powder is used in the reaction-bonding process, the nitride structure is altered, towards pure, β-Si_3N_4, and oxidation is increased.

Pure aluminium corrosion of the standard ceramic showed different behaviour, dependent upon the oxidation history of the ceramic. Aluminium reacts slightly with unoxidized ceramic, forming a carbide. For the oxidized material, the aluminium wets, and reacts with, the silica. This gives extensive degradation, mainly in the nitride phase. These observations are in good agreement with predictions based on the thermodynamics of the system.

ACKNOWLEDGEMENT

This work is supported by British Gas (Midlands Research Station) and the Department of Education (Northern Ireland)

REFERENCES

1. P. Burke and P.J. Wedge, Foundry Trade Jnl.,162 (3363),1988, p.92.

2. JANAF Thermochemical Tables, 2nd Edn., Eds. D. Stull and H. Prophet, NSRDS-NBS37 (1971).

3. S.C. Singhal, J. Mat. Sci., 11, 1976, p.1246.

4. H. Cappelen, K.H. Johansen and K. Motzfeldt, Acta Chem. Scand., A35, 1981, p. 247.

5. J.B. Washburton, J.E. Antill and R.H.W. Haws, J. Am. Ceram. Soc., 67, 1978, p. 67.

6. R.W. Davidge, A.G. Evans, D. Gilling and P.R. Wilyman, in "Special Ceramics: 5", Ed. P. Popper, Brit. Ceram. Res. Assoc., Stoke-on-Trent, 1972, p. 329.

7. F. Porz and F. Thümmler, J. Mat. Sci., 19, 1984, p. 1285.

8. A Hendry, in "Nitrogen Ceramics", Ed. F.L. Riley, Noordhoff, Netherlands, 1977, p. 183.

9. T. Iseki, T. Kameda and T. Maruyama, J. Mat. Sci., 19, 1984, p. 1692.

10. M.W. Johnston and J.A. Little, J. Mat. Sci., 25,1990, p. 5284.

11. D.J. Lloyd, H. Lagace, A. McLeod aud P.L. Morris, Mat. Sci. and Eng., A107, 1989, p. 73.

12. M. Naka, M. Kubo and I. Okamoto, J. Mat. Sci. Lett., 6, 1987, p. 965.

37 SIMS STUDIES OF THE TRANSPORT OF OXYGEN TRACER IN GROWING NICKEL OXIDE SCALES FORMED AT 600-900°C

A. W. Harris, A. Atkinson, D. P. Moon and S. Mountfort

Materials Development Division, Harwell Laboratory, Didcot, Oxon OX11 0RA, UK.

ABSTRACT

The transport of oxygen in nickel oxide scales during the oxidation of high purity nickel has been studied using ^{18}O tracer in combination with depth-profiling and transverse section imaging SIMS. Depth-profiling demonstrates that oxygen is transported to the scale-metal interface at all temperatures, the quantity of oxide formed at the interface increasing as the oxidation temperature is reduced. The relative volume of oxide formed at the scale-metal interface represents up to about 17% of the total oxide volume in the scale produced at 600°C.

Imaging of transverse sections through the scales formed at 600 and 700°C shows that oxygen is transported via a large number of fissures through the existing oxide. The fissures are healed by the growth of new oxide within the pre-existing scale. The distribution and spacing of the healed fissures suggests that they correspond to grain boundaries in the scale. The oxide grown at the scale-metal interface is laterally continuous for scales grown at 600°C but this is not so for higher oxidation temperatures.

Re-oxidation at either 700 or 900°C of scales formed at initial temperatures of 600-900°C shows that the oxygen transport is controlled by the initial oxidation temperature for the relatively short re-oxidation times studied here (total scale thickness less than about 3 μm). It is suggested that the enhancement of oxygen transport observed for the lower initial oxidation temperatures is due to the nature of the scale-metal interface formed at the different temperatures.

INTRODUCTION

The formation of an oxide scale on a metallic component during exposure to a corrosive atmosphere can protect the metal from rapid reaction with the gas and hence endow a useful service lifetime. The loss, or spallation, of part of a scale and the consequent exposure of unoxidised metal may lead to a more rapid reduction in component thickness and a lowering of service lifetime. The spallation of an oxide scale is controlled by several factors including stress in the scale and the scale-metal adhesion. These properties are determined by the scale microstructure and the nature of the scale-metal interface. Improved understanding of the fundamentals of the development of scale microstructures may allow closer control of the stress and adhesion and hence decrease spallation.

The oxidation of many transition metals, such as nickel, occurs principally by the transport of metal cations to the scale-gas interface and subsequent reaction to form new oxide [1]. This results in the arrival of cation vacancies at the scale-metal interface and these vacancies are important in determining the scale microstructure and the scale-metal adhesion [2]. In particular, it has been observed that additional new oxide units may be formed at the interface as a result of the accumulation of voidage [3]. The required through-scale transport of oxygen cannot be accounted for by a mechanism based on solid-state diffusion. It has been proposed that oxygen is transported as a molecular species via cracks in the scale [3,4].

The present work investigates the through-scale transport of oxygen during the oxidation of high-purity nickel using the combination of oxygen tracers and secondary ion mass spectrometry (SIMS) analysis. SIMS is used for both sputter depth-profiling and imaging of transverse sections of scales [5]. The effect of the scale microstructure on oxygen transport is determined by re-oxidising pre-existing scales at different temperatures. The data are used to confirm and refine existing models of oxygen transport in growing scales [3,4].

EXPERIMENTAL METHODS

The material used in these experiments was Puratronic-grade nickel containing less than 30 ppm total metallic impurities. The nickel was cold-rolled to a 0.5 mm thick sheet and slit into coupons of 1 cm^2 area. The coupons were diamond polished to a 1 μm grit finish and subsequently annealed for 1 hour at 900 °C in flowing dry hydrogen to remove any remaining cold-work and to de-carburise the nickel. Finally, each coupon was given a light polish on a 1 μm diamond lap to introduce a controlled amount of surface damage to encourage uniform scale growth.

The coupons were oxidised sequentially in $^{16}O_2$ (natural isotopic composition) and $^{18}O_2$ (enriched to 97% $^{18}O_2$) at a pressure of 21 kPa in a high-vacuum furnace system [3]. The coupons were maintained at temperature during the gas changeover in the isothermal oxidations. When the temperature of the $^{18}O_2$ tracer oxidation differed from that used initially, the coupon was allowed to cool and weighed prior to the re-oxidation to establish the thickness of the initial oxide scale. The coupons were heated to the oxidation temperature in a period of approximately 5 minutes in the oxidising gas. The temperature was subsequently maintained to better than ±5 °C.

SIMS depth profiles were obtained using a Cameca IMS-3F ion microscope employing a magnetic sector spectrometer. Primary excitation was a 15 keV Ar^+ ion beam rastered over a 250 x 250 μm area. The secondary signal was recorded from the central 60 x 60 μm region of the raster for ions of masses 16, 18, 28, 32, 34, 36, 52 and 62. Transverse section images were obtained from metallographically prepared sections perpendicular to the scale surface. The images were obtained from a modified VG Scientific MA 500 equipped with a high transmission quadrupole spectrometer and a high-brightness 28 keV Ga^+ primary ion probe rastered under computer control. The primary beam was focussed to a spot size of approximately 50 nm, giving sub-micron lateral resolution in the final images [5]. Images were obtained for ions of masses 16, 18, 32, 34, and 36 using the full resolution available on the instrument.

EXPERIMENTAL RESULTS

Depth Profiles of Isothermally Oxidised Scales

Figure 1 shows SIMS depth profiles through a series of scales produced by isothermal oxidation at temperatures of 600, 700 and 800°C in oxygen at 21 kPa pressure. The scale-metal interface is located at the position of a 50% decrease in the mass 62 depth profile. The nickel ion signal level is reduced in the metal compared to that in the scale. Sequential oxidation in $^{16}O_2$ and $^{18}O_2$ has led to an obvious increment of $Ni^{18}O$ at the scale-gas interface, as demonstrated by the mass 18 and 36 depth profiles. The ratio of the depth of the $Ni^{18}O$ region to that of the $Ni^{16}O$, shown in the mass 16 depth profile, is consistent with the ratio of the times of exposure to the isotopically distinct gases and assuming parabolic growth kinetics.

Both the mass 18 and 36 signal levels show a second "peak" in the region of the scale-metal interface. This indicates that new oxide units have grown in this region, as has been observed previously for oxidation at 700°C [3]. Table 1 gives the proportion of the total amount of $^{18}O_2$ incorporated into the scale which is present at the scale-metal interface for the scales shown in Fig. 1, derived from the areas beneath the depth profiles. The relative amount of the oxygen tracer transported to the scale-metal interface is enhanced as the oxidation temperature is decreased from 800 to 600°C.

The oxygen tracer has also been incorporated into the $Ni^{16}O$ regions of the scales. The mass 18 and 36 signal levels are significantly greater than predicted from the natural background expected in $^{16}O_2$. A comparison of the signal levels for the various monatomic and diatomic oxygen ions, masses 16, 18, 32 (not shown in Fig. 1), 34 and 36, provides information on the distribution of the incorporated oxygen. In particular, two models of the distribution of incorporated tracer have been proposed, based on either a randomly distribution of the ^{18}O atoms within the $Ni^{16}O$ or the formation of discrete regions of $Ni^{18}O$ [3]. The ratios of predicted mass 36 signal level for the two models to that actually measured for the scales illustrated in Fig. 1 are also included in Table 1. Exact agreement would give a ratio of unity. The table shows that neither model exactly describes the distribution of the tracer. Figure 2 illustrates schematically a refined model of the incorporation of the tracer into the existing scale.

SIMS Images of Isothermal Scales

Figure 3 shows a composite of SIMS images taken from a transverse section through a scale of 4.60 μm thickness produced at 700°C. The distinct regions of $Ni^{16}O$ and $Ni^{18}O$ are clearly demonstrated by the mass 16 and 18 images shown in Figs. 3e and 3a respectively. Enhancement of the mass 18 image using logarithmic contrast (Fig. 3b) shows the distribution

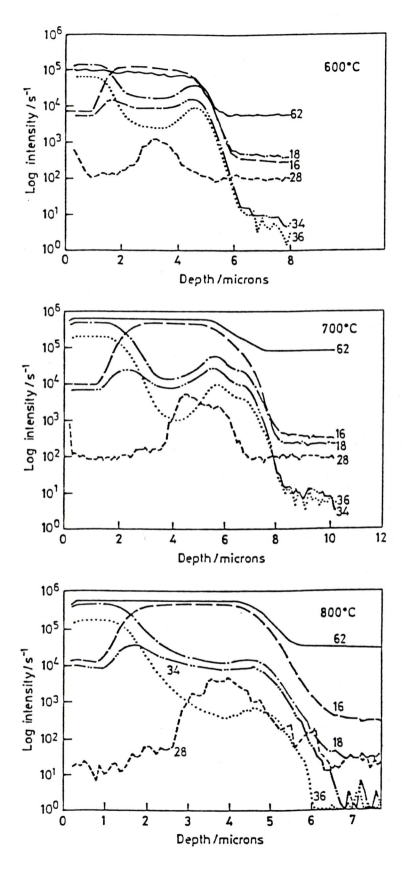

Figure 1: SIMS composition-depth profiles through isothermal oxide scales sequentially oxidised in $^{16}O_2$ and $^{18}O_2$. Oxidation temperatures as shown. Values within figure indicate mass number of appropriate ion species.

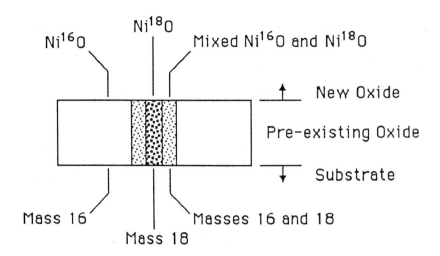

Figure 2: Model of the microstructure of pre-existing Ni^{16}O region. Mass numbers refer to the monatomic oxygen signals produced from each region.

Temp/°C	x/μm	^{18}O-incl.	Mass-36 ratio	
			Random	Separate
600	3.53	0.122	0.14	3.21
700	4.60	0.065	0.07	3.88
800	4.32	0.023	0.22	8.11

Table 1. Measured thickness, amount of oxygen tracer included at scale-metal interface as a fraction of total oxygen and comparison of predictions of the two models of ^{18}O distribution within the mid-part of the scale (ratio of predicted to measured mass 36 signal levels) for isothermal oxidation

of the Ni^{18}O incorporated into the existing Ni^{16}O and the presence of a distinct region of Ni^{18}O at the scale-metal interface. The incorporated oxide occurs as a series of branching "ribbons" about 50-100 nm wide with a separation of the order of 0.5 μm.

Images obtained from a scale formed at 600°C resembled those shown in Fig. 3, although the Ni^{18}O at the scale-metal interface was more distinct. The oxide formed at the scale-metal interface was laterally continuous at 600°C but this was not found to be the case for the 700°C scale. In the 700°C scale, the distribution of the ribbons of Ni^{18}O within the Ni^{16}O was also non-uniform. The ribbons were concentrated in the regions where the scale-metal interface oxide was present. Hence, the ribbons correspond to areas of the scale which allowed substantial oxygen access to the scale-metal interface and are assumed to be the result of the healing of the oxygen access pathways. The spacing of the Ni^{18}O ribbons in the scale produced at 600°C is of the order of 0.25 - 0.5 μm, approximately corresponding to the expected grain size in the scale [6]. As can be seen in Fig. 3, this spacing is simlar to that in the 700°C scale.

Comparison of the logarithmic contrast mass 18 image with the mass 34 images for the 700°C scale (Figs. 3c and 3d) shows that both the Ni^{18}O ribbons and the scale-metal interface oxide are not purely Ni^{18}O. A substantial mass 34 signal will be produced from regions of mixed Ni^{16}O and Ni^{18}O. This agrees with the observations made on the depth profiles discussed above.

Re-oxidised Scales

SIMS depth profiles were obtained for scales pre-oxidised at temperatures of 600-900°C in ^{16}O$_2$ and subsequently re-oxidised at 700 and 900°C using ^{18}O$_2$ tracer at 21 kPa pressure. Figure 4 shows a depth profile for the specimen pre-

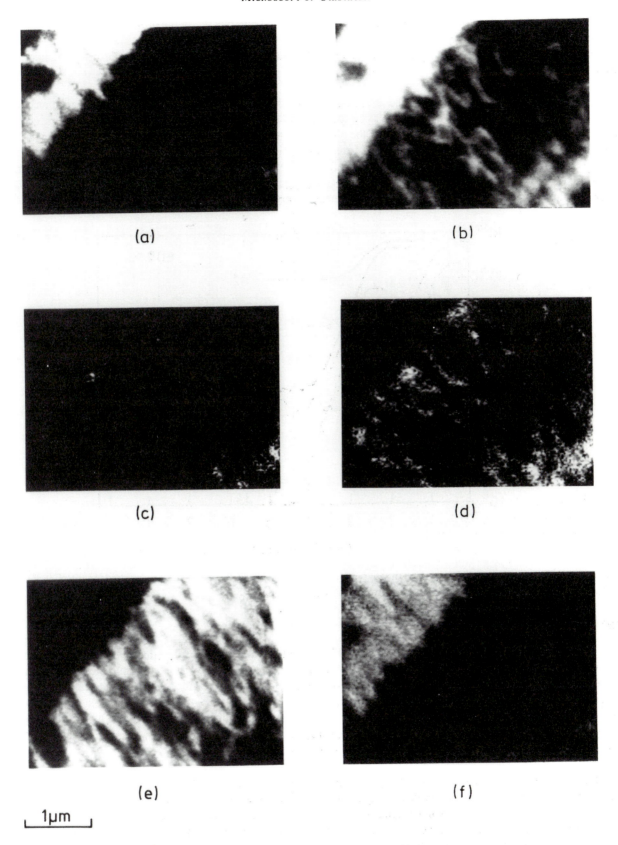

(a)

(b)

(c)

(d)

(e)

(f)

1μm

Figure 3: SIMS images of 4.60 μm oxide scale sequentially oxidised at 700°C. Substrate lies to bottom right in all images.
(a) Mass 18, linear intensity, (b) Mass 18, log, (c) Mass 34, linear, (d) Mass 34, log, (e) Mass 16, linear, (f) Mass 36, log.

oxidised at 600°C and re-oxidised at 900°C. Comparison with Fig. 1 shows that this depth profile closely resembles that for isothermal oxidation at 600°C. The relative amount of oxygen tracer at the scale-metal interface for re-oxidation at 900°C substantially exceeds that obtained from isothermal oxidation at 800°C (no depth profile was available for isothermal oxidation at 900°C).

The experimental results showed that, for the scales investigated, the relative amount of through-scale oxygen transport did not depend on the re-oxidation temperature or the time of exposure during re-oxidation. In contrast, increasing either the pre-oxidation temperature or the thickness of the initial scale gave a reduction in the amount of oxide formed at the interface. Changing the pre-oxidation temperature from 600 to 900°C deceased the relative amount of oxide formed at the interface from 13% to 2% for re-oxidation at 900°C for scales of 2.5 μm total thickness. The corresponding value for isothermal oxidation at 600°C is 17% from Fig. 1. Fracture sections of the re-oxidised scales show that the scale microstructures are single-phase and the grain size exceeds that expected for the initial oxidation temperature.

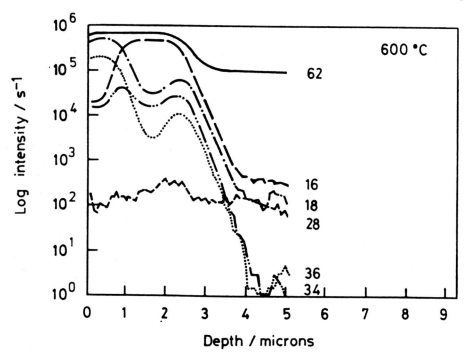

Figure 4: SIMS composition-depth profiles through scale pre-oxidised at 600°C in $^{16}O_2$ and re-oxidised at 900°C in $^{18}O_2$. Values within figure indicate mass number of appropriate ion species.

DISCUSSION

The SIMS depth profiles demonstrate that the growth of oxide at the scale-metal interface during the oxidation of pure nickel is controlled by the oxidation temperature. At lower temperatures the relative amount of interface oxide formed can be substantial. At 600°C, up to about 20% of the oxygen tracer is incorporated at the scale-metal interface. The inner part of a duplex scale grown on a nickel-0.1% chromium alloy was approximately 45% of the total scale thickness and showed a corresponding amount of oxygen incorporation [3]. Hence, whilst the scales grown on the high purity nickel do not exhibit as much interface oxide as "true" duplex scales, oxygen transport can still be substantial. It appears that all nickel oxide scales allow through-scale oxygen transport at lower temperatures.

The formation of oxide at the scale-metal interface requires the creation of space for the new oxide [3,4]. This occurs by the aggregation of nickel vacancies formed during the growth of the oxide by the outward diffusion of nickel. The total amount of space created depends on the balance between the vacancy flux, which increases with oxidation temperature as the oxidation rate rises, and the process of vacancy annihilation. The reduction in the amount of oxide formed with increasing temperature indicates that the effective activation energy for the annihilation of vacancies must exceed that for oxidation. The absolute values for the two processes must be such that at low temperatures the flux exceeds the annihilation rate.

Oxygen access to the interface occurs via a large number of pathways. These pathways are apparently healed during the oxidation. This suggests that each individual pathway must be open for only a relatively short period of time; the pathways are intermittent. The spacing of the healed pathways shown in the SIMS images appears consistent with the grain size of the scales and it is proposed that the pathways coincide with existing grain boundaries within a scale [3].

The utilisation of grain boundaries as oxygen access pathways suggests that the scale microstructure may have an influence on the through-scale oxygen transport process. The measurements of oxygen transport during the re-oxidation of existing scales investigated this possibility. It has been shown that the initial scale determines the degree of oxygen transport for the scales used in this work. However, the initial scale microstructure is not retained throughout the re-oxidation, instead the microstructure approaches that appropriate to the re-oxidation temperature and scale thickness. The relative amount of interface oxide does not depend on the re-oxidation time and hence the degree of deviation from the initial scale microstructure does not affect the oxygen transport. Overall, it appears that the microstruture of the initial scale cannot be controlling oxygen transport as this microstructure is substantially altered during re-oxidation.

It is proposed that the nature of the scale-metal interface, formed during the pre-oxidation, controls the relative amount of oxide formation at the interface. In particular, the vacancy annihilation processes may be affected. The nature of this interface is established during the early stages of oxidation and persists during subsequent exposure to different temperatures (at least for the time periods investigated here).

CONCLUSION

SIMS has been used for both depth profiling and transverse section imaging of oxide scales grown on high purity nickel to investigate through-scale oxygen transport and the growth of oxide at the scale-metal interface. It has been demonstrated that the amount of oxide grown at the interface is inversely dependent on oxidation temperature. At 600°C, the amount of oxide formed at the interface is almost half that observed for a "duplex" scale grown on a nickel-based alloy. The distinction between "single-layered" scales grown on pure nickel and "duplex scales" grown on alloys is therefore one of degree and not of kind.

SIMS imaging of transverse sections of scales demonstrates that oxygen transport occurs via a large number of intermittent pathways which probably coincide with grain boundaries in the scale. These pathways are healed during oxidation. The amount of oxide growth at the scale-metal interface is controlled by vacancy processes at the interface. In particular, the balance between the vacancy flux and the rate of vacancy annihilation determines the amount of space available for the growth of oxide at the interface.

Re-oxidation of pre-existing scales demonstrates that the amount of oxide formed at the scale-metal interface is determined by the temperature at which the interface structure is initially formed. The re-oxidation temperature and time have no significant effect on the oxygen transport. Annealing of this structure at higher temperatures does not affect the rate of oxygen transport.

ACKNOWLEDGEMENTS

We wish to thank Jason Brown for carrying out the re-oxidation experiments. This work forms part of the longer term research within the Underlying Programme of AEA Technology.

REFERENCES

1. A. Atkinson, Rev. Mod. Phys. 57, 1985, p.437.

2. R. Hales and A.C. Hill, Corros. Sci. 12, 1972, p.843.

3. A. Atkinson and D.W. Smart, J. Electrochem. Soc. 11, 1988, p.2886.

4. J. Robertson and M.I. Manning, Mat. Sci. Technol. 4, 1988, p.1064.

5. D.P. Moon, A.W. Harris, P.R. Chalker and S. Mountfort, Mat. Sci. Technol. 4, 1988, p.1101.

6. H.V. Atkinson, Oxid. Met. 28, 1987, p.353.

38 EFFECT OF DISPERSED OXIDES UPON OXIDATION OF NICKEL

N. Zhang, R. J.Bishop and R. E. Smallman

School of Metallurgy and Materials, The University of Birmingham, P.O.Box 363, Birmingham B15 2TT, UK.

ABSTRACT

Knowledge of the role of oxide dispersoids in ODS alloys during high-temperature oxidation in air is uncertain. The metal/oxide interface, across which matter must be transferred if oxidation is to continue, may be the preferred 'sink' for vacancies under all circumstances. Alternatively, the interfaces between dispersed oxide particles and the metallic matrix may provide favourable sites for vacancy annihilation, reduce the extent of condensation at the main metal/oxide interface and thereby improve the overall resistance to oxidation.

By comparing the oxidation behaviour of nickel, with and without oxide particles, it has been shown that the rate of oxidation for Ni-$0.5Y_2O_3$ samples is lower than that of commercially-pure Ni (99.9%). Metallographic evidence is provided to show that the coherent oxide scales formed on Ni-Y_2O_3 samples are associated with void formation at Y_2O_3 particles.

INTRODUCTION

Cavity formation in nickel at elevated temperatures has been a most important and popular research topic. Nickel, in common with most metals, forms a metal-deficient oxide. NiO is a p-type cation-deficient semiconductor and, therefore, the cations will migrate with electrons from the scale/metal interface towards the scale/gas interface during oxidation. The outward diffusion of cations must be counterbalanced by an inward flux of metal vacancies, and if these are not eliminated at the metal/oxide interface, they will be 'injected' into the underlying metal.

It is now well established that the oxidation of nickel gives rise to the formation of prolific voiding, particularly along grain boundaries. Vacancies injected into the substrate on oxidation are annihilated by forming voids, a process facilitated by the presence of carbon. In particular, oxygen diffuses along grain boundaries, produces a $C \rightarrow CO$ reaction and the resultant gas is accommodated in voids which also tend to acquire vacancies. Results for de-carburized nickel [1] show that vacancies which do not react with carbon atoms or collect at grain boundaries are able to condense at the metal/oxide interface, causing oxide decohesion and an increased oxidation rate.

In ODS alloys, oxide particle interfaces may well provide favourable sites for vacancy annihilation and therefore prevent the vacancies condensing at the metal/oxide interface, leading to a reduction in the oxidation rate. On the other hand, oxide particles in the grain boundaries may inhibit the ability of these alloys to annihilate vacancies, reducing their effectiveness as vacancy sinks; this would eventually lead to the vacancies condensing at the oxide/metal interface, thereby increasing rather than decreasing the oxidation rate. The role of oxide dispersoids in ODS alloys during oxidation is therefore by no means clear.

The present paper describes an investigation of the oxidation behaviour of nickel, with and without oxide dispersoids. Thermogravimetric work has been carried out in order to compare the oxidation behaviour of different alloys and metallographic studies have been made of alloy samples before and after oxidation.

EXPERIMENTAL WORK

Three alloy specimens of the oxide-dispersion-strengthened type have been investigated. A reference sample of commercially-pure nickel has been used. Table 1 shows the compositions of the materials used and their methods of manufacture.

The thermogravimetrical measurements of wafer specimens were made with a Stanton STA 1000 thermobalance at a temperature of 900°C, in static air. Specimens were ultrasonically cleaned in acetone before testing. A Polyvac E1000 mass spectrometer was used to determine the chemical composition of the specimens.

Specimens for morphological studies were mounted in conducting bakelite. After final polishing with 0.5 μm α-Al$_2$O$_3$ powders, specimens were etched in a 93: 2: 5 hydrochloric acid-nitric acid-sulphuric acid solution. The preparation of discs for cross-sectional TEM observations was based on a method developed by Manning and Rowlands [2] and later modified by Newcomb et al. [3].

The scanning electron microscopy was carried out using an ISI 100A instrument with LINK 860 EDX. Analytical electron microscopy carried out with a Philips EM 400T with EDX 9100 and also with a JEOL 4000 FX with LINK AN 10000 EDX.

RESULTS AND DISCUSSION

Pure Ni

Figure 1 shows that the weight-gain curves of all the specimens obtained at 900°C for 120 hours; pure nickel was used as a reference sample (N). A coarse, columnar NiO scale layer formed on this pure Ni substrate, without oxide dispersoids, at a temperature of 1200°C in 3 hours (Fig.2). The voids extended from the oxide/metal interface to the substrate along grain boundaries and appeared inside grains, in agreement with previous work [1]. These internal pores in nickel could result from the oxidation of carbon, even when its content is less than 0.003wt% [4]. It appears that oxygen atoms diffuse inwards along grain boundaries and react with carbon to produce CO bubbles; these bubbles provide convenient condensation sites for injected vacancies.

Ni-Al$_2$O$_3$

In contrast to the pure nickel sample, Ni-Al$_2$O$_3$ (A), which had been made by a mechanical alloying (MA) method, formed on oxide scale of duplex structure under the same oxidising conditions as those for pure Ni (1200°C, 3 h). Figure 3 shows that oxide scale grew parallel to the extrusion direction of the Ni-Al$_2$O$_3$ substrate. In common with previously-reported observations, the NiO scale consisted of two layers, the outer being compact and the inner porous. A distinct gap separated the two layers. The poor adhesion of this duplex scale caused it to spall easily, either at the inner/outer interface or at the oxide/metal interface. Because of its duplex structure, NiO was unable to deform with the contracting metal core and contact was soon lost at the metal/oxide interface, causing an increase in the rate of oxidation of the substrate (Fig.1).

The transfer of cations of nickel into the scale creates residual vacancies at the oxide/metal interface. Because ions continue to be driven through the scale by the chemical potential gradient, the oxide separated from the metal dissociates

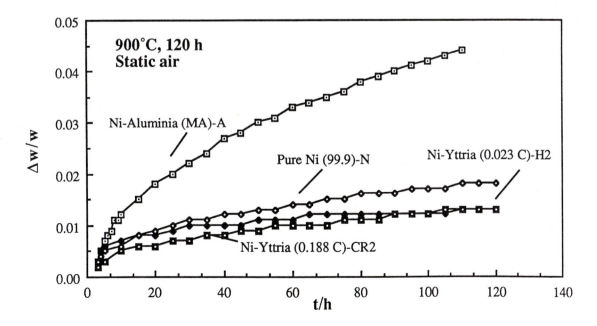

Figure 1

Table 1. Materials used for oxidation studies

Sample	Original Powder Size/μm	Process	Source	Reference
Ni-1Al₂O₃	Ni ~3, Al₂O₃ ~0.03	Mechanical Alloying	NIRM (Japan)	A
Ni-0.5Y₂O₃	'123' 99.99% Ni ~5 Y₂O₃ ~0.5	Hot isostatic pressing,1100°C 100MPa, 8 h	INCO(Hereford)	H2
Ni-0.5Y₂O₃	'123' 99.99% Ni ~5 Y₂O₃ ~0.5	Hot isostatic pressing, 1200°C 2000bar, 4 h	Cranfield Inst. of Technology	CR2
Ni	99.9% purity	Casting	Ventron (F.R.G.)	N

Table 2. Compositions (wt%) of Ni, Ni-Al$_2$O$_3$ and Ni-Y$_2$O$_3$ specimens using mass spectrometer

Sample	C	Si	Al	Co	Cr	Fe	Ni
Ni (N)	0.000	0.007	0.041	0.020	0.002	0.015	Rem.
Ni-Al₂O₃ (A)	0.014	0.033	0.574	0.083	0.116	0.418	Rem.
Ni-Y₂O₃ (CR2)	0.188	0.019	0.010	0.000	0.103	0.003	Rem.
Ni-Y₂O₃ (H2)	0.023	0.007	0.000	0.000	0.001	0.023	Rem.

and oxygen is transported across the gap and is able to oxidize the underlying exposed metal. The cross-sectional TEM micrograph (Fig.4) illustrates these ideas. It shows the interface of Ni-Al$_2$O$_3$ after oxidation at 1100°C for 6 hours. The inner layer in the immediate vicinity of the substrate surface exhibited a porous columnar structure. A visible gap was present at the oxide/metal interface. External oxygen could penetrate via micropaths in the oxide scale and arrive at the oxide/metal interface to form new NiO, thereby increasing the overall rate of oxidation.

Long straight dislocations can be seen in TEM specimens of oxidised Ni-Al$_2$O$_3$ (1100°C, 6h) close to the oxide/metal interface (Fig.5b). However, short and tangled dislocations were observed in the same sample before oxidation (Fig. 5a). It may be inferred that there was a movement of metal vacancies through the matrix near the oxide/metal interface. However, although metal vacancies are able to migrate into the substrate, it appears that the duplex scale, with its porous inner layer, is more likely to act as a sink for vacancies than features such as dislocations, grain boundaries and the surfaces of oxide dispersoids.

TEM examination of oxide particles and surrounding metallic matrix, before and after oxidation, showed that no significant changes had taken place (Fig. 6). In particular, no voids had formed.

Table 2 shows that impurity elements were present in these alloys, e.g. Fe, Cr. Since these elements are either trivalent or tetravalent, they will, when dissolved in the oxide, increase the mobility of cations in NiO and promote its rate of formation. The formation of a duplex structure is undesirable since it ultimately leads to exfoliation of the oxide layer, thereby reducing the oxidation resistance of the nickel. In order to eliminate impurities and reduce the grain boundary contribution, samples of a second OD alloy, Ni-Y$_2$O$_3$, were produced by hot isostatic pressing.

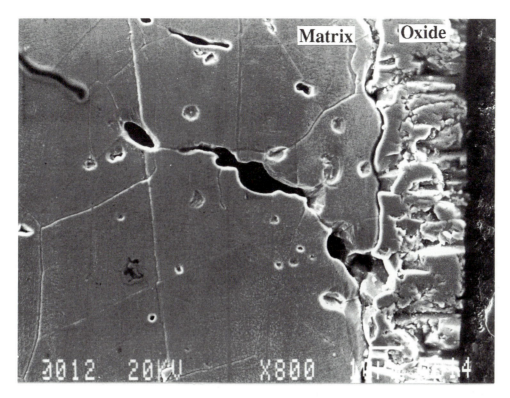

SEM

Figure 2

Ni-Y$_2$O$_3$

Two types of Ni-Y$_2$O$_3$ samples (CR2 and H2) were made by HIPing, as shown in Table 1. Both H2 and CR2 samples exhibited lower rates of oxidation than pure Ni when oxidised at 900°C for 120 hours (Fig. 1). After oxidation at 1200°C for 3 hours, a single and coherent oxide scale formed on both H2 or CR2 samples (Fig.7b and Fig.9b). Some voids were detected in the underlying substrate of both samples. It is interesting to compare the morphology of the voids in these samples before and after oxidation, using a common magnification. In sample H2, oxidation encouraged void formation in the matrix (Fig.7a,b). It appeared that some of the holes, produced by the detachment of Y$_2$O$_3$ particles during etching, became larger during oxidation.

In addition, some new voids formed along grain boundaries, probably as a result of the oxidation of carbon, as described above. Figure 8 shows that some voids were associated with particles, either at grain boundaries or inside grains. Energy-dispersive X-ray analysis was used to determine the composition of these particles. They were found to be yttrium-rich phases, i.e. stable Y$_2$O$_3$. In this instance, the oxide particles did not affect the ability of the boundary to act as a sink. The preferred regions for void annihilation are grain boundaries and surfaces of Y$_2$O$_3$ dispersoids rather than the main oxide/metal (NiO/Ni) interface. Oxidation of carbon and the presence of the Y$_2$O$_3$ dispersion both affected cavity formation during the oxidation of sample H2.

In sample CR2, oxidation caused an enlargement of the voids (Fig. 9a,b). However, no significant voids were found at grain boundaries in this sample after oxidation. Most of the voids formed around Y$_2$O$_3$ particles rather than at the NiO/Ni interface (Fig.10). EDX analysis showed that the particles associated with voids in sample CR2 were yttrium-rich.

Both Ni-Y$_2$O$_3$ samples (H2 and CR2) showed that oxidation produced an increase in the number of voids within grain interiors. This effect is attributed to 'vacancy injection', particularly as these samples exhibited a coherent oxide scale. In other words, vacancies were not destroyed at the main oxide/metal interface. The voids derived from vacancy injection formed at the surfaces of Y$_2$O$_3$ dispersoids as well as at grain boundaries. As the oxide layer grows by the outward diffusion of cations, a flow of cation vacancies equivalent to the outward flow of metal moves inwards. Any vacancies which are not annihilated at the oxide/metal interface are injected into the metal where they either diffuse to sinks on Y$_2$O$_3$ particle sites or to grain boundaries. In both cases, they nucleate voids.

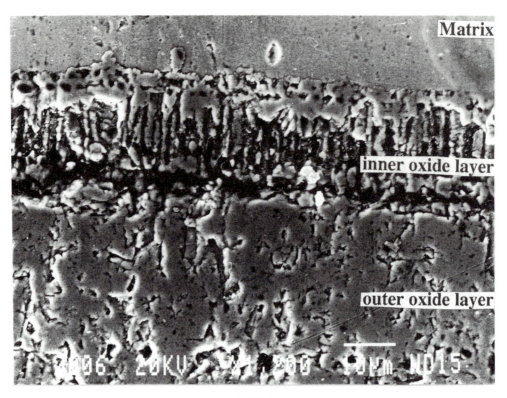

SEM

Figure 3

It is arguable that the interfaces of the Y_2O_3 dispersoids are likely to be particularly favourable sinks for vacancies during oxidation with no vacancies being destroyed at the main oxide/metal interfaces. Preservation of an adherent oxide scale helps to protect the nickel substrate from further oxidation. In this way, one can explain why the samples of Ni-Y_2O_3 oxidised more slowly than commercially-pure Ni samples when tested at 900°C for 120 hours.

Comparison of the oxidation curves for the two Ni-Y_2O_3 samples (H2 and CR2), reveals that the rate of oxidation of sample CR2 was slightly lower than that for sample H2 during the first 90 hours (Fig. 1). Subsequently, they oxidised at the same rate. Sample CR2 initially contained more carbon (0.188wt.%) than sample H2 (0.023wt.%), as shown in Table 2, and it appears that carbon was responsible for the differences in behaviour between these two samples. In accordance with previous work [1, 4] on the effect of carbon upon cavity formation, it thus appears that carbon in sample CR2, which was eight times more plentiful than in sample H2, reacted with oxygen which had penetrated along micro-channels in the oxide scale. CO or CO_2 then formed bubbles in the substrate. These bubbles may also have provided sites for vacancy annihilation in the substrate. Vacancies were thus less likely to be destroyed at the main oxide/metal interface. The tendency for gas bubble formation increases with the carbon content of the alloy. The Ni-Y_2O_3 matrix of sample CR2 was initially high in carbon (0.188 wt.%) but the amount of available carbon was steadily depleted during 90 hours of oxidation. This sample oxidised at a slightly lower rate than sample H2 (0.023wt.% C) during the first 90 hours of oxidation. Eventually, after 90 hours, the oxidation curves for sample CR2 (0.1 88wt.%) and sample H2 (0.023wt.%) coincide.

CONCLUSIONS

1. The rate of oxidation for Ni-0.5%Y_2O_3 (HIPed) is lower than that of commercially-pure Ni (99.99%).
2. Coherent oxide scales were formed on Ni-Y_2O_3 samples and voids formed at Y_2O_3 particles. Y_2O_3 dispersoids in a nickel substrate provided favourable vacancy annihilation sites and therefore prevented condensation of vacancies at the oxide scale/metal interface. This result provides general support for the concept of vacancy injection.
3. Differences in carbon content between Ni-0.5%Y_2O_3 samples affected their oxidation behaviour at 900°C over the first 90 hours of oxidation.

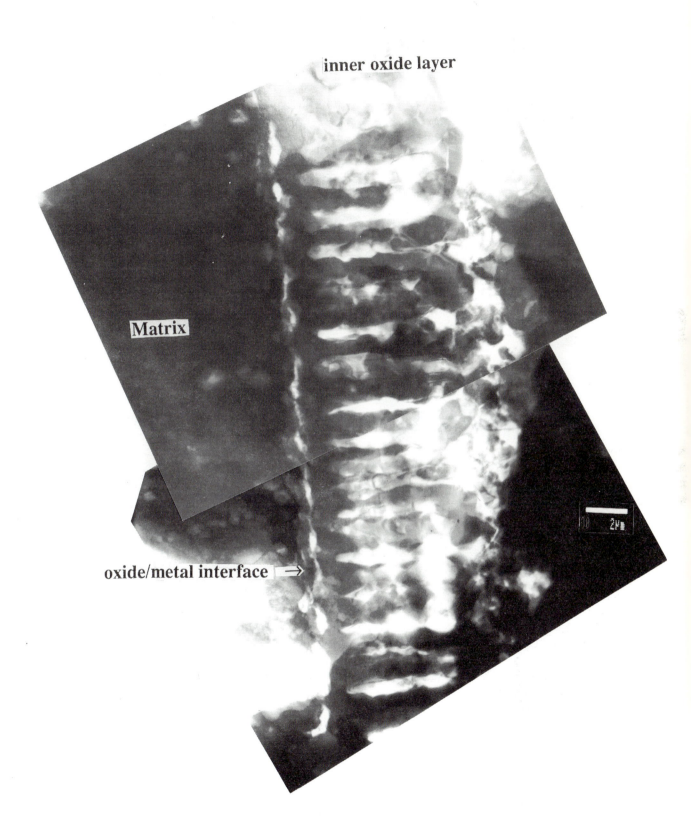

Figure 4

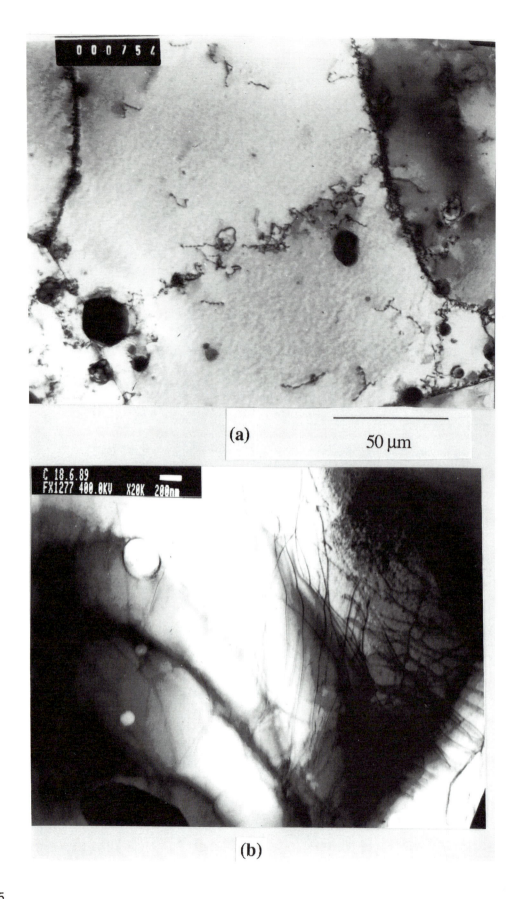

(a)

50 µm

(b)

Figure 5

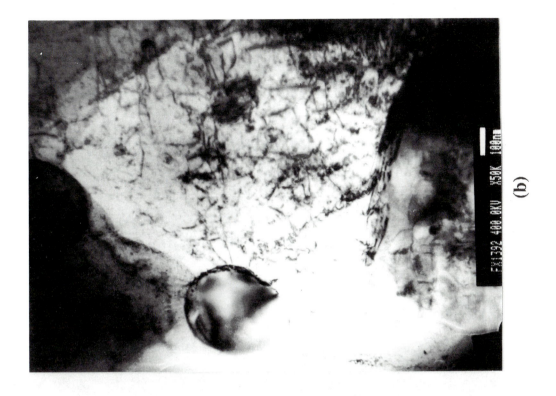

Figure 6

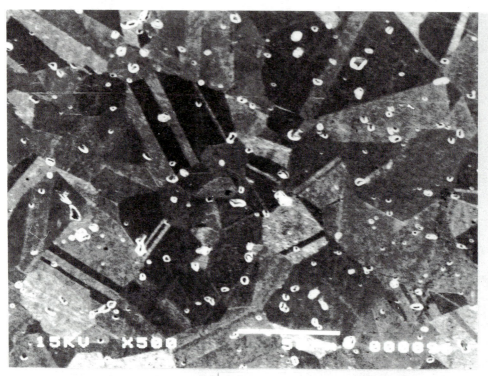

(a) SEM

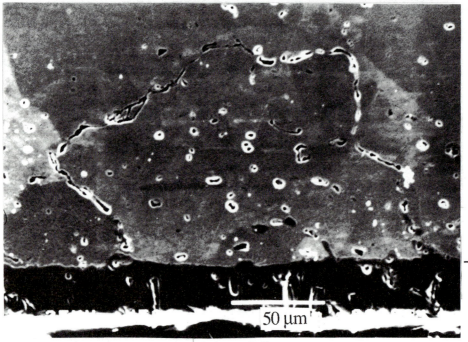

Matrix

Oxide

50 μm

(b) SEM

Figure 7

4. A brittle oxide scale with duplex structure was formed on Ni-Al$_2$O$_3$ made by mechanical alloying. The presence of impurities (e.g. Fe and Cr), probably due to mill erosion, increased its rate of oxidation.

ACKNOWLEDGEMENTS

The authors wish to thank INCO Alloys Ltd. (Hereford), The Cranfield Institute of Technology and the National Research Institute of Metals (Japan) for supplying the materials.

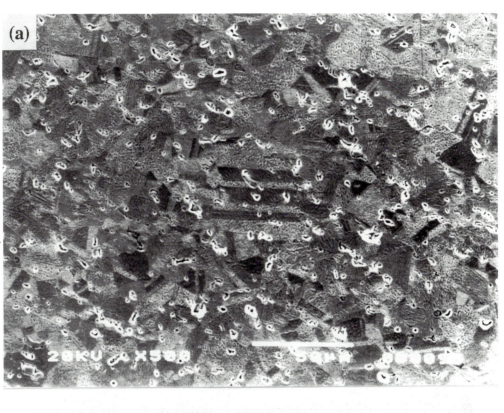

Matrix

Oxide

Figure 8

SEM

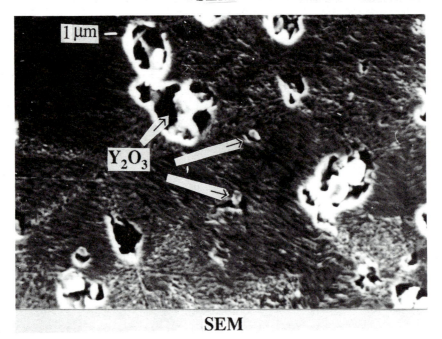

SEM

Figure 9

REFERENCES

1. C. Deacon, M. H.Loretto and R. E. Smallman, Mat. Sci. Technol. 1, 1985, p. 344.

2. M. I. Manning and P. C. Rowland, Brit. Corros. J. 15, 1980, p.184.

3. S. B. Newcomb, C. B. Boothroyd and W. M. Stobbs, J. Microsc. 140, 1985, p.195.

4. D. R. Caplan, J. Hussey, G. I. Sproule and M. J. Graham, Oxid. Met. 14(4),1980, p. 279.

5. P. Hancock and R. Fletcher, ATB Metall. 6, 1966, p. 1.

6. J. E. Harris, Materials Science and Technology 4, 1988, p. 457.

39 COMPETITIVE INTERFACE REACTIONS IN THE OXIDATION OF Ni-Cr ALLOYS

S.B. Newcomb and L.W. Hobbs*

*Department of Materials Science and Metallurgy,
Cambridge University, Pembroke Street, Cambridge CB2 3QZ, UK.*

**Department of Materials Science and Engineering,
Massachusetts Institute of Technology, Cambridge MA 02139, USA.*

ABSTRACT

Presented here are the preliminary TEM results aimed at characterizing the oxidation behaviour of Ni-Cr alloys with nominal compositions Ni-5wt%Cr and Ni-20wt%Cr. The results highlight the manner in which scaling processes come to be dominated by either oxygen or chromium diffusion and lead to the development of structurally distinct scales.

INTRODUCTION

Transmission electron microscopy and associated analytical techniques have proved particularly useful in following the microstructural and compositional development of high temperature oxidation scales. Their application to the study of complex systems, such as Fe-Ni-Cr alloys [1-3] for example, has particularly emphasized the inhomogeneity of the different scaling processes but has led to the identification of a number of competing oxidation mechanisms which are not readily predictable. Equally, TEM has been used to study apparently simple elemental systems, such as pure Ni, and has again revealed a surprising complexity of oxidation behaviours [4-6]. Here are described some preliminary TEM results for two Ni-Cr alloys which represent a half-way house between elemental metals and complex alloys, drawing on accumulated knowledge of both. The primary aim was to characterize the different scales formed for two different simple alloy regimes, with nominal compositions Ni-5wt%Cr and Ni-20wt%Cr, to see what effect a higher Cr content would have on the different types of oxides that these alloys can potentially form and thus on the overall balance of scaling reactions that are occurring. While it was initially thought that these differences might be most accentuated in the vicinities of alloy grain boundaries where accelerated oxidation is known to occur [7], it turns out that the two alloys exhibit more obvious intrinsic scaling differences where growth of each scale is controlled by a competitive balance in the diffusivities of oxidant and metal ions.

EXPERIMENTAL METHODS

The compositions of the two alloys investigated were Ni-5.3wt%Cr (hereafter Ni-5Cr) and Ni17.6wt%Cr (hereafter Ni-20Cr). Coupons were annealed in argon for 2 hours at 1373K, ground to 1200-grit SiC paper, ultrasonically cleaned in methanol and then oxidised for 100 hours in air at 1273K. Cross-sectional specimens for TEM examination were made using standard techniques [8]. Light microscopy of cross-sectioned coupons was also used to provide grosser details of overall microstructure and of the heterogeneities that developed in the scales.

LIGHT MICROSCOPY

Light microscopy established that the overall oxidation morphologies reported by earlier investigators [9] had been obtained. The oxide formed on the Ni-5Cr alloy spalled rather badly when the specimen was removed from the furnace, so that only the lower region (~10μm) of the scale may be seen in the light micrograph shown in Fig. 1a. Here can be seen a zone of precipitation to a depth of some 15 μm beneath the scale, while the metal/oxide interface is rather flat and not particularly suggestive of inward oxidation having taken place. Equally, the overall regularity of the scale, as well as of the internal oxidation zone, indicated that preferential oxidation at metal grain boundaries does not generally occur in this alloy. By comparison, the scale formed on Ni-20Cr (Fig. 1b) was much less regular and showed both gross undulations at the metal/oxide interface and regular intrusions into alloy grain boundaries to a depth up to 50 μm. The scale itself was found to be layered, though the irregular morphology of the layering did not suggest simple duplex scaling.

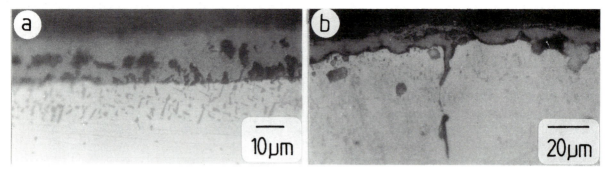

Figure 1: Light optical micrographs of the scales formed on (a) Ni-5Cr, and (b) Ni-20Cr.

TRANSMISSION ELECTRON MICROSCOPY

Ni-5Cr

The bulk of the alloy examined in the TEM was found to be coarse grained with a low dislocation content typical of annealed microstructures. Immediately beneath the scale and to a depth of ~12μm, there is an internal oxide precipitation zone. Here the oxide was found to be Cr_2O_3 (containing less than about 1wt% Ni) and to exhibit both coarse 'needle'- and 'block'-type morphologies. At the base of the zone, the Cr_2O_3 tended to be more needle-like; Fig. 2a shows a typical example where the oxide is approximately 2 μm in length. The alloy surrounding these precipitates was dislocated, as shown in Fig. 2b, reflecting the volume changes associated with the internal oxidation. Nearer the metal/oxide interface, the Cr_2O_3 precipitates had a more 'blocky' appearance (Fig. 2c), dark field micrographs suggesting that the surrounding alloy was rather less dislocated than when the precipitates had a needle morphology. Precipitates of both types were oriented with respect to the alloy, with both $(100)Ni//(0111)Cr_2O_3$ and $(100)Ni//(1211)Cr_2O_3$ orientation relationships exhibited by each precipitate type. XEDS indicated that the metal within the internal oxidation zone was totally denuded in Cr and contained almost entirely Ni, with no indication of compositional variation across the precipitation band within either the metal or the oxide particles themselves. Cr depletion of the metal did not, however, extend beyond the precipitation zone, with the alloy less than 1 μm from the zone exhibiting the bulk composition (5.3wt%). This

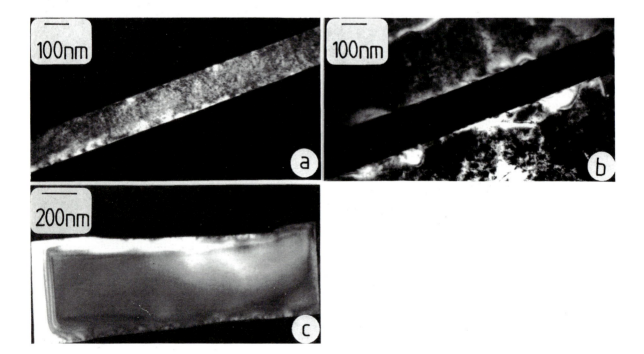

Figure 2: TEM micrographs showing Cr_2O_3 internal oxide in Ni-5Cr (a) needle-like oxide morphology, (b) dislocations in the alloy surrounding the oxide, and (c) block-type Cr_2O_3.

observation, together with that of the coarse oxide precipitates in the metal, implies that the scaling process is dominated by an inward flux of oxygen rather than by the diffusion of Cr from the metal to the scaling zone. Given the tendency of the scale formed on this alloy to spall, it is probably not unreasonable to suggest that gross lateral cracking of the oxide may occur at a temperature which would then facilitate the ingress of oxygen to the lower regions of the scale. While the apparent ease with which oxygen can diffuse into the Cr depleted alloy is perhaps surprising, it should be emphasized that Ni itself continues to diffuse upwards into the scale, as evidenced by the continued growth of Ni-rich oxides there.

The region of the scale examined in the TEM was found to be mostly NiO. Relatively coarse grains (~2 μm) of low defect-content NiO were seen at the metal/oxide interface, as shown in Fig. 3a. An area was also seen where one of the Cr_2O_3 precipitates had been incorporated into the scale, the metal/oxide interface presumably migrating into the alloy during scaling. A dark field micrograph of a Cr_2O_3 precipitate, now situated some 0.25 μm into the scale, is shown in Fig. 3b. The Cr_2O_3 here, although coarse (~1 μm), does not have the 'blocky' morphology seen before, suggesting that it has partially dissociated during scaling. That it had dissociated was confirmed by examining the oxide grains immediately adjacent to it, where it was found that Ni-Cr spinel had formed (see (110) inset diffraction pattern, Fig. 3c). Figure 3c shows the typical morphology of a ~200nm spinel grain which, by comparison with the NiO seen away from such areas, is comparatively defective. Rather smaller spinel grains are also seen decorating the rounded Cr_2O_3 grains, as arrowed in Fig. 3b. Beyond the spinel oxide, again are found coarse grains of NiO (Fig. 3d), which had tended to undercut the Cr_2O_3 (again suggesting that an inward oxidation process is being observed), although the dislocation content of the NiO was found to be significantly higher than in the same oxide even further away (Fig. 3a). The defective NiO again contained <1% Cr, whereas the rounded Cr_2O_3 grains described above contained 4wt% Ni and the Ni-Cr spinel only some 4-5wt% Cr.

While at first sight the formation of the spinel oxide is rather surprising, given the rapid inward flux of oxygen and the outward diffusion of Ni, its formation is energetically favoured and can occur through the simple reaction between NiO and Cr_2O_3. Alternatively, this could be the remnants of the very initial stage of oxidation when all three types of oxide would be expected to form before growth of NiO comes to dominate the scaling process. The data has, however, emphasized the apparent ease with which oxygen diffuses through the scale formed on this alloy, and it would thus be reasonable to assume that the ensuing high oxygen partial pressure can promote, to at least some degree, formation of new

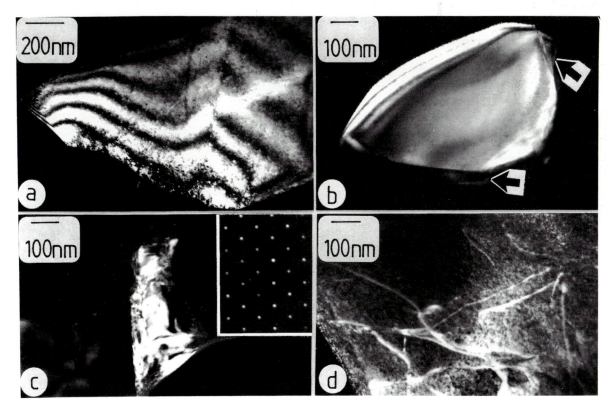

Figure 3: TEM micrographs of the scale formed on Ni-5Cr showing (a) NiO, and (b) Cr_2O_3 within the scale. The oxide adjacent to the Cr_2O_3 was (c) Ni-Cr spinel, and (d) defective NiO.

oxide at the base of the scale. The inference that NiO does indeed migrate into the alloy is supported by the behaviour of Ni-1%Cr studied using microlithographic markers.[10] This new growth then leads to the incorporation of Cr_2O_3 into the scale and to the formation of the Ni-Cr spinel, both of which would be expected to cause a reduction in the overall inward flux of oxygen after long-time oxidation and thus to a significant reduction in the rate of scaling. Equally, the gross Cr depletion seen in the precipitation zone beneath the scale could also play a significant role in promoting accelerated oxidation should oxide spallation occur at the metal/scale interface.

Ni-20Cr

Regions of the metal/oxide interface where enhanced scaling had taken place at an alloy grain boundary were examined and compared to those away from such a zone. This has enabled an investigation into how the delicate balance of scaling reactions is modified by a higher alloy Cr content and whether this then affects the way in which oxygen is apparently able to diffuse readily through the scales formed.

The metal/oxide interface region formed away from an alloy grain boundary was found to contain coarse grains of Cr_2O_3. A typical area is shown in Fig. 4a, where the grain size of most of the oxide is ~0.7µm, although other somewhat finer grains (200nm) may also be seen. While the light micrographs shown in Fig. 1 suggest that at least some of the scale has grown into the alloy, the observation here of relatively coarse grains is somewhat surprising, given the way in which inward growing oxides are so typically fine grained, particularly in dilute alloy systems.[2,10] Furthermore, the morphology of the metal/oxide interface itself, where the absence of any oxide cusping is particularly noticeable, is not suggestive of an active growth front. Nevertheless, the obvious similarity in the form of the grains seen here and at alloy grain boundaries (where it is inescapable that inward oxidation has occurred) simply emphasizes how grain size alone cannot be used to determine whether or not inward oxidation has occurred and at the same time suggests that the oxide seen in Fig. 4a has grown into the alloy. It should be emphasized that the oxide seen here was apparently generally compact and did not contain voids. The metal immediately beneath the scale showed some dislocation activity (Fig. 4b), suggesting that the alloy had undergone local creep deformation. Isolated oxide precipitates were also found within the sub-scale metal, a typical example being shown in Fig. 4c where 300nm grains of Cr_2O_3 can be seen. This precipitate has a rather similar morphology to that of the internal oxide seen in the M-5Cr alloy, though a significantly lower volume fraction of such precipitates was seen here.

Figure 4: TEM micrographs of the scale seen on Ni-20Cr showing (a) the metal/oxide interface, (b) dislocation activity beneath the oxide, and (c) Cr_2O_3 precipitation beneath the scale.

In regions of the scale where accelerated oxidation has occurred at an alloy grain boundary, it was found that the oxide had a very similar morphology to that of the Cr_2O_3 described above. An area where coarse Cr_2O_3 grains may be seen some 2.5 μm beneath the main scale is shown in Fig. 5a, the alloy grain boundary also being marked. While it is clear that no gross change in the oxidation process has taken place in this region of enhanced diffusivity, an alloy sub-grain boundary may be seen adjacent to the 'boundary oxide'. This is shown in Fig. 5b (as arrowed), where dislocation activity may also be seen in the surrounding alloy. Separation of the metal and oxide at the interface itself was observed (see Fig.5b), though it is not clear whether void coalescence at the interface or cracks that were formed during thin foil preparation are being seen.

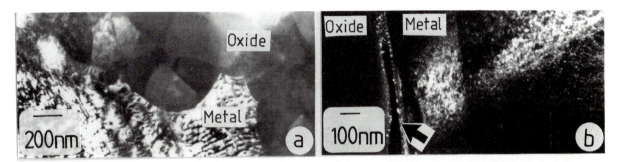

Figure 5: A region of accelerated oxidation at an alloy grain boundary showing (a) coarse Cr_2O_3 grains, and (b) sub-grain formation in the Ni-20Cr alloy.

The oxide found immediately above the metal here, where the scale had a total thickness of ~26μm, was again coarse grained Cr_2O_3 which contained few defects and had a cation content of ~98wt% Cr. The metal beneath the scale was, however, depleted in Cr to approximately 7.9wt% and to a depth of ~25μm. The way in which the metal here is depleted (to a depth equal to that of the scale thickness above) is quite different to the behaviour seen for the Ni-5Cr alloy, where depletion occurred only within the precipitation zone. What is apparently being seeing is a change in the oxidation process from one dominated by the diffusion of oxygen through the scale to one where the diffusion of Cr in the alloy then controls the scaling process. The data for the Ni-5Cr alloy does, however, suggest that coarse oxide precipitation may well have taken place at an earlier stage of the oxidation process which, as has been seen, leads to only localized Cr depletion in the alloy. Whereas in Ni-5Cr the Cr content of the alloy eventually falls to zero, the higher (though depleted) Cr level of the Ni-20Cr alloy examined here would favour the continued formation of Cr_2O_3 internal oxide as well as the coarsening of pre-formed precipitates within the internal oxidation zone. Continuation of this process would then naturally result in the formation of a complete layer of Cr_2O_3, but beneath the pre-oxidation surface of the alloy. While oxygen may well still be able to rapidly diffuse through the outer part of the scale, its diffusivity in Cr_2O_3 is slow, even at 1273K.[11] Evidence that oxygen is not at this point readily diffusing into the alloy comes from the way in which internal oxide precipitates are only rarely seen and the fact that the metal/oxide interface away from grain boundary regions does not look particularly active. It can now be seen how there has been a change in the dominant diffusing species once the ingress of oxygen is inhibited by the formation of a continuous Cr_2O_3 layer.

The 26μm scale examined above the grain boundary zone described was found to contain Cr_2O_3, Ni-Cr spinel and NiO. Typically the generally coarse grained Cr_2O_3, which extended 15 μm from the metal/oxide interface, exhibited a cation content of 93-98wt% Cr. Most of the oxide seen in a 7.5μm band above the Cr_2O_3 was found to be Ni-Cr spinel, though isolated grains of NiO were observed in both the Cr_2O_3 and the Ni-Cr spinel layers. A typical region where a 400nm NiO grain lies adjacent to the spinel is shown in Fig. 6a, while the morphology of the spinel is shown in Fig. 6b. The observation of the spinel is significant because it again reflects a high oxygen and Cr activity within the scale, as evidenced by the higher Cr cation content (64wt%) of the spinel here compared to that formed on the Ni-5Cr alloy. Just as NiO forms on Ni-5Cr, it would be expected that the same oxide would be formed here and that it would continue to grow until the formation of the complete layer of low Ni-content Cr_2O_3 inhibits the continued diffusion of Ni into the scale. When this occurs, as already seen, Cr diffusion becomes dominant, though oxygen will still be diffused into the scale through the NiO formed above. The diffusion of both species must then promote the reaction of NiO and Cr_2O_3 leading to the formation of the Ni-Cr spinel. The presence of isolated grains of NiO within the zone suggests that the process is both active and incomplete.

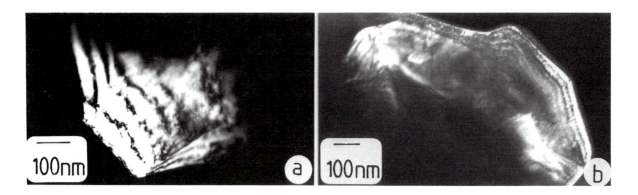

Figure 6: TEM micrographs of the mixed oxide layer formed on Ni-20Cr showing (a) an isolated NiO grain lying adjacent to (b) Ni-Cr spinel.

Equally, the surprisingly thin outer layer (~3.5μm) of NiO which was observed may be the consequence of continuing spinel formation, promoted by the upward diffusion of Cr.

Further evidence that the scale has been structurally modified comes from a significant observation of the NiO outer layer. Unlike the NiO seen on the Ni-5Cr alloy, the NiO formed on Ni-20Cr exhibited a defect superlattice structure reminiscent of wüstite $(Fe_{1-x}O)$[12] as deduced from the diffraction pattern (Fig. 7a). A dark field micrograph of the NiO is shown in Fig. 7b which demonstrates how the thickness fringes are disrupted by an apparently ordered 'phase' within the oxide. Figure 7c shows a micrograph imaged using a (220) superlattice reflection and indicates that precipitate-like clustering is being seen. While it is clear that the domains extend through the thickness of the thin foil, they do not appear to be an artifact of ion beam thinning or electron beam irradiation. It would not be unreasonable to suggest that the type of non-stoichiometry seen here in the NiO has been promoted by the presence of trivalent Cr^{3+} ions, which can occupy interstitial tetrahedral sites in NiO, with the accompanying formation of octahedral cation vacancies in clusters in the same way as occurs for Fe^{3+} ions in wüstite. Equally it should be emphasized that such non-stoichiometry will also increase the diffusivity of this part of the scale and is yet a further example of how diverse the scaling of these 'simple' alloys can be.

CONCLUSIONS

The TEM results have highlighted the disparate oxidation behaviours of Ni-5Cr and Ni-20Cr. It has been shown that on the one hand scaling in the Ni-5Cr alloy is dominated by an inward flux of oxygen and on the other how this oxidation process can change in the Ni-20Cr alloy to one dominated by Cr diffusion. This preliminary investigation underscores how important is a detailed knowledge of the composition and microstructural disposition of each of the various phases present to the understanding of an oxidising system.

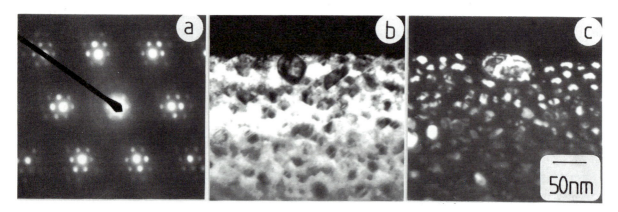

Figure 7: (a) (110) NiO diffraction pattern exhibiting superlattice type reflections. Dark field micrographs using (b) NiO and (c) superlattice reflections.

ACKNOWLEDGEMENTS

We thank Professor D. Hull FRS for the provision of laboratory facilities, the SERC [SBN] and NATO (Grant No. CRG 891024) [SBN and LWH] for financial support of research, and the Massachusetts Institute of Technology for award of sabbatical leave [LWH].

REFERENCES

1. S.B. Newcomb, W.M. Stobbs and E. Metcalfe, Phil. Trans Roy. Soc. (Lond.) A319, 1986, p.191.

2. W.M. Stobbs, S.B. Newcomb and E. Metcalfe, Phil. Trans Roy. Soc. (Lond.) A319, 1986, p.219.

3. S.B. Newcomb and W.M. Stobbs, Mat. Sci. and Technol. 4, 1988, p.384.

4. L.W. Hobbs, H.T. Sawhill and M.T. Tinker, Trans. Jap. Instit. Met. JIMIS-3, 1983, p.115.

5. L.W. Hobbs, H.T. Sawhill and M.T. Tinker, Rad. Effects 74, 1983, p.229.

6. H.T. Sawhill and L.W. Hobbs, Proc. Int. Congr. on Metallic Corros. (NRC Canada) 1, 1984, p.21.

7. G.C. Wood and F.H. Stott, Mat. Sci. Technol. 3, 1987, p.519.

8. S.B. Newcomb, C.B. Boothroyd and W.M. Stobbs, J.Microsc. 140, 1985, p.195.

9. B. Chattopadhyay and G.C. Wood, Oxid. Met. 2, 1970, p.373.

10. C.K. Kim, S-K. Fan and L.W. Hobbs, this volume ,1991.

11. P. Kofstad, Non-stoichiometry, Diffusion and Electrical Conductivity in Binary Metal Oxides, New York, John Wiley, 1972.

12. C. Lebreton and L.W. Hobbs. Rad. Effects 74, 1983, p.219.

40 OXIDATION IN A TEMPERATURE GRADIENT

S. Malik and A.V. Chadwick

University Chemical Laboratory
University of Kent
Canterbury CT2 7NH, U.K.

ABSTRACT

In many industrial situations metallic materials are subjected to a large temperature gradient and hence there is a large heat flux through the material. Typical examples are the alloy tubes in heat exchangers and boilers in the power industry. The temperatures of operation are usually high and therefore the metal is undergoing slow oxidation. Nearly all basic studies of corrosion are undertaken in isothermal conditions and there are very few data to indicate the effect of an imposed thermal gradient. There has been one attempt to model the effect of a heat flux on oxidation and this shows that under certain conditions "catastrophic" oxidation can occur, i.e. the oxidation rate increases with oxide thickness.

During a fundamental study of the effects of a heat flux on the oxidation of pure metals, the authors have developed the thermodynamics of the process in a more complete form than in earlier work. The results of an experimental study of the process are reported here.

1. INTRODUCTION

The effect of imposing a temperature gradient on a system is to induce a flow of heat, the heat flux. An additional effect of the application of a temperature gradient is the induction of a flow of matter, usually termed **thermomigration**. In a crystalline solid thermomigration can cause preferential diffusion of atoms either up or down the temperature gradient and the phenomenon has been the subject of several reviews [1-4]. Thermomigration can be of considerable technological importance when materials are subjected to high temperatures and large temperature gradients, for example in nuclear reactors, furnaces and heat exchangers. A major cause for concern is the effect of the thermomigration causing compositional changes in the material, for example changing the stoichiometry across a nuclear oxide fuel in a reactor or the "unmixing" of a metal alloy. Another potential problem, which has received very little attention, is the effect of thermomigration on the oxidation rates of metal components at high temperatures. It has been proposed that the effect of the temperature gradient is to alter the corrosion kinetics and there is experimental evidence to indicate that corrosion rates are different (usually faster) in heat exchangers. However, in real-life situations it is not easy to distinguish between several effects that could alter the oxidation rate, e.g. erosion by dust, attack by acid fumes, etc. There has been a theoretical modelling of this problem by Glover [5] which suggests that under certain conditions the effect could be dramatic and lead to "catastrophic" oxidation, i.e. the oxidation rate increases with oxide thickness. The experimental data for oxidation in a temperature gradient are very sparse [6, 7] and are not sufficient to test the model. Recently the authors initiated a programme of research to study the effect of a temperature gradient on metal oxidation with the aims of developing the theoretical background and providing reliable experimental data based on laboratory measurements for simple metals. It is the initial results of this work, particularly the development of the experimental procedures and some preliminary results, that are presented in this paper.

The study of oxidation in a temperature gradient presents problems from both the theoretical and experimental viewpoints. The modelling of the kinetics is complicated by having to deal with two driving forces on the diffusion of ions in the oxide layer; the gradients of oxygen partial pressure and temperature. This problem was addressed by Glover and the basic procedures were developed in his paper. However, he neglected the fact that the point defects in the oxide have **effective** charges and a diffusion electric potential is developed in the layer. Here the same procedures used by Glover have been followed but the diffusion potential has been included. The final general expressions for the flow of ions in the oxide layer and the time dependence of the layer thickness have been obtained [8] and solved for specific experimental systems. The major experimental difficulty in studying the oxidation in a temperature gradient is the measurement of the oxide thickness on the sample surface. Most isothermal studies of oxidation use a metal foil and use thermogravimetric methods which involve determination of the oxide thickness on **all** exposed surfaces. For temperature gradient work the system must be precisely defined and the oxide layer must be measured with a known direction of the temperature gradient. Several possible procedures including stripping off the layer and ultrasonic measurements have been examined.

The preferred method is to use electron microscopy measurements of sections through the oxide layer; this method is described in detail in a later section.

THEORETICAL MODELLING

Thermomigration in a crystalline solid is a subtle phenomenon and can be considered as arising from two contributions. The first contribution is the gradient of point defects that are generated across a sample due to the two ends of a sample being at different temperatures. This results in a flow of defects from the hot end (higher defect concentration) to the cold end (lower defect concentration). The second contribution is due to the temperature gradient biassing of the diffusive jumps of the atoms via the defect. This is governed by the **heat of transport,** Q^* of the atom. Q^* is an energy term and represents the heat carried from the initial site to the final site when an atom makes a jump in an isothermal system. The available data [1-4] indicate that for vacancy diffusion Q^* is positive and is roughly equal in magnitude to the enthalpy of migration of the atom, h_m.

Glover [5] coupled the classical Wagner model for isothermal oxidation (see, for example, Birks and Meir [9]) with the models of thermomigration to treat oxidation in a temperature gradient. This is not a trivial problem since it requires consideration of two driving forces for diffusion, the differences in oxygen partial pressure and temperature at the two surfaces of the oxide film.

The results that he obtained showed that the oxidation kinetics depend on the nature of the non-stoichiometry of the oxide, i.e. whether it is n or p- type oxide, and the sign and magnitude of Q^*. A careful examination of the Glover model shows that he neglected the **effective** charge on the diffusing point defects (vacancies or interstitial ions) in the oxide film and the resulting diffusion potential. This adds a further complication to the modelling and the final expression for the flux, J_i, of a species, i, in the oxide is given by

$$J_i = \frac{-Nc_iD_i}{kT} \left\{ (1-t_i) \left[\frac{du_i}{dx} + \frac{Q^*_i}{T}\frac{dT}{dx} \right] - q_i \left[\sum \frac{t_k}{q_k} \left(\frac{du_k}{dx} + \frac{Q^*_k}{T}\frac{dT}{dx} \right) \right] \right\} \qquad (1)$$

Here D_i, c_i, u_i, q_i, Q^*_i and t_i are the diffusion coefficient, concentration, chemical potential, effective charge, heat of transport and transport number of species i, respectively. N is the number of lattice sites available to i per unit volume, k is the Boltzmann constant, T the temperature and x the distance into the oxide film normal to the metal surface, using x=O as the oxygen/metal oxide interface. It is the last term on the right hand side of Equation (1) that is missing from the expression derived by Glover. The integration of $J_i dx$ across the film can be pursued in the same manner adopted by Glover. However, it is now more complex and is only possible by making specific assumptions about the nature of the diffusing ions, their transport mechanism and the nature of the defect structure of the oxide, i.e. whether it is oxygen excess or deficient, the types of defect and the transport numbers of all the defects (both point defects and electronic defects). We have performed these calculations for some welldocumented systems and the detailed results will be presented in a subsequent papers. Focus has been placed on two basic systems, metals whose oxide layers are n and p- type oxides and with zinc and nickel oxidation as the systems for the experimental test-beds of the modelling.

EXPERIMENTAL

The experimental procedure was basically simple and a schematic diagram of the oxidation rig is shown in Figure 1. A polished metal sample (35 mm diameter x 5mm thick) was firmly contacted to an air-cooled stainless steel block and placed in an alumina sheath. The sheath was air-tight and allowed control of the atmosphere surrounding the sample. A high temperature furnace surrounded the sheath. The temperature of the atmosphere, and the temperatures at two positions near the front and back faces of the sample were monitored with thermocouples. The readings from these latter two thermocouples enabled the heat flux through the sample to be determined. The sample was oxidised in the temperature gradient for a known time and temperature, and the oxide layer thickness monitored. However, usual methods of thickness determination could not be employed i.e. weighing the sample etc. Thus direct measurements of the oxide layer have been developed to overcome this problem. A small groove was cut at an angle of 45 degrees into the surface with a 3mm diameter diamond burr, as shown in Figure 2. The section of the oxide layer in the groove was observed in a scanning electron microscope (SEM) and a Polaroid photograph taken of the image. It was found that coating the oxide surface with a graphite layer, using a colloidal graphite paint, in the region of the drill hole improved the contrast. The true thickness of the oxide layer, x, was obtained from the relation x= 1.sin45°, where 1 is the observed thickness of the layer on the Polaroid. The estimated error on x is ±10%, which mainly arises from the measurement of the Polaroid film. A typical micrograph is shown in Figure 3. The advantages of this approach are that it is fairly straightforward and the

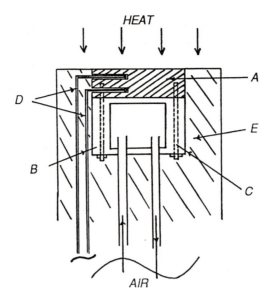

Figure 1: Schematic diagram of the heat flux oxidation experiment. A=sample; B=air-cooled stainless steel block; C=stainless steel studs; D=Pt/Pt-13%Rh thermocouples; E=ceramic wool thermal insulation.

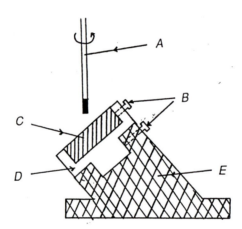

Figure 2: Schematic diagram of the jig for drilling probe holes in the oxide layer. A=diamond burr; B=locking screws; C=sample; D=sample holder; E=jig to hold sample at 45°.

resolution depends only on the fineness of the diamond grit and the magnification of the SEM. In addition, the hole is so small that the sample can be used several times and the measurements made effectively in a continuous manner (i.e. thickness with time) for a single sample.

The main study has been on oxidation of pure nickel. The samples were obtained from Johnson Matthey and were of 4N+ purity. The oxidation was studied in the temperature range 600 to 900°C with heat fluxes of 0 to 240 kWm^{-2} in a pure oxygen.

RESULTS AND DISCUSSION

The primary quantitative results from this work are presented in Table 1. Also included in this table are some literature data for the oxidation of pure nickel films.

The results in the temperature range 600 to 800°C suggest that for the heat fluxes that have been applied there is no significant affect on the oxidation kinetics of nickel within the limits of the experimental error. This can be seen by comparing the measured film thicknesses with runs where no temperature gradient has been applied and by comparison with the literature data for isothermal oxidation. In some respects this is disappointing and may originate for two reasons. The effects may be too small to detect with the magnitudes of the heat flux which have been employed. The second reason

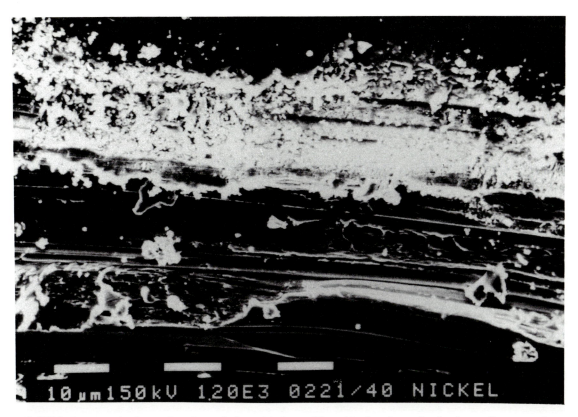

Figure 3: SEM photograph of a section through the oxide layer. The white band running horizontally across the centre is the oxide layer.

could be due to the nature of the oxidation process at these temperatures. Very elegant studies of the oxidation of nickel and diffusion in nickel oxide by Atkinson *et al.* suggest that at these temperatures the diffusion of nickel ions is predominantly through the bulk defects in the film rather than point defects. The current theoretical model of Glover [5] and that developed here is for diffusion via point defects and may be inapplicable to these experiments. It is worth noting that the theories of thermomigration in solids are only for point defect diffusion and the problem of boundary and dislocation thermomigration has not been addressed.

The temperature gradient oxidation at 900°C produced a remarkably thick oxide film, approximately twice that predicted from the literature isothermal data. This clearly needs to be verified and isothermal experiments are being currently performed in our experimental system. The isothermal result obtained at 855°C in our rig also produced an unusually thick oxide layer and as yet we cannot be certain of the role of the temperature gradient in oxidation at these higher temperatures.

Our work has revealed some unusual surface structures of the oxide films produced in a temperature gradient. Visual inspection reveals differences in the layers. The isothermally produced oxide layers have an even, light green appearance. Those formed in a temperature gradient are much darker and have a mottled appearance. In addition the grain sizes of the oxide are different, being much larger in the layers produced in a temperature gradient which is the origin of the mottled appearance. This is demonstrated in Figures 4a and 4b which show SEM pictures of the oxide surfaces produced at 700°C isothermally and in a temperature gradient, respectively. The cause of these differences is not yet clear. Although there is a large temperature gradient across the oxide layer the actual temperature difference is not very large due to the thinness of the film. Across a 10µm oxide layer in the temperature gradient experiments it is estimated that the temperature difference will be the order of 1°C, and it is probably not just an effect of temperature of the film. However, to produce the gradient the gas temperature has to be very high, typically 300-400°C higher than the temperature of the metal sample, and this could be causing the observed effect.

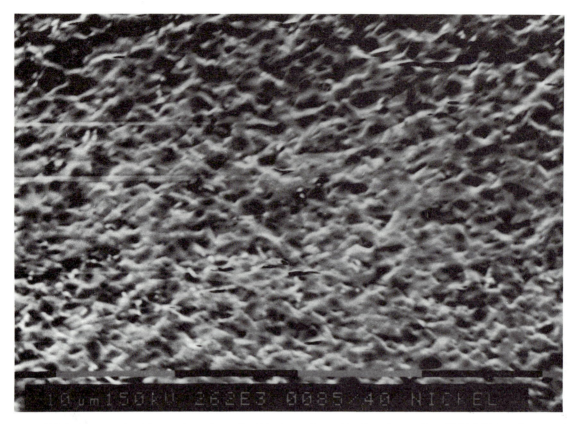

Figure 4a: SEM photograph of the oxide surface of a nickel sample oxidised isothermally at 700°C. This is for sample used in Run 8 of Table 1.

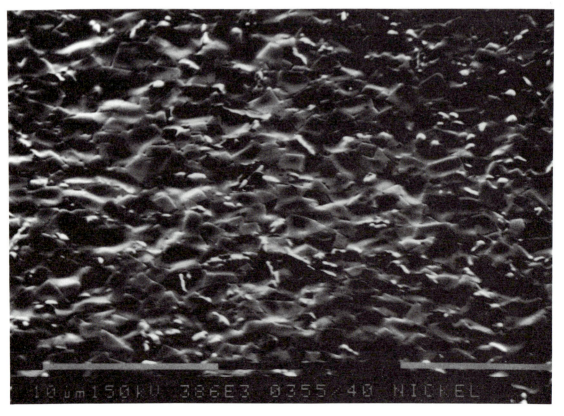

Figure 4b: SEM photograph of the oxide surface of a nickel sample oxidised in a heat flux at 700°C. This is for sample used in Run 7 of Table 1.

Table 1. Results for the oxidation of nickel

Run	Temperature °C	Heat flux kWm^{-2}	Time hours	x μm	Lit*. x μm
1	600	180	113.5	3.5	6.4
2	600	120	336	12.0	12.0
3	600	0	114	6.4	6.0
4	600	0	312	12.0	9.9
5a	700	60	1	1.5	1.9
5b	700	60	6	4.7	3.8
5c	700	60	34	5.1	8.9
5d	700	60	75	11.7	12.4
5e	700	60	115	16.7	14.7
6	700	180	114.25	13.0	13.3
7	700	216	352.5	24.8	23.4
8	700	0	476	32.5	27.2
9	800	216	144	28.3	29.3
10	855	0	120	45.3	28.7
11	900	240	480	141.4	57.2

* Literature data for the isothermal oxidation of nickel taken
   from the work of Atkinson et al[11]

CONCLUSIONS

The present experiments suggest that for heat fluxes up to 200kWm^{-2} there is no effect on the oxidation rate of pure nickel. The preliminary results at 900°C, using similar heat fluxes, are not consistent with isothermal oxidation data. More work is in progress at this high temperature regime.

ACKNOWLEDGEMENTS

We wish to acknowledge the SERC and the CEGB for the award of a CASE studentship to S.M. We also wish to thank Dr. Barry Meadowcroft for help, advice and encouragement throughout this work.

7. REFERENCES

1. R.E. Howard and A.B. Lidiard, Rep. Prog. Phys., 27, 1964, p. 161.

2. A.R. Allnatt and A.V. Chadwick, Chem. Rev., 67, 1967, p. 681.

3. M.J.Gillan, 1983, in "Mass Transport in Solids", Eds. F. Beniere and C.R.A. Catlow, Plenum Press, New York, p.227.

4. J. Philibert, "Diffusion et transport de matiere dans les solides", Les Editions de Physique, Paris, 1985, ch. 8.

5. D.M. Glover, Corros. Sci., 20, 1980, p. 1185.

6. J.C. Griess, J.H. De Van and W.A.Maxwell, Materials Performance, 17, 1978, p. 9.

7. J.C. Griess, J.H. De Van and W.A. Maxwell, Materials Performance, 19, 1980, p. 46.

8. A.R. Allnatt, S. Malik and A.V. Chadwick, to be published.

9. N. Birks and G.H. Meir,"Introduction to High Temperature Oxidation of Metals", Arnold, London, 1983.

10. A. Atkinson, R.I. Taylor and A.E. Hughes, Phil. Mag. A45, 1982, p. 823.

41 OXIDATION BEHAVIOUR AND PROTECTION OF CARBON-CARBON COMPOSITES

F. J. Buchanan and J. A. Little

Department of Materials Science and Metallurgy, University of Cambridge, Pembroke Street, Cambridge CB2 3QZ, UK.

ABSTRACT

The oxidation behaviour of both unprotected and protected carbon-carbon (C/C) composites has been studied in a temperature range of 500°C to 1100°C. Protective systems investigated have included SiC coatings, either alone, or combined with a crack sealing glass material.

Chemical Vapour Deposition has been used to apply the SiC coatings to the C/C composites. Scanning and transmission electron microscopy have been important in the characterisation of microstructures related to coating parameters and in the investigation of crack patterns due to thermal expansion mismatches between the SiC coating and the C/C substrate. A boric oxide glass sealant has been investigated, important considerations being viscosity, oxygen diffusivity, effects of moisture and volatility of the glass at high temperature.

INTRODUCTION

Carbon-Carbon (C/C) composites appear to be ideal high temperature materials in many respects. They maintain very high strength and stiffness properties above 2000°C whilst being lightweight and creep resistant. Current applications include rocket propulsion components, re-entry thermal protection and aircraft brakes, making use of the low density and shock resistance of the material [1, 2]. These are not primary load bearing applications. Possible future applications such as aero engine components, hypersonic vehicle airframe structures and space structures would take advantage of the attractive mechanical properties obtainable from C/C composites [3]. However, oxidation of carbon occurs at temperatures as low as 500°C [4] and many of the anticipated applications require operation at temperatures of up to 2000°C [2].

Several approaches are being considered to protect carbon from oxidising. Inhibitors within the matrix have been used to slow down the oxidation rate and protective coatings have been applied to prevent oxygen from reaching the underlying carbon [5].

Unfortunately, due to thermal expansion differences ceramic coatings such as silicon carbide applied by chemical vapour deposition develop cracks during thermal cycling and thus do not form a satisfactory oxygen barrier [4]. A sealant material is required to provide self-healing capability by flowing into thermal stress cracks formed in the ceramic layer [2].

A glassy film of boron oxide when applied over all exposed surfaces of C/C composites has been effective in protection against oxidation in air at temperatures of up to 1000°C [4]. However the use of this glass is limited, mainly due to moisture sensitivity and volatility at higher temperatures. Thus there is a requirement for an improved glass sealant material, important properties being viscosity, oxygen diffusivity, effects of moisture and volatility of the glass [3].

The present investigation studies the oxidation behaviour of a C/C composite material, initially unprotected and subsequently after the chemical vapour deposition of a silicon carbide coating. The microstructures of the SiC coating are examined and would seem to be deposited in a layer-by-layer manner. The limiting factors in the use of boron oxide as a sealant material, namely moisture sensitivity and volatility are also investigated.

EXPERIMENTAL PROCEDURE

The C/C composite materials ('K-Karb') used contained rayon based fibres and a carbonised phenolic resin impregnate matrix in a 2D fibre lay-up (BP Research, Sunbury).

Silicon carbide coatings were deposited onto 1cm sided cubes of the composite, using a deposition temperature of 1125°C and a pressure of 17mbar. Methyltrichlorosilane (MTS) was used as the source of SiC with hydrogen as the carrier gas.

A deposition rate of 5 to 12 μm/hour was achieved depending on sample positioning within the chamber. In order to ensure total sample coverage a double layer coating was applied with the samples being inverted between applications. Single layer coating thicknesses ranged from 25 to 50 μm.

A borate glass coating was applied to both SiC coated and uncoated samples. This was achieved by heating the sample, surrounded with boric oxide powder, to 800°C, holding for 2 hours and cooling to room temperature. Excess glass was allowed to flow off the sample while at temperature to leave a coating ranging in thickness from 0.15mm to 0.25mm. Moisture absorption experiments were carried out in room temperature flowing air of 25% relative humidity (flow rate 100ml/min).

Measurements of oxidation kinetics of all samples were carried out in a vertically mounted furnace in dry flowing air (21% O_2, 79% N_2, flow rate 50 - 150 ml/min), at temperatures between 500°C and 1100°C. The specimens were supported by Ni-Cr wire from a Stanton-Redcroft mass balance and mass changes were measured as a function of time. The furnace was either held at a constant temperature for each test run, or ramped at 100°C intervals, holding for 20 minutes at each temperature.

Specimens were characterised using reflected light, scanning and transmission electron microscopy. TEM sample preparation involved polishing 3mm discs to a 6 μm finish followed by ion beam thinning.

RESULTS AND DISCUSSION

Oxidation of Uncoated C/C Composite

A measurable weight loss occurred at temperatures as low as 475°C. Above this temperature the oxidation rate increases with temperature up to approximately 750°C. In this low temperature region the rate is controlled by the chemical reaction at the gas/substrate interface. At higher temperatures the rate is controlled by gas phase diffusion across the boundary layer and increases with air flow rate rather than temperature.

Weight loss was found to be linear at constant temperature, indicating both fibres and matrix were oxidised simultaneously. This was confirmed by scanning electron microscopy of the oxidised composite surface. The surface appeared highly porous and showed degradation of both the fibre and matrix components (Fig.2) compared with the unoxidised composite (Fig.3). Occasional isolated large matrix particles were not as severely degraded as the surrounding fibres and smaller matrix particles due to the lesser exposed surface area (Fig.4).

Silicon Carbide Coated C/C Composite

SiC coated C/C samples, examined by optical and scanning electron microscopy, contained cracks after cooling from the chemical vapour deposition temperature of 1125°C due to tensile stresses within the coating. Figures 5 and 6 show a typical cracked surface and Fig. 7 shows the fracture surface within a crack.

SEM examination of the coating surface morphology revealed rounded hillocks, with each hillock being comprised of small faceted grains (Figs.5, 6 and 8). The structure is similar to that described by Cheng [6] for low deposition temperatures. It was suggested that a surface kinetics controlled mechanism produced this type of morphology, whereas, at higher deposition temperatures, a mass transport controlled mechanism produces larger, columnar grains [6].

When examined by TEM the SiC coating was found to have a cubic (β) type structure, confirmed by measurement of the electron diffraction pattern (Fig.9). This structure is common for SiC CVD deposits [6 and 7]. A layered structure was observed within some of the SiC grains (Figs.10,11 and 12). The layers lay perpendicular to the growth direction. This type of columnar growth has been previously reported as that of one-dimensionally disordered crystals. Each crystal is marked with dense parallel striations indicative of very heavy faulting [7].

Oxidation of SiC coated C/C composite

The rate of oxidation of this system varied with temperature as shown in Fig. 1. As temperature increases the cracks begin to close, due to the higher thermal expansion of the SiC. This causes a reduction in the amount of oxygen able to diffuse down the cracks and attack the substrate. This is balanced by the increased reactivity of the carbon at higher temperatures. Weight loss increases to a maximum at approximately 900°C, then drops considerably up to 11 00°C at which point the cracks should be virtually closed.

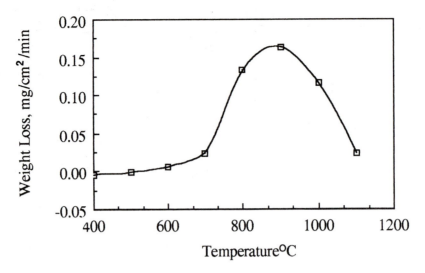

Figure 1: Oxidation of SiC coated C/C composite.

Glass Sealant Materials

Borate glass coatings applied alone to the C/C composite provided only a few hours protection at 900°C and less at higher temperatures. The combined low viscosity and high volatility of the glass at these temperatures caused thinning of the glass to a point where oxygen could rapidly penetrate the composite at the sample cube corners, causing severe degradation of the substrate (Fig.13).

Borate glass applied to SiC coated C/C sample provided good oxidation protection. However, at higher temperatures the glass volatility became significant with a linear weight loss of approximately 0.25mg/cm^2/hour occurring at 1100°C in dry air. Volatility in moist air was considerably enhanced as also found by McKee [8]. The borate glass was found to absorb moisture at room temperature.

Clearly a glass composition with superior moisture sensitivity and volatility characteristics is required, with other important properties being oxygen diffusivity and viscosity.

CONCLUSIONS

Oxidation of unprotected C/C composite occurs at temperatures as low as 475°C. The rate increases with temperature provided sufficient oxygen is available.

The SiC CVD coating on the C/C composite contained cracks permitting oxygen to penetrate through to the underlying substrate. The rate of oxidation of the substrate is balanced by the increase in both the carbon reactivity and the extent of coating crack closure as the temperature is raised.

The SiC coating structure consisted of a pattern of rounded hillocks, with each hillock containing fine faceted grains. The grains had a cubic (β) type structure with dense parallel striations being observed within some of the grains.

Pure borate glass absorbs moisture at room temperature when exposed to a moist environment. Moisture sensitivity at higher temperatures causes a notable increase in the volatility of the glass.

Borate glass does not sufficiently protect C/C composites from oxidation at 900°C and above. However, when applied to a SiC coated C/C composite oxidation protection is greatly improved. Volatility of the glass becomes significant at higher temperatures especially in the presence of water vapour.

A glass composition is required which will protected SiC coated C/C composites from oxidation in the temperature range 500 - 1100°C whilst having good moisture sensitivity and volatility characteristics.

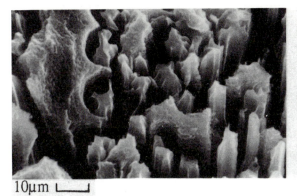

10μm ⌐___⌐

Figure 2: Composite surface after oxidation for 60 min at 800°C (SEM).

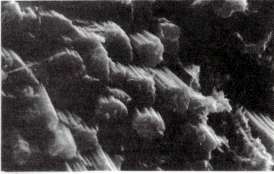

10μm ⌐___⌐

Figure 3: Unoxidised C/C composite surface (SEM).

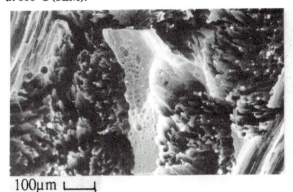

100μm ⌐___⌐

Figure 4: Composite surface after oxidation for 85 min at 1100°C (SEM).

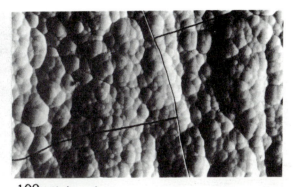

100μm ⌐___⌐

Figure 5: SiC coating showing typical 'rounded hillock' morphology (SEM).

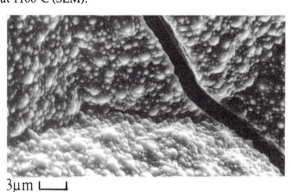

3μm ⌐___⌐

Figure 6: SiC coating showing thermal mismatch cracking (SEM).

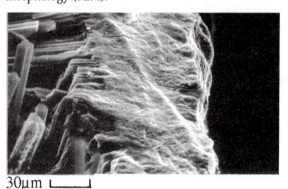

30μm ⌐___⌐

Figure 7: Typical fracture surface of a SiC coating crack (SEM).

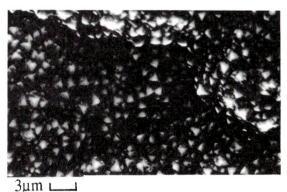

3μm ⌐___⌐

Figure 8: Fine faceted grains (SiC oxidised for 30 min at 1100°C) (SEM).

Figure 9: Typical cubic (β) structure of SiC (TEM).

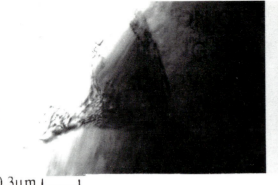

0.3μm ⊢————⊣

Figure 10: SiC showing layered deposition perpendicular to growth direction (TEM).

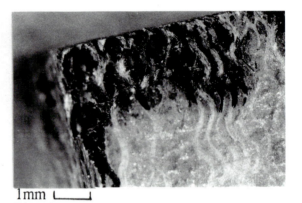

Figure 11: Diffraction pattern of central grain in Fig.10.

0.3μm ⊢————⊣

Figure 12: Dense parallel striations within a SiC columnar grain (TEM).

1mm ⊢————⊣

Figure 13: Boron oxide coated composite oxidised for 120 min at 1000°C (optical).

REFERENCES

1. J.D. Buckley, Am. Ceram. Soc. Bul. 67, 1988, p.364.

2. G. Savage, Metals and Materials 4, 1988, p.544.

3. J.E. Sheehan, Carbon 27, 1989, p.709.

4. D.W. McKee, Carbon 25, 1987, p.551.

5. K.L. Luthra, Carbon 26, 1988, p.217.

6. D.J. Cheng, J. Electrochem. Soc. 134, 1987, p.3145.

7. S.S. Shinizaki and S. J. Hiroshi, Am. Ceram. Soc. 61, 1978, p.425.

8. D.W. McKee, Extended abstracts of 18th Bienniel Conf. on Carbon, Worcester, M.A., Am Ceram Soc, Uni. Pk.,P.A., 1987, p.448.

42 ISOCHRONAL OXIDATION EXPERIMENTS ON BORIDE AND CARBIDE POWDERS

M. Stewart

National Physical Laboratory
Teddington, Middlesex TW11 0LW, UK.

ABSTRACT

The isochronal oxidation of WC, iron-neodymium-boron ($Fe_{14}Nd_2B$) and TiB_2 powders have been investigated to establish an experimental basis for quantitative particle size and distribution determination. The chemical reactions involved have been established using a range of surface analytical techniques, including optical and scanning electron microscopy and X-ray diffraction.

INTRODUCTION

In the production of parts by the powder metallurgy route it is useful to know the particle size distribution of the starting powder since this will control the sintering behaviour and thus the final properties. Previous work on the isothermal oxidation of WC powders has shown that the particle size and size distribution can be determined from oxidation rate [1].

The basis of this technique is simply that finer powders will oxidise more quickly due to their larger surface area to mass ratio. In the sizing experiments the controlled oxidation of powders is carried out in a thermogravimetric analyser (TGA) under isochronal conditions, i.e. a linear increase in temperature.

In order to perform the analysis it is necessary to understand the kinetics of the oxidation process and its temperature dependence, i.e. a rate equation and an activation energy. To this end the oxide scale formation on these particles was examined at various points during the experiment using optical microscopy, scanning electron microscopy (SEM) and X-ray diffraction (XRD).

EXPERIMENTAL

The oxidation in air of 20mg batches of powder was carried out in a Stanton Redcroft 780 TGA, and were subjected to a linear increase in temperature of between 2.5°C/min and 20°C/min over the range 25-1100°C. Earlier studies have shown that oxidation behaviour outside these limits is dependent on the sample mass and heating rate [2]. For heating rates greater than 20°C/min the reaction becomes pyrophoric and the furnace cannot maintain a linear heating rate. Also for large sample masses there is an abrupt decrease in the oxidation rate, probably due to the formation of a protective crust inhibiting air reaching the unoxidised powder.

Three different materials were studied in the sizing experiments: tungsten carbide, a tool material; iron-neodymium-boron, a high strength permanent magnet material; and titanium diboride, a wear resistant ceramic. Two grades of ball milled WC were examined, a fine (Fisher sub-sieve size 0.83μm, Ledermann GmbH) and a coarse (FSSS 5.59μm, GTE Sylvania). The iron-neodymium-boron examined was a hydrogen decrepitated and ball milled stoichiometric $Fe_{14}Nd_2B$, and the titanium diboride a stoichiometric TiB_2 (H.C.Starck).

RESULTS AND DISCUSSION

Tungsten Carbide

The tungsten carbide powders transform to oxide by the reaction -

$$2WC + 5O_2 \rightarrow 2WO_3 + 2CO_2 \qquad (1)$$

Since the carbon dioxide is lost, the mass increase when the WC has completely transformed should be equal to the ratio of molecular weight of the carbide to the oxide. All the experiments taken to completion agreed with the predicted weight

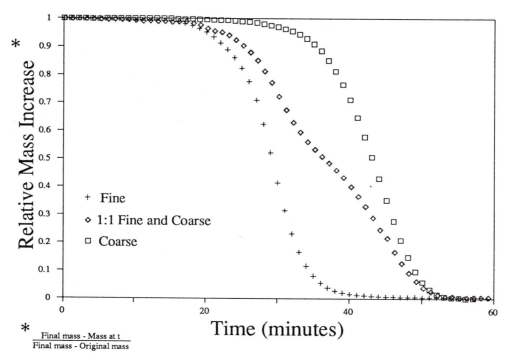

Figure 1: Normalised oxidation curves for three WC powders; a coarse, a fine, and a 1:1 mixture of the coarse and fine. Starting temperature 300°C, heating rate 10°C/min.

gain of 18.3%. In order to separate the effects of slightly differing sample sizes the curves are normalised with respect to the overall weight gain. Figure 1 shows the normalised oxidation curves for three WC powders with contrasting size distributions. This shows that the technique can be used easily for qualitatively sizing powders. The method of quantitatively sizing using this method has been discussed elsewhere [2,3], but the requirements for the analysis are the oxidation rate law, the activation energy and the pre-exponential factor from the Arrhenius equation. For WC this information is available in the literature, but in to order confirm that only reaction (1) is taking place the oxidation products were examined. Figure 2b shows the partially transformed coarse WC powder after heating at 10°C/min from room temperature to 670°C, and compares to the as received powder (Fig. 2a). As the oxide forms it spalls off giving rosette like particles with fresh WC at the centre. This behaviour equates to a linear rate law, which for the purposes of sizing is ideal since the completion of the reaction is more rapid. X-ray diffraction confirmed that WO_3 was the only solid reaction product.

Iron-Neodymium-Boron

Figure 3b shows the fully oxidised $Fe_{14}Nd_2B$ powder after heating at 10°C/min from room temperature to 1100°C, and compares to the as received powder (Fig. 3a). In contrast with WC the oxide on $Fe_{14}Nd_2B$ is adherent and the particles have grown in size. There is also some evidence that sintering of the particles has occurred. This is undesirable for the sizing experiments as this reduces the surface area, although as it is only apparent at the asperities the effect on sizing will not be pronounced.

Examination of the partially oxidised samples by X-ray diffraction and metallography of the polished cross sections reveals that the process is two-stage. At temperatures below 450°C the $Fe_{14}Nd_2B$ transforms by the inward growth of an oxide layer. X-ray diffraction of the product at this stage shows only a broad α-iron peak. This is consistent with TEM studies that have found a finely dispersed hexagonal Nd_2O_3 within a highly disturbed α-iron [4]. The weight gain at this stage is also consistent with the formation of Nd_2O_3. Above 450°C iron oxides begin to form. The presence of Fe_2O_3 and Fe_3O_4 was detected by XRD, however no FeO was discovered. The bulk of the oxide consisted of Fe_2O_3, and the total weight gain of approximately 35% agrees with the formation of oxides with a metal cation ratio of 2:3. Figure 4 shows the energy dispersive X-ray maps for oxygen, neodymium, and iron of a cross-sectioned powder heated at 10°C/min from room temperature to 840°C. This shows that the formation of the iron oxide is by outward diffusion of iron, leaving a neodymium rich centre which marks the size of the original particle. The oxidation is completed when the supply of iron from within the particle has been used.

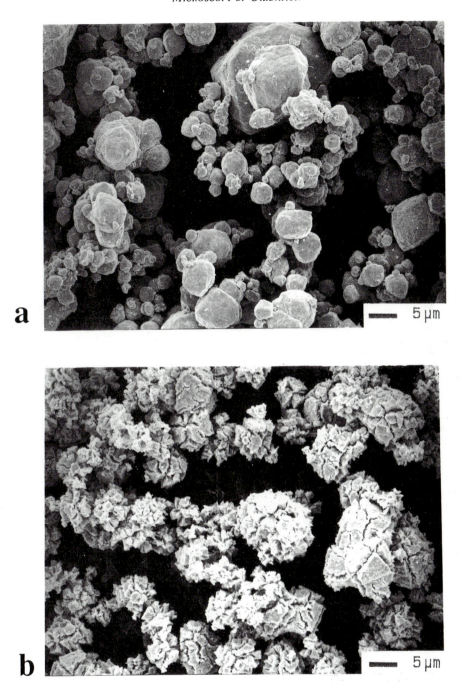

Figure 2: (a) As received coarse grained tungsten carbide powder. (b) Partially oxidised WC powder, 10°C/min to 670°C. The WO_3 formed spalls off giving rosette like particles with fresh WC in the centre.

The implications of this two-stage behaviour on the particle sizing of $Fe_{14}Nd_2B$ is that the oxidation rate law for both oxides are needed, and the respective Arrhenius equations, and also the ratio of molecular weights of the oxides.

Titanium Diboride

Preliminary studies on TiB_2 have shown that the oxidation behaviour is dependent on sample mass. Since the material is less dense than either WC or $Fe_{14}Nd_2B$ a sample mass of 20 mg fills the platinum crucible to a greater depth.

Figure 3: (a) As received hydrogen decrepitated and ball milled stoichiometric $Fe_{14}Nd_2B$ powder. (b) Fully oxidised $Fe_{14}Nd_2B$ powder after 10°C/min to 1100°C.

During heating at 10°C/min to 1100°C the sample mass gains over 75%. If the material oxidised stoichiometrically to TiO_2 and B_2O_3 there would be a weight gain of 115%. After oxidation the powder has sintered to a compact and can only be removed from the crucible by immersion in 10% HF:HCl. Examination of this compact reveals that underneath the white oxide there is still a considerable amount of grey unoxidised TiB_2. It was only by reducing the sample mass to 1mg that all the powder could be oxidised. X-ray diffraction of the reaction products reveals rutile, and a glassy phase. The weight gain data shows an abrupt change in slope at about 500°C. In experiments on TiC, which exhibit similar weight gain behaviour, this has been attributed to the change in oxide being formed from anatase to rutile [5]. However anatase was not detected in this case, and it would seem that the change in oxidation rate is due to the formation of a boron rich glassy phase, probably B_2O_3. Figures 5 and 6 show the as received and oxidised TiB_2 powders. It can be seen that the

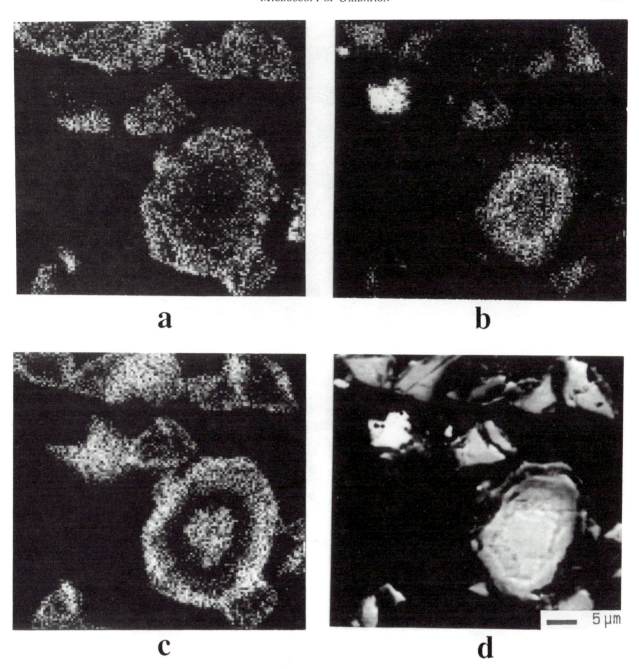

Figure 4: X-ray maps of cross-sectioned $Fe_{14}Nd_2B$ powder after 10°C/min to 840°C. (a) Oxygen Kα, (b) Neodymium Lα, (c) iron Kα, (d) Backscattered electron image.

rutile formed on the surface is much finer than the original powder. At higher magnification, Fig. 7, the glass phase can be seen between the fine grained rutile. This glass phase causes the particles to sinter at relatively low temperatures, creating an impervious crust on the top surface of the large sample masses inhibiting the complete oxidation.

Although there is a large weight gain in this system, the sintering enhanced by the liquid phase is not ideal for powder sizing. The sintering can be reduced by decreasing the sample mass, and thus the 'packing' density, however this is at the expense of experimental sensitivity. Further work is needed on this material to determine whether a temperature regime exists where the formation of TiO_2 is dominant, or to size using a different gaseous reaction.

Figure 5: As received titanium diboride powder.

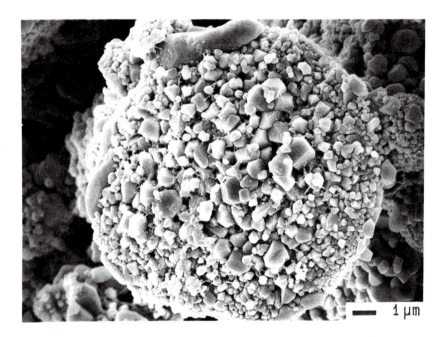

Figure 6: Titanium diboride particle oxidised at 10°C/min to 1100°C. The original TiB$_2$ particle oxidises to form a fine grained rutile on the surface.

CONCLUSIONS

The nature of oxide growth during isochronal oxidation experiments has been investigated for three materials, in order to establish an experimental basis for the quantitative sizing.

For WC the formation of WO$_3$ obeys a linear rate law and it is simple to predict the weight gain assuming spherical particles. In the case of Fe$_{14}$Nd$_2$B the behaviour was found to be two stage, below 450°C the oxide grows inwards, and above this temperature the particles grow by the outward diffusion of iron. Although the two stage process doubles the

Figure 7: Titanium diboride particle oxidised at 10°C/min to 1100°C. A glassy phase, probably B_2O_3, is formed between the fine grained rutile.

number of unknowns needed for the quantification this material can still be sized by this method. However in the TiB_2 system the formation of a liquid phase during oxidation makes sizing this material difficult. Further work is needed on this material to find a reaction that inhibits the formation of the liquid phase.

ACKNOWLEDGEMENTS

The stoichiometric $Fe_{14}Nd_2B$ powder was provided by Professor I.R. Harris of the University of Birmingham.

REFERENCES

1. B. Roebuck, E.G. Bennett, E.A. Almond and M.G. Gee, J.Mat.Sci. 21, 1986, p.2033.

2. B. Roebuck, Int. J. Refrac. Hard Met. 7, 1988, p.41.

3. M. Stewart, M.G. Geeand B. Roebuck, to be published in J. Mat. Sci.

4. P. Schrey, IEEE Trans. Magnetics, 22-5, 1986, p.913.

5. B. Roebuck, E.A. Almond and J.L.F. Kellie, Horizons of Powder Metallurgy, Eds. W.A.Kaysser and W.J. Huppmann, 1986, p.123.

SECTION SEVEN
Newer techniques

43 THE INTERNAL ATTACK OF 20%Cr/25%Ni/Nb STAINLESS STEEL REVEALED BY SELECTIVE ETCHING

J. A. Desport and M. J. Bennett

*Materials and Manufacturing Technology Division, Building 393, Harwell Laboratory,
AEA Technology, Didcot, Oxon OX11 0RA, UK.*

ABSTRACT

Improved procedures have been developed, by combining mechanical abrasion with selective etching, for the preparation of specimens to enable specific aspects of the attack of alloys to be examined more extensively by conventional surface analytical techniques, such as scanning electron microscopy and X-ray diffraction. The techniques have been utilised for the microstructural characterisation of the internal attack of 20Cr/25Ni/Nb stainless steel in CO_2-based environments at 900-1000°C.

INTRODUCTION

Further understanding of the complexities of the high temperature corrosion of metals is dependent critically upon elucidation of the microstructure and composition of the attack. No single technique can reveal everything and the overall picture has to be built up from a composite of information derived from the most appropriate approaches available. However, more comprehensive microstructural insight could emerge from the development of improved sample preparation procedures aimed at the defined exposure of specific aspects of the corrosive attack for subsequent examination by standard surface analytical techniques, such as scanning electron microscopy (SEM), energy dispersive analysis (EDX) and X-ray diffraction (XRD).

Although the procedure of selective etching has been utilised successfully by many workers over several decades to reveal corrosion products by dissolving away the surrounding metal, its potential has not been exploited fully. This paper will report the development of a series of simple preparation procedures, which by combining mechanical abrasion with appropriate chemical and electrolytic etching techniques, enables complete flexibility in the way the external scale and internal attack may be exposed for surface analysis. These developments have been employed to characterise the nature of the attack, particularly that occurring internally, during the exposure of 20Cr/25Ni/Nb stabilised stainless steel to carbon dioxide-based environments at 900-1000°C.

EXPERIMENTAL

Material

The 20Cr/25Ni/Nb stainless steel, with the analysis (wt.%) 20.9 Cr, 25.0 Ni, 0.59 Nb, 0.75 Mn, 0.61 Si, 0.026 C and 0.005 N was tested in the standard 930°C, dry hydrogen annealed metallurgical condition. The specimens were sections of advanced gas cooled nuclear reactor fuel cans and were oxidised for either 3088 or 8094 h in CO_2 + 2%CO + 350 vpm CH_4 + 200 vpm H_2O at 900°C or for 120 h in CO_2 at 1000°C.

Mechanical Abrasion

A small EMCO Unimat 3 lathe was converted into a lapping machine by mounting a mandrel cradle on the cross slide with a micrometer attachment at the rear in line with the mandrel axis (Fig. 1). A small variable control stepper motor was coupled by a belt drive to the mandrel to permit rotation of the mandrel in a direction counter to that of the rotating lap.

For polishing, a small specimen section, typically 3.5 mm long x 2.5 mm wide, was mounted using a low melting point wax on a porous ceramic optical flat forming part of the mandrel head. Grinding was started on a 600 grit SiC lap and was continued with successively finer diamond laps to reduce the surface finish to 1 μm. The sample was then both released from the mandrel and cleaned in methylated spirits. For further reduction in thickness below 100 μm the specimen

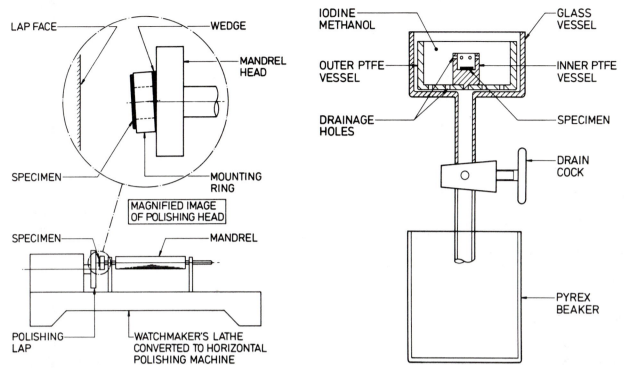

Figure 1: Technique for mechanical preparation of oxidised specimen section prior to selective etching.

Figure 2: Apparatus for extraction of internal precipitates following complete metal dissolution in iodine-methanol.

had to be more firmly secured. This was achieved using the support of a stainless steel ring (4 mm o.d., 2.8 mm i.d., 0.25 mm thick), whose surfaces were lapped and polished as described. The specimen was placed in the centre of the ring and welded at the four corners using a micro-tweezer welder. Further reduction of the specimen thickness could be continued as required. In certain instances it was desirable to produce a tapered wedge shaped profile across the specimen surface. This was accomplished by inserting a small piece of metal shim, ~ 50 μm thick, underneath one side of the ring so that the specimen assembly was fixed in a tilted position on the porous ceramic. Again lapping and polishing was continued until intermittent checking with an optical microscope revealed the emergence at the polished surface of the type of attack to be examined.

Selective Etching

Metal grains surrounding oxidised regions were dissolved chemically by several procedures. If desired, only a small portion of the surface could be etched by carefully masking the remainder with 'Lacomet' lacquer. For electropolishing, since constant monitoring was necessary, a special unit was constructed for this purpose. This consisted of a PTFE base block into which two small reservoirs were located, one to contain the electrolyte, Lenoirs solution (78 g CrO_3, 310 ml H_3PO_4, 67 ml H_2SO_4 and 120 ml H_2O) and the other, distilled water. A vertical PTFE column was also mounted onto the block with a moveable head through which was fitted a retractable arm attached to platinum tipped tweezers to hold the specimen for insertion into the electrolyte at an angle of about 30° to its surface. The tweezer arm allowed specimen movement also from one bath to the other. The reservoir containing the electrolyte was fitted with a platinum electrode screwed to the base block through which electrical contact was made. Power was supplied by a POLIPOWER unit. Since the electrolyte was partially opaque the specimen had to be held just beneath the surface so that visibility was unimpaired and this could be controlled with the tweezer arm. The etching process was carried out at 3 V with a 50 mA current typically for ~ 45 seconds and could be observed directly with a low power, optical microscope. At the completion of etching the specimen was transferred quickly to the distilled water bath. The masking lacquer was removed with acetone and the specimen was cleaned finally in distilled water and methylated spirits.

For SEM examination, a standard aluminium alloy stub was adapted to accommodate a ring assembly, thus avoiding the use of colloidal graphite normally employed for both fixing a specimen and providing electrical contact. For this purpose a recess was machined in the stub centre to accommodate the ring. The recess depth ensured that the uppermost surface stood proud of that of the stub. The ring was secured by two small screws and washers into the stub, one on either side of the ring.

Complete dissolution of metal to expose internal precipitates was effected using an iodine-methanol solution. The main problem was the small mass of the precipitates and the necessity of containing them within a small and defined volume. This was achieved with the apparatus shown in Fig. 2. The oxidised specimen, from which the outer scale and silica intrusions on both major faces had been removed by mechanical abrasion, was placed in the small inner PTFE vessel having a row of small drainage holes beneath the top rim. This vessel had a spiggot protruding from its underside for location within a larger PTFE vessel with bottom drainage holes and this in turn was loaded into a glass vessel fitted with a drainage cock. The glass and PTFE vessels were filled with iodine-methanol solution. After 24 h the solution was drained and depending on whether or not metal dissolution was complete, iodine-methanol solution was replenished or residual iodine was removed by flushing a number of times with methanol. After cleaning the precipitates were transferred using a fine paintbrush to a silicon wafer for XRD, these being held in position by a drop of cyclo-hexane and for SEM to a standard stub covered with a self-adhesive label with the adhesive side uppermost.

RESULTS AND DISCUSSION

Optical microscopy and fractography [1,2] has revealed that the attack of 20Cr/25Ni/Nb stainless steel in CO_2 based environments at 900-1000°C is characterised by the formation of an external scale consisting of an outer FeCrMn spinel and an inner Cr_2O_3 layer, beneath which are silica intergranular instrusions into the steel and often internal precipitates. The attack is shown schematically in Fig. 3. To reveal specific aspects of the microstructure and composition of the different areas of the attack specimen preparation procedures were developed to examine:

(a) the underside of the scale and the intrusions; (b) a shallow taper section through the intrusions; (c) the distribution of internal precipitates through the alloy, and (d) the morphology and composition of the internal precipitates.

For examination of the underside of the scale the oxidised 20Cr/25Ni/Nb stainless steel was abraded from one major face to just above the silica intrusions on the face to be examined. This was achieved by support on a stainless steel ring, as described. Only a small area of the specimen was electrochemically etched in Lenoirs solution, so that the surrounding unattacked steel supported the oxide scale window exposed. A scanning electron micrograph of this region (Fig. 4) shows at the top left-hand corner the edge of unetched steel. The silica intrusions proved to be platelets, which grew from the

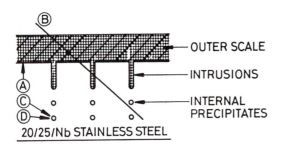

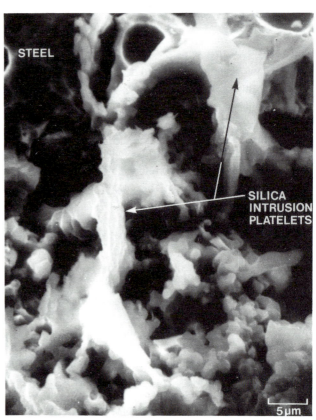

Figure 3: Specific aspects of the microstructure and composition of the attack of 20/25/Nb steel to be revealed.

Figure 4: Scanning electron micrograph of the underside of scale formed on 20/25/Nb steel following 120 h oxidation in CO_2 at 1000°C.

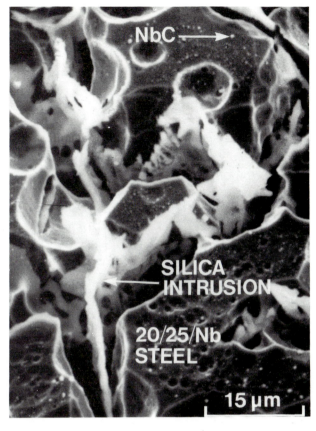

Figure 5: Silica intrusions formed during 120 h oxidation at 1000°C at 20/25/Nb steel grain boundaries.

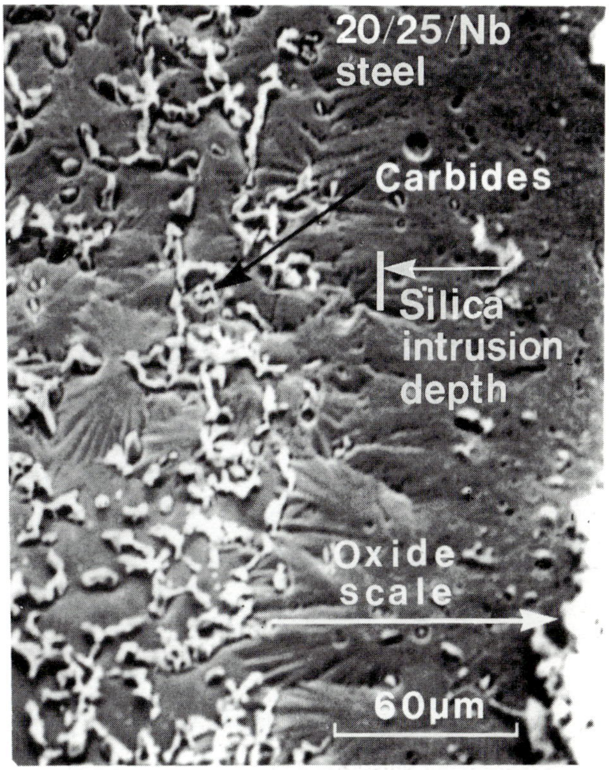

Figure 6: Distribution of precipitates in 20/25/Nb steel beneath the silica intrusions following 8094 h oxidation in a CO_2 based environment at 900°C.

bottom-most layer of the external scale with a globular structure. EDX suggested these could be essentially coated particles consisting of silica overlaying chromia.

It was possible that a finer and more fragile structure at the intrusion extremities could have been detached during electrochemical etching. To facilitate examination of the intrusion structure throughout its depth the thickness of this region was magnified by preparing a taper section. The external scale was first removed from one side of the oxidised specimen, which was then welded onto a support ring with the intact oxidised surface underneath. Using the thin wedge, as described, lapping and polishing was continued until a taper section fully exposed the external scale and intergranular silica intrusions. A short electropolish removed sufficient steel at the intrusion interface to reveal the feather-like structure of their extremities (Fig. 5). The small white particles revealed were Nb rich and probably NbC.

The distribution of the internal precipitates was exposed by etching in aqua regia a transverse cross-section of the oxidised steel. A scanning electron micrograph of a region from the base of the silica intrusions is shown in Fig. 6. Small Nb rich particles, varying in size, were apparent in the steel grains at about the same density across the section. Larger Cr rich particles formed a continuous network in the grain boundaries of the steel underlying the silica intrusions.

Extraction of the internal precipitates was achieved firstly by abrasion from both specimen faces of the external scale and metal to the depth of the silica intrusions followed by complete dissolution of the remaining steel in iodine-methanol. The X-ray diffraction pattern of the particles indicated that they were $Cr_{23}C_6$ and NbC. A scanning electron micrograph (Fig. 7) revealed that the former had an angular and often well defined crystalline structure, with smaller NbC particles attached to the outer surfaces.

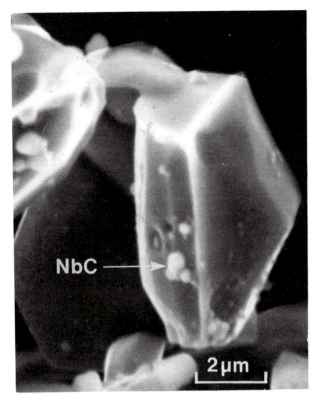

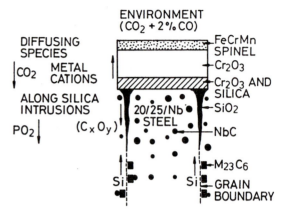

Figure 7: Scanning electron micrograph of Cr23C6 and NbC particles formed intergranularly within 20/25/Nb steel during 3088 h oxidation in a CO_2 based environment at 900°C.

Figure 8: Mechanism of internal oxidation of 20/25 Nb steel in CO_2 based environments at 900-1000°C.

The microstructural evidence would suggest that under these oxidation conditions internal attack with the formation of silica intrusions and carbide precipitates involves inward movement of the oxidant, CO_2, in addition to the outward movement of cations. Taken in conjunction with other observations [3,4], a probable mechanism for the internal attack is shown schematically in Fig. 8. The oxygen partial pressure (p_{O_2}) decreases with increasing depth so that the oxides form in order of increasing thermodynamic stability. As in crevice-type corrosion, as the p_{O_2} diminished along the silica intrusions so the associated carbon activity (a_c) permitted carbon formation, dissolution into the steel, and precipitation as carbides. Reaction would have occurred initially with extraneous niobium (i.e. that not already combined with carbon and nitrogen). Over the depth of the silica intrusions the steel composition would have been depleted in chromium by an extent decreasing with depth from the underside of the external scale. Only beneath the intrusions was the chromium activity (a_{Cr}) sufficiently high to react with extraneous carbon to form $C_{23}C_6$. The precipitation of these carbides in the steel grain boundaries would suggest that these were the primary transport paths for carbon ingress.

CONCLUSIONS

1. Simple procedures, based on a combination of mechanical abrasion and selective etching, provide complete flexibility for specimen preparation for the physical and chemical characterisation of internal attack of alloys by standard surface analysis techniques such as SEM, EDX and XRD.

2. Use of these procedures has provided a mechanistic understanding of the internal oxidation of 20Cr/25Ni/Nb stainless steel in CO_2-based environments at 900-1000°C.

ACKNOWLEDGEMENT

This study was part of the longer term research carried out within the Underlying Programme of the United Kingdom Atomic Energy Authority. We are grateful to Dr. A.M. Emsley (National Power - Leatherhead) for providing some of the oxidised specimens and to our colleagues B.A. Bellamy and M.D. Fones for the XRD.

REFERENCES

1. J.A. Desport and M.J. Bennett, Oxid. Met., 29,1988, p. 327.

2. R.C. Lobb and H.E. Evans, Materials for nuclear reactor core applications, British Nuclear Energy Society, London, 1987, p. 335.

3. M.J. Bennett, Proc. of the 10th International Congress on Metallic Corrosion, Oxford and IBH Publishing Co. PVT Ltd., New Delhi, 4, 1987, p. 3761.

4. M.J. Bennett, A.C. Roberts, M.W. Spindler and D.H. Wells, UKAEA Harwell Laboratory Report AERE-R13685 (1989).

44 OBSERVATION OF OXIDE/METAL INTERFACE USING SCANNING AUGER MICROSCOPY WITH AN *IN SITU* SCRATCH TECHNIQUE

P. Y. Hou and J. Stringer*

Materials and Chemical Sciences Division, Lawrence Berkeley Laboratory, Berkeley CA 94720, USA.
*Electric Power Research Institute, Palo Alto CA 94304, USA.

ABSTRACT

The chemistry of the scale/metal interface of non-adherent oxides has been investigated and related to its microstructure. Al_2O_3 or Cr_2O_3 scales were formed at elevated temperatures in oxygen on Fe-18wt%Cr-5wt%Al and Ni-25wt%Cr alloys. Newly exposed interfaces were made by scratching the oxide with a diamond stylus located in the vacuum chamber of a scanning Auger microprobe. The scratching operation often caused the scales to fracture and exfoliate at the scale/metal interface. On both the Al_2O_3- and the Cr_2O_3-forming alloys, sulfur was found to be present everywhere on the metal surface. This sulfur coverage was no more than 3 monolayers thick. On areas where the scale had lost contact with the metal at temperature, the sulfur was embedded in a thin oxide layer formed during cooling of the specimen.

INTRODUCTION

The protection of high temperature alloys and coatings against oxidation is provided by the formation of slow-growing oxide films which often contain Al_2O_3 and/or Cr_2O_3. One major limiting factor in the protection is the tendency of the oxides to spall under thermal cycling conditions. The degree of spallation depends on a large number of factors, for example: the thermal expansion mismatch of the alloy and the oxide, the oxide thickness, the cooling rate, the mechanical properties of the oxide and the alloy, the growth processes of the oxide scale, and the strength of the oxide/alloy interface. It has often been assumed that the oxide/metal interface found on Al_2O_3- and Cr_2O_3-forming alloys is weak because spallation often occurs there in practical situations. Funkenbusch et al. [1-2] however, were the first to suggest that the interface is in fact intrinsically strong, but is weakened by an impurity such as sulfur which segregates from the bulk of the alloy to the interface. The effect of reactive element addition [3], which greatly improves the scale adhesion in the alloy, was suggested to prevent such segregation.

Although segregation of sulfur to alloy surfaces has been repeatedly demonstrated with in situ heating in a vacuum chamber [1,2,4,5], a clear indication that sulfur or other impurities are present at the scale/metal interface of non-adherent oxide scales is lacking. Smeggil et al. have reported [2,6] slight sulfur enrichment at the scale/metal interface using electron microprobe analysis after extensive searching on metallographically polished cross-sections. However, if the sulfur, which segregated from the bulk of the alloy to the interface, was anticipated to be present as chemically adsorbed layers, one would not expect the resolution of the microprobe to detect such segregation. More likely then, the microprobe analysis detected sulfide particles which were located at or near the interface. Smeggil and Peterson [5] using energy-dispersive X-ray analysis (EDX) on stripped Al_2O_3 scales have found some presence of sulfur on the underside of the oxide, but no sulfur was detected on the metal side. Kim et al. [7] have detected a high sulfur concentration at the interface between Cr_2O_3 and Fe-25wt%Cr alloy using analytical electron microscopy. However, much of this sulfur may have migrated through the oxide grain boundaries from the oxidizing atmosphere which contained a small amount of H_2S.

In this study, the chemistry of the scale/metal interface of non-adherent oxides has been investigated and related to its microstructure. The oxide scales, Al_2O_3 or Cr_2O_3, were formed at elevated temperatures in oxygen on Fe-18Cr-5Al and Ni-25Cr alloys. Newly exposed interfaces were made by scratching the oxide with a diamond stylus located in the vacuum chamber of a scanning Auger microprobe.

EXPERIMENTAL METHODS

Both the Fe-18Cr-5Al and the Ni-25Cr alloys were made by induction melting and casting from high purity materials

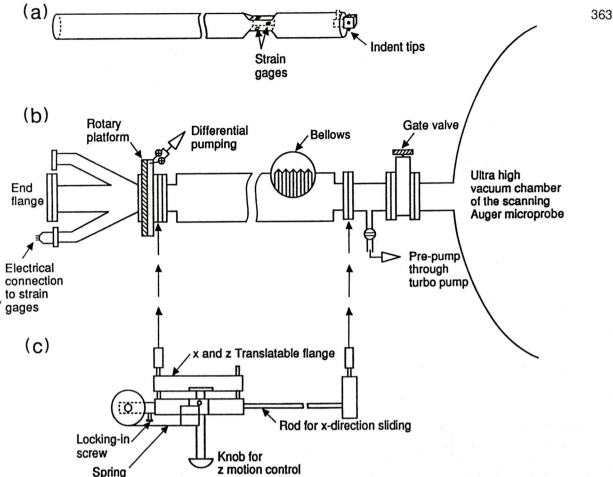

(a) Strain gages · Indent tips

(b) Rotary platform · Differential pumping · Bellows · Gate valve · Ultra high vacuum chamber of the scanning Auger microprobe · End flange · Electrical connection to strain gages · Pre-pump through turbo pump

(c) x and z Translatable flange · Rod for x-direction sliding · Locking-in screw · Spring · Knob for z motion control

Figure 1: Schematic illustration of the Friction, Hardness and Adhesion Tester within the Scanning Auger Microprobe (FHATSAM).

under an argon atmosphere. Subsequently the ingots were homogenized in vacuum at 1100°C for 24 hours. The average grain size for the Ni-25Cr was 1.5 mm x 0.4 mm x 0.3 mm and for the Fe-18Cr-SAl was 0.83 mm x 0.48 mm x 0.48 mm. Specimens 15 mm x 10 mm x 1 mm were cut from the center of the ingots. Prior to the oxidation treatments the specimens were polished to 600 SiC grit and ultrasonically cleaned.

Oxidation tests were carried out in dry oxygen at 1 atm total pressure in a horizontal tube furnace. Specimens in alumina crucibles were placed at one end of a closed mullite reaction tube while the system was being equilibrated with flowing oxygen. Oxidation was initiated by pushing the crucibles into the hot zone of the furnace. After a desired time, the crucibles were quickly removed from the furnace and capped, and the specimens were allowed to cool to room temperature in air. If cooling had caused any noticeable spallation, the specimen surface was lightly gold-coated prior to its observation under the scanning Auger microprobe (SAM). This procedure separated the interfaces exposed during cooling from those uncovered by the in-situ scratching technique.

The apparatus used for the in-situ scratching experiments was a device designed by the research group of Prof. G. A. Somorjai of UC Berkeley, and fabricated at LBL. This equipment, called the FHATSAM (Friction, Hardness and Adhesion Tester within the Scanning Auger Microprobe), can be used to measure mechanical properties in ultra high vacuum (UHV), and to expose interfaces for subsequent Auger and SEM (scanning electron microscopy) analysis. A schematic diagram of the apparatus is illustrate in Fig. 1. In Fig. 1(a) is shown the body of the FHATSAM which is a half-inch rod with a narrowed neck region designed to hold up to a maximum of 2 kg. Two pairs of strain gages are mounted normal to each other on this neck region. Located at the front end of the rod is a multiple tip holder with a Knoop, a Rockwell and a Vickers tip. During a scratch motion, the strain on the rod in the horizontal and the vertical directions can be measured. The usual noise level is about 50 g, mainly due to vibration of the rod. The strain gages are calibrated by comparing indentations made on polycrystalline Cu with those made using a commercial micro-hardness tester.

The vacuum assembly housing the rod is shown in Fig. 1(b). The end of the rod is fastened at the end flange. Five inches away from the end flange the rod is clamped to a rotary platform. The entire end assembly, from the end flange to the rotary platform, can be rotated with respect to the SAM. The seal between this rotating end and the rest of the assembly can be pumped differentially during operation. The entire system is initially isolated from the UHV through the gate valve. After pre-pumping through a turbo pump the gate valve opens the FHATSAM to the UHV. The rod with the indenting tips can then be pushed in toward the specimen surface by compressing the bellows. A translatable flange unit (Fig. 1(c)) is fixed at the ends of the bellows. This unit provides a mechanism by which the FHATSAM can be locked in position at any point along the x-direction. It also provides some degree of control in the z-direction, so that the indenter can be positioned closer or further away from the specimen surface. The translatable flange is spring-loaded to protect the FHATSAM from any accidental forward movement which may bring damage to itself or to the SAM. Once the FHATSAM is in position, subsequent motion during indentation and/or scratching is achieved by manipulating the specimen stage control of the SAM. Normal chamber pressure during a FHATSAM operation is in the range of 10^{-9} torr. Within the UHV chamber is a PHI 660 SAM equipped with a secondary ion mass spectrometer and an ion gun for sputtering.

Loads ranging from 200 g to 1000 g were usually used on the oxidized specimens using the Vickers tip. After positioning the tip, the specimen was raised manually using the z-control of the sample stage of the SAM until the desired load is reached. The specimen was then held at this load for 10 seconds before relief. If a scratch was intended, the load would be held for 5 seconds, and the specimen moved manually 0.6 mm in the y-direction within 5 seconds, held for another 5 seconds, then released. The entire process of achieving and measuring the loading of the FHATSAM is presently being improved. However, for the purpose of mainly exposing the scale/metal interface using the device, the present set-up is sufficient.

RESULTS AND DISCUSSIONS

Figure 2(a) shows a typical scratch mark made on the surface of a Fe-18Cr-5Al alloy. The FeCrAl specimen was oxidized at 1100°C isothermally for 50 hours. More than half of the Al_2O_3 scale spalled upon cooling. Scratching through a piece of remaining oxide usually cause it to exfoliate at the scale/metal interface. The surface from which the scale had spalled during cooling appears light because of the presence of the coated gold layer; the darker region corresponds to the surface from which the scale had been removed by the scratching operation. A magnified view of the interfacial area is shown in Fig. 2(b). As with most Al_2O_3/metal interfaces, it consists of voids (smooth regions), which are believed to correspond to areas where the oxide had lost contact with the metal at temperature, and regions of oxide imprint where there was oxide-to-metal contact at temperature. Auger analyses of the void and the imprinted areas are presented in Figs. 2 (c) and (d) respectively. On neither surface was Au present, whose major Auger excitations are at 2024 and 2111 eV. This indicates that the surface being analyzed was indeed freshly exposed in the UHV. A strong oxygen signal and some carbon were present on both areas, which may have arisen from the residual gases in the UHV. S and small amounts of Cl were found on the imprinted area, but not on the void area. The imprinted area also showed the presence of Cr and Al, but the void area did not.

The difference in chemistry of the void area and the imprinted area was studied further using SAM, and the results are shown in Fig. 3. As seen in Fig. 3(c), depth profiles through one of the imprinted areas (3) show a concentrated S peak at the metal surface. The thickness of this S layer is only about 12 A, which is no more than 3 monolayers. On the void area (2) as well as the second imprinted area (1) the surface was found to be covered with an oxide layer between 160-200 A thick. This oxide consisted of an outer layer containing only Fe and an inner layer containing Fe, Cr and Al. Within the inner layer and immediately ahead of the metal surface a broad S peak was detected. Analyses of a number of areas at the interface showed that this type of oxide layer exists on all of the void areas and on 1/3 of the oxide-imprinted areas. The remaining 2/3 of the imprinted areas were true metal surfaces with a sharp S peak present on its surface; only a few of these surfaces were found to also contain a small amount of Cl.

The thin, two-layered oxide is believed to be formed during cooling of the specimen. Since the voids were areas where the scale had lost contact with the underlying metal at temperature, the scales above the voids would crack more easily under the imposed thermal stress, thus allowing the surfaces of the voids to be slightly oxidized. Evidently, on some of the areas where the scale was in contact with the metal at temperature it also became detached during cooling, allowing the same kind of slight oxidation to take place on some of the oxide-imprinted areas. Once this thin oxide is removed, sulfur is found everywhere at the scale/metal interface, and the amount is the same on the imprinted areas and on the void surfaces. A thickness of no more than 3 monolayer was found. This shows that the sulfur was present as a chemisorbed layer rather than as a sulfide. When the oxidation of the metal surface took place during cooling, the sulfur on the surface instead of reacting with atmospheric oxygen to form volatile compounds, became embedded in the thin oxide layer ahead of the metal surface.

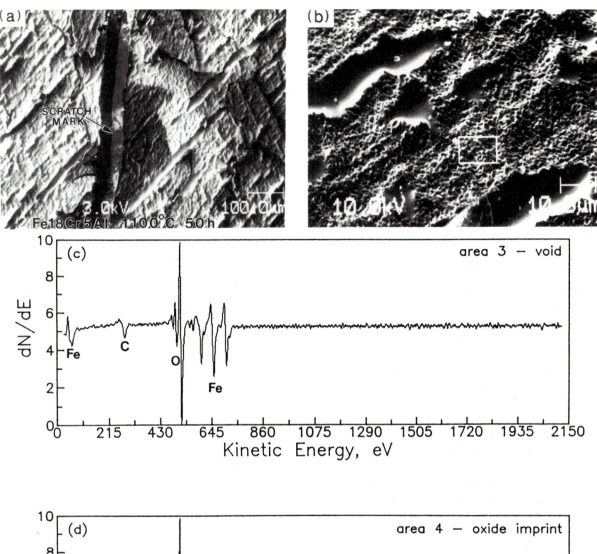

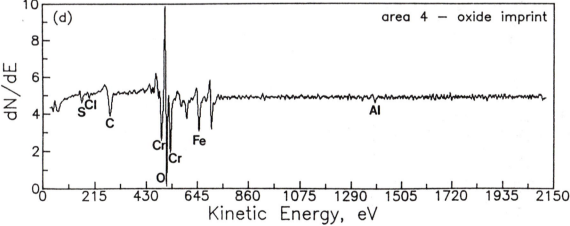

Figure 2: (a) SEM micrograph of an 800 g scratch made on a Fe-18Cr-5Al alloy oxidized at 1100°C for 50 hours. (b) Newly exposed interfacial area after oxide exfoliation caused by the scratch. (c)-(d) AES analysis of areas 3 and 4 on (b).

The presence of sulfur on metal/oxide interfaces was also found on the Cr_2O_3-forming Ni-25Cr alloy after a short time of oxidation. At the lower left hand corner of Fig. 4(a) is shown part of an indentation mark made by a Vickers indenter under 350 g load. The Ni-25Cr specimen was oxidized at 1000°C for only 20 minutes. A complete layer of Cr_2O_3 formed under these oxidation conditions, and the scale showed no noticeable spallation upon cooling. As seen in Fig. 4(a), the indentation caused some scale fractures adjacent to the indentation mark. SAM mappings of S, Cr and Ni are shown in Figs. 4(b-d). It is seen that the fractures occurred at the scale/metal interface, exposing the Ni substrate; furthermore, the presence of S is closely related to where the scale/metal interface had been exposed.

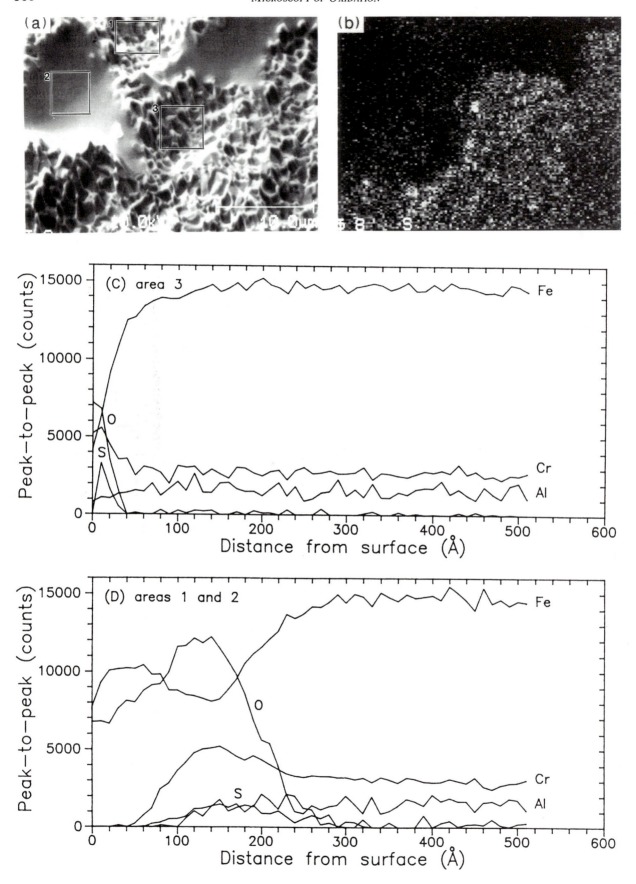

Figure 3: (a) Magnified SEM image of the metal surface shown in Fig. 2(b). (b) SAM sulfur map of the area shown in (a). (c)-(d) AES depth profiles of areas 1-3 on (a).

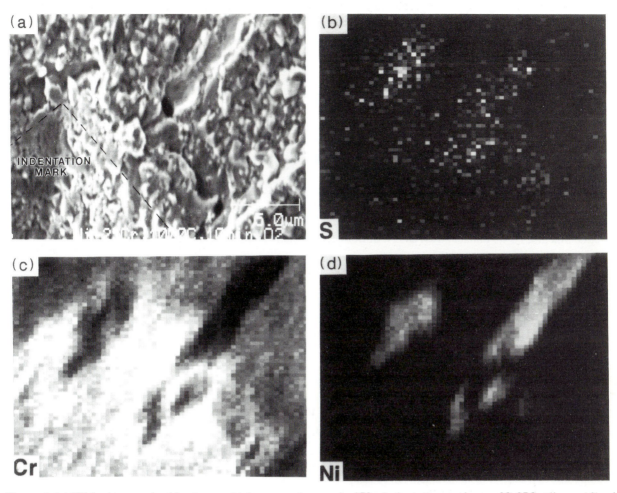

Figure 4: (a) SEM micrograph of fractures which occurred around a 350 g indentation made on a Ni-25Cr alloy oxidized at 1000°C for 20 minutes. (b)-(d) SAM mappings of S, Cr and Ni respectively.

The present results have clearly demonstrated the presence of sulfur on the metal/scale interface of non-adherent Al_2O_3 and Cr_2O_3 scales, and the source of this sulfur is the indigenous sulfur in the alloy instead of surface contamination. However, further study of the chemistry at the scale/alloy interface of adherent scales is needed in order to determine whether sulfur at the interface does indeed weaken it, and if the role of reactive elements is to prevent such segregation to provide a stronger interface. Hussey et al[8] have used SIMS to study the composition of oxide scales formed on Fe-20wt%Cr and Fe-20wt%Cr-0.078wt%Ce alloys, and have found the presence of sulfur near the scale/metal interface in both alloys. The sulfur was found to be concentrated at the interfacial region of the Ce-containing alloy, but was dispersed throughout the Cr_2O_3 scale on the Fe-20Cr alloy. The total amount of sulfur increased slightly with oxidation time and its concentration was only slightly lower for the Ce-containing alloy. Both alloys developed adherent scales regardless of the presence of S near the interface. These results suggest that (i) the presence of S at the scale/metal interface is not a strong factor in determining spallation behavior of the scale, and (ii) the presence of reactive element in the alloy did not significantly eliminate sulfur segregation to the scale/alloy interface. Work is presently being conducted to study the chemistry of the scale/metal interface of adherent Al_2O_3 and Cr_2O_3 scales, particularly those formed on alloys with reactive element additions.

CONCLUSIONS

Indentation or scratches made on non-adherent Al_2O_3 or Cr_2O_3 scales often causes the scales to fracture and exfoliate. The fracture is usually located at the scale/metal interface. In-situ scratching under ultra high vacuum and subsequent studies of the interface revealed much about the interfacial chemistry. On both the Al_2O_3- and Cr_2O_3-forming alloys, sulfur was found to be everywhere at the interface. The sulfur coverage was no more than 3 monolayers thick. A few areas were also found to contain chlorine, but its distribution was not as widespread as that of sulfur. Surfaces which appeared "smooth" under the SEM often did not represent the true metal surface. An oxide layer approximately 200 A thick which was formed during specimen cooling was present on all the void areas, where the scale had lost contact with the metal at temperature, and on some of the oxide-imprinted areas, where the scale was in contact with the metal at temperature. Future work needs to concentrate on the chemistry of adherent scales in order to truly understand the role of sulfur in scale adhesion and the role of reactive elements in relation to sulfur segregation at scale/metal interfaces.

ACKNOWLEDGEMENTS

The authors would like to thank Mr. Gilroy Vandentop for his instructions in operating the FHATSAM equipment. This work was supported by the Electric Power Institute under contract No. RP 2261-1, through an agreement with the U. S. Department of Energy under Contract No. DE-AC03-76SF00098.

REFERENCES

1. A. W. Funkenbusch, J. G. Smeggil and N. S. Bornstein, Metall. Trans., Vol. 16A, 1985, p. 1164.

2. J. G. Smegill, A. W. Funkenbusch and N. S. Bornstein, Metall. Trans., Vol. 17A, 1986, p. 923.

3. D. P. Whittle, and J. Stringer, Phil. Trans. Roy. Soc. London, Vol. A27, 1979, p. 309.

4. Krishan L. Luthra and Clyde L. Briant, Oxid. of Metals, Vol. 26, 1986, p. 397.

5. J. G. Smeggil and G. G. Peterson, Oxid. of Metals, Vol. 29, 1988, p. 103.

6. J. G. Smeggil, "Corrosion and Particle Erosion at High Temperatures" Eds. V. Srinivasan and K. Vedula, TMS pub., 1989, p. 403.

7. Y.-K. Kim, K. Przybylski and G. J. Yurek, Proc. Symp. Fundemental Aspects of High Temp. Corr.-II, p. 259, Eds. D. A. Shores and G. J. Yurek, Electrochem. Society, Boston, May 4-9, (1986).

8. R. J. Hussey, P. Papaiacovou, J. Shen, D. F. Mitchell and M. J. Graham, "Corrosion and Particle Erosion at High Temperatures", Eds. V. Srinivasan and K. Vedula, TMS pub., 1989, p. 567.

45 A COMPARISON OF *IN SITU* AND *EX SITU* OXIDATION PROCESSES FOR ZnTe AND CdTe SURFACES AS STUDIED BY HIGH-RESOLUTION ELECTRON MICROSCOPY

Ping Lu and D. J. Smith

Center for Solid State Science and Department of Physics, Arizona State University, Tempe AZ 85287, USA.

ABSTRACT

The *in situ* and *ex situ* oxidation of ZnTe and CdTe surface has been studied by high-resolution electron microscopy. The *in situ* oxidation occurred as a result of electron irradiation within the electron microscope and was found to involve an initial amorphization of the surface of the bulk crystal, followed by the formation of small crystallites of the cation oxide. It appeared that the anion species were being desorbed from the surface by a non-thermal electron-stimulated process. The *ex situ* oxidation simply involved annealing in air, with subsequent observations by electron microscopy. Differences between the materials were seen in the *ex situ* studies. In the case of ZnTe, a layered sequence of ZnTe/Te/ZnO was observed, with the large Te crystals(up to 1000Å across) invariably in an epitaxial relationship with the bulk material, and the smaller ZnO crystal(~50Å) at the surface in random orientations. The formation of an epitaxial TeO_2 layer immediately adjacent to the bulk CdTe crystal was, however, observed in the *ex situ* CdTe oxidation studies.

INTRODUCTION

The growth of an oxide layer on the surfaces of a semiconductor can easily occur during device fabrication, and these layers can have a strong influence on the optical and electrical behavior of the semiconductor device. The semiconductor materials CdTe and ZnTe have important applications such as thin-film solar cells and IR detectors. Surface oxidation of these materials has previously been studied using various techniques such as low-energy electron-loss spectroscopy and photoemission spectroscopy [1-4], Auger electron spectroscopy [4], as well as high-resolution electron microscopy (HREM) [5-6]. In these studies, the oxide films were produced by thermal oxidation of semiconductors in oxygen or air under a variety of temperature and pressure combinations.

Under the high electron current density conditions which are normally employed for high-resolution imaging(~5-20A/ cm^2), reaction of the electron beam with the exposed specimen surfaces is often found to take place. The growth of surface oxide layers on several In-compound semiconductors inside the electron microscope during exposure to the electron beam has been previously described [7]. In the present study, the surface oxidation of ZnTe and CdTe *in situ* has been induced by electron beam irradiation within the electron microscope and studied by high-resolution electron microscopy. The results of the oxidation are compared with those which resulted from the *ex situ* oxidation processes which are induced by heating the crystal in air. Further details of the oxidation studies for ZnTe have been reported elsewhere [8].

EXPERIMENTAL

Small amounts of ZnTe and CdTe crystals were crushed under purified methanol to form a fine powder which was then deposited on holey carbon support films. In order to avoid, or at least to minimize, surface contamination, those specimens of ZnTe and CdTe intended for *in situ* oxidation were immediately transferred to the microscope airlock where they were dried out during evacuation. Samples of ZnTe and CdTe for the *ex situ* oxidation studies were heated in air at 260°C for various periods of time before transferral into the microscope for observation. The microscope used for observations was a JEM-4000EX which had a vacuum pressure of ~10^{-7} torr in the vicinity of the specimen. This microscope had an interpretable resolution of better than 1.7Å when operated at 400keV. The image pickup and viewing system attached to the microscope made it possible to observe dynamic surface processes in real time. Optical diffractograms (ODMs) from the electron micrographs were used to identify the phases present in the surface region, with the lattice spacing of the bulk crystal as an internal reference.

RESULTS AND DISCUSSION

In Situ Observations

The specimens of ZnTe and CdTe prepared for *in situ* studies had either clean, or nearly clean, surfaces (a thin amorphous layer of perhaps 5-10Å thickness). Oxidation of the surfaces usually took place following irradiation by an intense 400keV electron beam. Figures 1 and 2, respectively, show the ZnTe and CdTe crystals aligned with [110] parallel to the incident electron beam direction, after exposure for several minutes to a beam with a current density of about 20A/cm^2; the corresponding ODMs are included as the insets. The small crystals visible in the surface region of Fig.1 have been formed during the electron irradiation, and were identified as hexagonal ZnO by using ODMs from the small crystals and the nearby bulk crystal, with the lattice spacing of ZnTe as an internal reference. These oxide crystals, which have sizes ranging approximately from 20 to 100Å, clearly have no definite orientation relationship with the bulk crystal. The surfaces of the CdTe crystal shown in Fig. 2 were originally covered by a few layers of amorphous material. The small crystals which developed during the electron irradiation could be identified as cubic CdO, again by using the corresponding ODMs.

By using the image pick-up and viewing system attached to the microscope, the gradual development of the oxides (ZnO and CdO respectively) during the exposure of ZnTe and CdTe to the electron beam could be directly followed. It was observed that a similar oxidation process occurred for both materials. It appeared that the edges of the crystal were gradually amorphized under the electron beam and that small crystals of the corresponding cation oxide then nucleated and grew out of these areas. In the process, however, no Te or Te oxides were observed, the local crystal thickness has gradually decreased, and small holes were visible in some places. This result indicates that the Te species are being desorbed during the electron beam irradiation. Further observations with changes of the beam current density by more than a factor of three revealed that the oxidation rate had only a marginal dependence on the current density, suggesting that the oxidation process was basically non-thermal.

There are several possible mechanisms for removing the Te species from the ZnTe and CdTe crystals under electron beam irradiation. For example, the Te species could be removed by direct momentum transfer or electron-stimulated desorption processes [9]. Since the threshold energy for the direct atomic displacement of Te in crystalline ZnTe and CdTe is above 200keV [10,11], we would not expect any substantial loss of Te to occur at 200keV if indeed the loss was only due to direct moment transfer. Our observations revealed, however, that the same basic oxidation process took place even when the microscope was operated at 200kV. It therefore appears that the loss of Te is not due to a direct momentum transfer process but due to an electron-simulated desorption mechanism which would probably actually increase at lower energies. The *in situ* oxidation process seems to involve amorphization of the crystal, accompanied by the removal of the Te species from the crystal and followed by oxygen diffusion to form the cation oxides. The oxidation process seems to be very similar to that previously reported for In III-V compound semiconductors [7].

Ex Situ Oxidation

The samples of ZnTe and CdTe were heated at 260°C in air for various periods of time and subsequently observed in the microscope. In order to avoid any effects which might be induced by electron irradiation, images were usually recorded from areas without significant prior exposure to the electron beam. For ZnTe, it was observed that the extra phases formed during the annealing treatments were mostly crystals of ZnO and Te, as identified by the corresponding ODMs. The surface oxidation regions usually formed a layer structure in the sequence ZnTe/Te/ZnO, with Te metal always in direct contact with the bulk ZnTe and small ZnO crystallites near the crystal surface in random orientations. This layering of the oxidized region was observed for crystals heated for only 10 minutes as well as for crystals heated for more than an hour. Figure 3 shows a ZnTe crystal, heated for 100 minutes, which displays the typical layer structure. The Te metal visible in Fig. 3 has an epitaxial relationship with the bulk ZnTe such that $(111)[1,-1,0]_{ZnTe}//(011)[1,-1,1]_{Te}$.

In the case of the CdTe sample, heating at 260°C for less than 30 minutes usually resulted in oxide phases which could not be identified from their lattice spacings. Crystals of TeO$_2$, which has an orthorhombic structure [12], were, however, found for samples heated for a longer time. For example, Fig. 4 shows an crystal which had been heated for 200 minutes, together with its corresponding optical diffractogram. This crystal of TeO$_2$ is characterized by its two sets of {121} planes with spacings of 3.28Å and an angle of 72.5° between planes in different sets. The oxide is in a well-defined epitaxial relationship with the bulk CdTe as shown in the inset which is the optical diffractogram obtained near the interface. The epitaxial relation between oxide and bulk is $(111)[1,-1,0]_{CdTe}//(121)[0,1,-2]_{TeO_2}$. The 13% difference in the spacings ($d_{(111)CdTe}$=3.74Å, $d_{(121)TeO_2}$=3.28Å) is accommodated at the interface by adding atomic planes on the TeO$_2$ side.

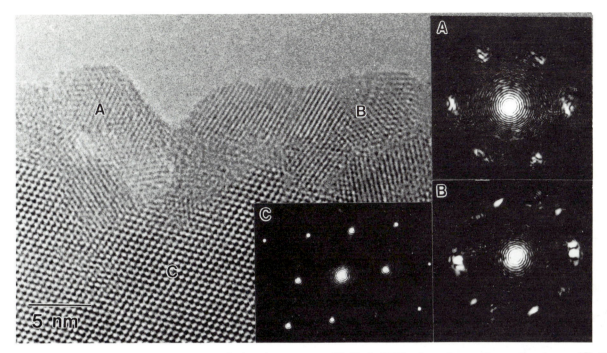

Figure 1: High-resolution electron micrograph showing an area of ZnTe, in [110] orientation, with a surface layer of ZnO (areas A, B) which has developed during irradiation by the incident electron beam for several minutes. The insets show ODMs from areas A, B and C(ZnTe) used for analysis of lattice fringe spacings.

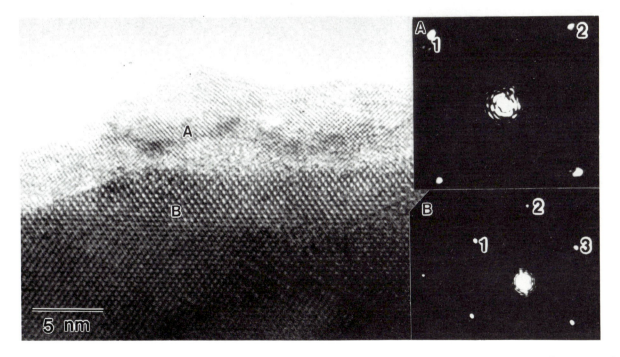

Figure 2: High-resolution electron micrograph showing an area of CdTe, in [110] orientation, with surface layer of CdO(area A) which developed during irradiation by the incident electron beam for several minutes. The insets show ODMs from areas A and B(CdTe) used for analysis of lattice fringe spacings.

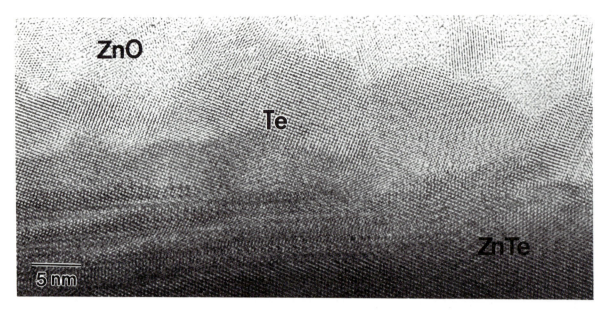

Figure 3: High-resolution electron micrograph showing part of ZnTe crystal which had been pre-heated in air at 260°C for 100 minutes. Note the layering of the oxide region (ZnTe/Te/ZnO) and the well-defined epitaxial relation of ZnTe and Te.

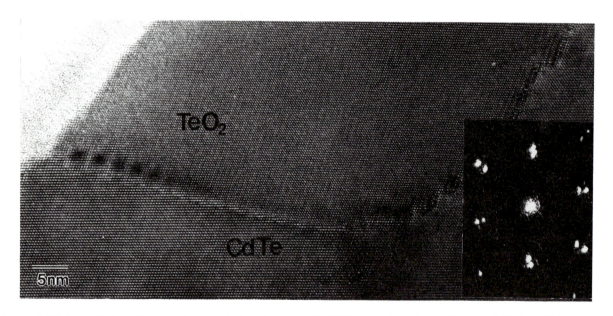

Figure 4: High-resolution electron micrograph showing part of a CdTe crystal pre-heated in air at 260°C for 200 minutes. The inset shows the corresponding ODM from the interface of the oxide (TeO$_2$) and the bulk CdTe. Note the well-defined epitaxial relationship between the oxide and the bulk CdTe.

The *ex situ* oxidation is strictly a thermal process. For ZnTe, the layer sequence of ZnTe/Te/ZnO observed in the surface region suggests that the oxidation process involves the diffusion of Zn atoms to the surface with subsequent formation of ZnO by reaction with oxygen. The oxidation process for CdTe involves the diffusion of Cd atoms to the surface while oxygen diffuses in through the oxide and substitutes for Cd in the sample. The formation of ZnO and Te but not TeO$_2$ in the case of ZnTe, and the formation of TeO$_2$ but not CdO in the case of CdTe is most likely due to the differences in the heats of formation of the respective oxides(for TeO$_2$, ZnO and CdO, these are -77.3Kcal/mol, -83.2Kcal/mol and -61.6Kcal, respectively [13], so that ZnO is more stable than TeO$_2$, and TeO$_2$ is more stable than CdO) although it may also be affected by the relative diffusion rates of metal cations through the surface. The surface layering observed here for the oxidized ZnTe crystals has also been reported in the oxidation of other compound semiconductors such as GaAs, GaSb [14], InAs and InSb [15,16], with the Group V species (As,Sb) located at the semiconductor-oxide boundaries.

CONCLUSION

High-resolution electron microscopy has been used to compare the initial oxidation of CdTe and ZnTe crystals as induced by electron irradiation inside the microscope with that due to annealing in air. The *ex situ* results indicate that the technique can provide useful insights into the development of layering as the oxidation reaction proceeds. However, it needs to be appreciated that the *in situ* studies, whilst interesting, are substantially different. Careful distinctions should thus always be made in high-resolution studies of *ex situ* and *in situ* oxidation reactions because of the likelihood that the latter could involve extraneous artifactual information.

ACKNOWLEDGEMENT

This work has been supported by NSF Grant DMR-8514583.

REFERENCES

1. A. Ebina, K. Asano, Y. Suda and T. Takahaski, J. Vac. Sci. Technol. 17, 1980, p.1074.

2. U. Solzbach and H.J. Richter, Surf. Sci. 97, 1980, p.191.

3. T.L. Chu, S.S. Chu and S.T. Ang, J. Appl. Phys. 58, 1985, p.3206.

4. F. Wang, A. Schwartzman, A.L. Fahrenbruch, R. Sinclair and R.H. Bube, J. Appl. Phys. 62, 1987, p.1469.

5. F.A. Ponce, R. Sinclair and R.H. Bube, Appl. Phys. Lett. 39, 1981, p.951.

6. C. Kaito, N. Nakamura and Y. Saito, Appl. Surf. Sci. 22/23, 1985, p.604.

7. A.K. Petford-Long and D.J. Smith, Phil. Mag. A54, 1986, p.837.

8. P. Lu and D.J. Smith, Phys. Stat. Sol. A107, 1988, p.681.

9. L.W. Hobbs, Quantitative Electron Microscopy, Eds. J.N. Chapman and A.J. Craven, (Scottish Universities Summer Schools in Physics, Edinburgh), 1984, p.437.

10. J.M. Meese and Y.S. Park, Radiation Damage and Defects in Semiconductors, Eds. J.E. Whitehouse (The Institute of Physics, London and Bristol), 1972, p.51.

11. F.J. Bryant and A.T. Baker, Phys. Lett. A35, 1971, p.457.

12. R.W.G. Wyckoff, Crystal Structures, 2nd edition, Interscience, New York, 1, 1963, p.254.

13. O. Kubaschewski, E.L. Evans and C.B. Alcock, Metallurgical Thermochemistry, 5th edition, Pergamon Press, New York, 1979, p.354.

14. G.P. Schwartz, G.J. Gualtieri, J.E. Griffiths, C.D. Thurmond and B. Schwartz, J. Electrochem. Soc. 127, 1981, p.2488.

15. R.L. Farrow, R.K. Chang, S. Mroczkowski and F.H. Pollak, Appl. Phys. Lett. 31, 1976, p.768.

16. T.P. Smirnova, A.N. Golubenko, N.F. Zacharchuk, N.F. Belyi, G.A. Kokovin and N.A. Valisheva, Thin Solid Films 76, 1981, p.11.

46 THE USE OF MICROLITHOGRAPHIC MARKERS IN ELECTRON MICROSCOPY OF HIGH-TEMPERATURE OXIDATION PROCESSES

C.K. Kim, S.-K.Fan* and L.W. Hobbs

Department of Materials Science and Engineering,
Massachusetts Institute of Technology, Cambridge MA 02139, USA.
*now at Texas Instruments, Dallas TX 75265, USA.

ABSTRACT

A method for photolithographic deposition of inert markers of submicrometre dimensions onto metal substrates prior to oxidation is described. The fine scale of the marker deposition avoids the difficulties encountered with conventional markers, whose size may exceed that of those microstructural features controlling oxidation. Their regularity provides a more reliable demarcation of original interface postitions than do other irregularly-deposited markers. The new technique has been applied to the high-temperature oxidation of pure Ni, Ni-1at%Al and Ni-1at%Cr alloys. Observation of the markers using cross-sectional transmission electron microscopy (XTEM), scanning electron microscopy (SEM) and scanning transmission electron microscopy (STEM) has revealed striking differences in the scale morphology, microstructures and oxidation mechanisms between pure Ni and the dilute alloys. In particular, the results suggest that small Al and Cr additions promote significant inward transport of oxygen. The presence of the markers appears to affect only marginally the oxidation kinetics and local oxide morphology.

INTRODUCTION

Identification of the diffusing species and their relative mobilities is central to the elucidation of high temperature oxidation mechanisms. The requisite transport measurements must, however, be performed on oxide scales in contact with their parent metals, since transport properties are sensitive to local microstructures, interface structures and chemical compositions. Relative motion of the scale/substrate interface during oxidation provides one source of in situ information about the rate controlling diffusing species, and inert markers have long been used to document interface position and motion. In oxidation experiments, Pt wires spot-welded [1] or wound tightly around substrate coupons [2] have served historically as markers, but the dimensions of such markers (and the resolution of the usual examination technique - light microscopy - used to locate them) are large compared to the microstructural features controlling or reflecting interface position and movement and often larger than individual layer thicknesses in multilayer scales. The markers in this case act as a significant barrier, impeding or modifying scale development.

With the advent of more microscopical examination techniques, such as EPMA, SEM and TEM, other methods have been adopted for generating markers: vacuum deposition of noble metals as very thin inert films [3] or through mesh grids [4], implantation of noble metal ions [5], and spraying metal droplets or ceramic particulates [6]. Locating such markers can be adventitious, and the reliability and utility of inert marker methods has continued to be questioned [5].

The spatial specificity and resolution of cross-sectional TEM (XTEM) and associated analytical techniques (AEM) have proven particularly powerful in characterizing the microstructural and compositional details of scale and interface development in elemental metals [7-11] and alloys [12-14] (see, for example, the study of the Ni-Cr model system by Newcomb and Hobbs [15] in this volume). The success of XTEM and AEM in revealing the microstructural origins of morphological features in semiconductor devices as well [16] suggested that micromarkers of appropriate dimensions for electron microscopy could be laid down using standard microlithographic deposition methods regularly employed in the manufacture of microcircuits. The facile control of size and spacing of markers provided by these methods ensures that markers spaced closely enough to be encountered frequently in TEM investigation but widely enough not to seriously interfere with oxidation kinetics or microstructural development can be generated[11]. The methods also provide sufficient freedom in the choice of marker chemistry that marker inertness can be achieved, even in the multiphase systems which characterize oxidation substrates and products, or at least the extent of the interaction can be verified using TEM and AEM probes.

A technique is described here for generating microscopic markers for use in high-temperature oxidation investigations. The technique has been applied to a comparative investigation of the transport mechanisms responsible for the development of duplex scales in the oxidation of pure Ni, Ni-1at%Al and Ni-1at%Cr.

SUBSTRATE PREPARATION AND OXIDATION

Ni (99.99%) was arc melted, with and without additions of Al and Cr (99.99%) in an argon atmosphere. After drop-casting, the metal was cold-rolled to 0.5mm thickness. Strips 20 x 2 x 0.5mm were cut from the resulting sheet and polished from 600-grit SiC paper down to 0.3μm alumina powder. After cleaning, samples were encapsulated in a vycor tube and annealed in argon at 1273K for several hours to achieve an average grain size of 200 μm. The substrates were then subjected to a final polish with 0.05μm alumina powder before deposition of marker materials, as described below. After deposition, samples were oxidized in air at 1273K and sectioned for SEM, XTEM and AEM observations.

MICROSCOPIC MARKER DEPOSITION

Considerable effort was made to optimize the photolithographic process due to the reactivity of the metal substrates compared to Si wafers and the greater difficulty in achieving flat, highly-polished surfaces. The sequence of procedures is schematized in Fig. 1.

Mask Fabrication

The mask used for photolithographic contact printing was prepared as a glass plate with a chromium coating on one side onto which the marker pattern is etched. For this study, the basic pattern consisted of an 8 x 8 array of 2.55 x 2.55mm squares each containing 256 parallel lines 1μm wide and spaced 10μm apart. This pattern was produced by a 10X reduction from an original 25.5 x 25.5mm reticle which was repeated 64 times to form the final mask. A pattern with a periodic array of parallel lines of this width and spacing was chosen to maximize the probability of finding the markers in XTEM with minimal disruption of the oxidation process.

Photolithography

After final mechanical polishing and cleaning, samples were coated with Shipley 1470SF photoresist and spun at 5000rpm to achieve a film thickness of about 1 μm. Coated samples were then softbaked in a convection oven at 363K for 30 minutes to remove the major portion of the solvent. Contact printing through the mask was performed using a Karl Suss Model 505 mask aligner and an illuminating wavelength of 365nm. Because the surface flatness varies from sample to sample, some over-exposure was required to guarantee an undercut profile in the photoresist. Typically, an exposure time of 6s with light intensity of 140J/m^2 produced good results. The exposed substrates were then developed in alkaline aqueous solution (Shipley 351 developer) for 60s, rinsed in hot water for an additional 60s, dried and stored for marker deposition.

Deposition and Lift-Off

Marker patterning was achieved by applying a lift-off technique. Composite markers comprising an initial 10nm-thick layer of Cr (for the Ni-1%Al) or Mo (for the Ni-1 %Cr) under a 150nm-thick layer of Pt marker material were Ar plasma-sputtered onto the substrates using an MRC 8620 sputtering system. The samples were then immersed in acetone to remove the remaining photoresist, leaving the marker pattern shown in Fig. 2. The SEM micrograph in Fig. 3 illustrates the high resolution achievable through this microphotolithographic process. The initial Cr or Mo layer in the composite marker was found necessary to act as a diffusion barrier precluding dissolution of the Pt marker into the Ni-base substrate during oxidation.

RESULTS

Oxidation in air at 1273K produced very different results for pure Ni and for Ni-1at%Al and Ni-1at%Cr alloys. In all three cases, however, the markers are seen to remain substantially intact and embedded in a scale thickness of 10-20μm. The regularity of the marker pattern eliminates any ambiguity in marker identification, a problem with other micromarker techniques [12].

For pure Ni, the markers were found very close to the substrate/scale interface (Fig. 4). A little oxide has formed immediately beneath each marker, indicating a minor local perturbation of the oxidation process by the markers, but the fact that the unperturbed portion of the scale/substrate interface lies along the line of the markers confirms that the

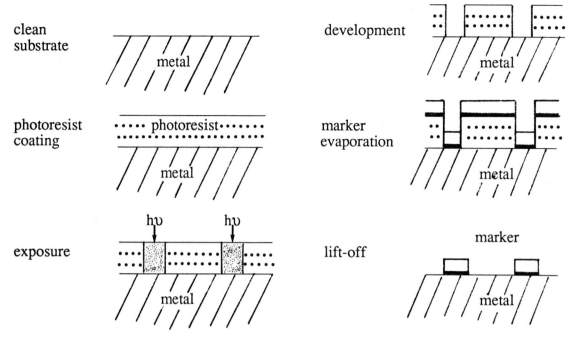

Figure 1: Schematic representation of processing sequence for photolithographic deposition of markers.

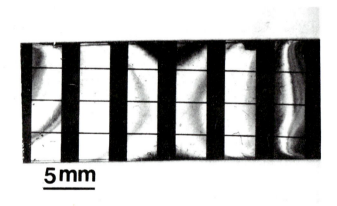

Figure 2: Light micrograph of the marker pattern generated on an alloy coupon before oxidation.

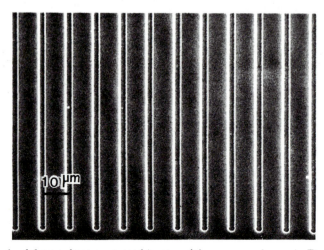

Figure 3: SEM micrograph of the markers generated in one of the squares shown in Fig. 2.

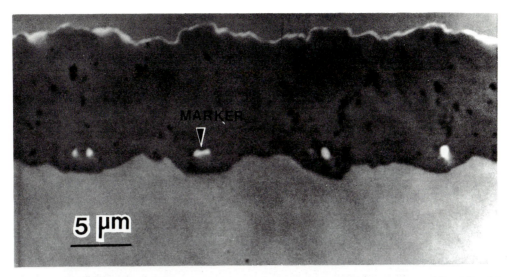

Figure 4: Transverse section SEM micrograph (back-scattered image) of oxide formed on pure Ni oxidized in air for 1 h at 1273 K, showing markers embedded close to oxide/metal interface.

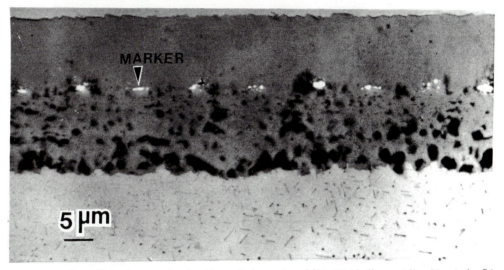

Figure 5: Transverse section SEM micrograph of oxide scale formed on Ni-1at%Al alloy oxidized in air for 5 h at 1273 K, showing markers located at the boundary between inner and outer scale layers.

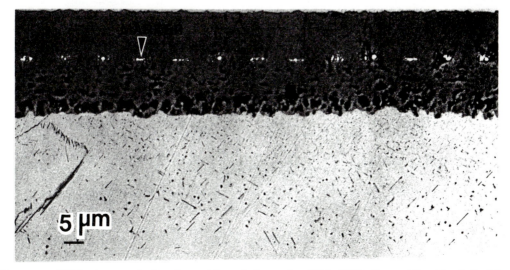

Figure 6: Transverse SEM micrograph showing oxide scale formed on Ni-1at%Cr alloy oxidized in air for 5 h at 1273 K, showing markers located at inner/outer scale interface, as for Ni-1at%Al.

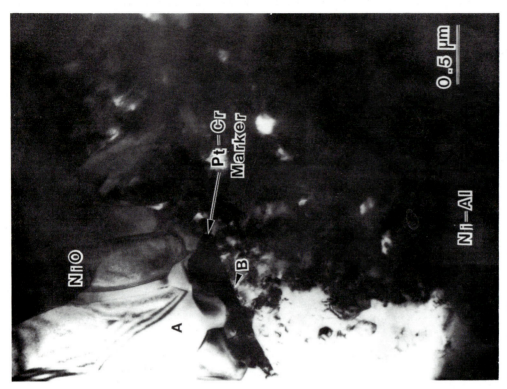

Figure 8: XTEM dark-field image of NiO duplex scale formed on Ni-1at%Al substrate oxidized 10 min at 1273 K in air, showing Pt-Cr marker embedded in scale at interface between fine-grained inner layer and largely columnar outer layer scales.

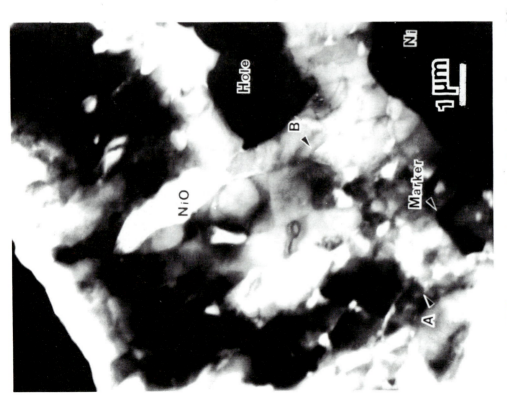

Figure 7: XTEM dark-field image of NiO scale formed on pure Ni substrate oxidized 30 min at 1273 K in air, showing Pt-Cr markers near substrate/scale interface, fine-grained inner layer, and columnar outer scale.

interface has not moved appreciably during scale growth. It is concluded that oxidation is taking place essentially solely at the oxide/gas interface, necessarily by migration of Ni cations through the scale. The XTEM micrograph (Fig. 7) indicates further that the markers are found below the ~1μm-thick inner layer scale of equiaxed oxide grains which lies beneath the outer columnar oxide [7,8]. This thin inner layer, the orientation of whose grains with respect to the underlying metal it is which governs the metal/oxide interface properties[17], does not therefore form from inward diffusion of oxygen but as a consequence of developments early in the scale's formation [10].

By contrast, for both Ni-1at%Al and Ni-1at%Cr, the markers are embedded well into the scale (Figs. 5 and 6) at the interface between two oxide layers, the lower of which appears to be highly porous. XTEM (Fig. 8) indeed reveals the markers at the juncture of these two layers, the upper of which is columnar oxide similar to that formed on pure Ni, the lower of which is extremely fine-grained but nevertheless compact. The porous appearance of this layer in SEM, often remarked upon in observations by light microscopy of duplex scales, is in fact an artefactual result of grain pull-out during metallographic polishing. The formation of this layer does not therefore occur by a dissociative mechanism. Both alloys also exhibit an internal oxidation zone below the oxide scale; in the Ni-Al alloy, the internal oxide particles appear to be γ-alumina and Ni-Al spinel, in the Ni-Cr alloy α-Cr_2O_3. The composition of the fine-grained inner layer is, however, single-phase NiO, which the marker position reveals in both cases to have formed by inward transport of oxygen. An inner NiO scale, advancing into the internal oxide zone of the alloy, will incorporate these internal oxides, as surmised for the case of Ni-5wt%Cr [15].

The precise role of the alloy additions in effecting this change in oxidation mechanism is not clear. STEM investigation revealed only a marginal presence of Al in the inner NiO layer for the Ni-Al alloy, but did detect Al segregation to columnar grain boundaries in the outer scale. For the Ni-Cr case, the inner NiO scale was enriched to 3% Cr cation content, as was also found by Atkinson,[18] while the outer scale contained only 0.09% Cr whose distribution has yet to be determined. Segregation of Al or Cr to columnar oxide grain boundaries could substantially alter oxygen transport in NiO, either by providing rapid grain-boundary transport paths or by affecting the creep behavior at such boundaries and promoting microcracks which permit oxygen ingress. No example of the latter has been found, but for the ratio of inner to outer scale thicknesses of about 1:1 in these experiments, ionic diffusion of oxygen along grain boundaries alone appears too slow to account for the observed magnitude of inner layer growth [19].

CONCLUSIONS

Micrometre-size markers, deposited by a photolithographic method, can be produced routinely and regularly enough to be useful for SEM and TEM investigations and are seen to reliably indicate the relative motion of the metal/oxide interface during oxidation. The addition of small amounts (1 at%) of Al and Cr into Ni alters the preferred oxidation mechanism in Ni from one dominated by outward cation diffusion to one where substantial oxygen ingress occurs.

ACKNOWLEDGEMENT

This work has been supported by the National Science Foundation.

REFERENCES

1. B.L. Gleeson, D.L. Douglass and F. Gesmundo, Oxid.Met. 31, 1989, p.209.

2. E.M. Fryt, G.C. Wood, F.H. Stott and D.P. Whittle, Oxid.Met. 23, 1985, p.77.

3. H.M. Hindam and W.W. Smeltzer, Oxid.Met. 14, 1980, p.337.

4. C.M. Cotell, K. Przybylski and G.J. Yurek, Fundamental Aspects of High Temperature Corrosion, Ed. D.A. Shores and G.J. Yurek, The Electrochemical Society, 2, 1986, p.103.

5. E.W.A. Young, H.E. Bishop and J.H. deWit, Surface Interface Anal. 9, 1986, p.163.

6. T.A. Ramanarayanan, R. Ayer, R. Petkovic-Luton and D.P. Leta, Oxid.Met. 29, 1988, p.445.

7. L.W. Hobbs, H.T. Sawhill and M.T. Tinker, Trans.Japan.Inst.Metals JIMIS 3 Suppl.,1983, p.115.

8. L.W. Hobbs, H.T. Sawhill and M.T. Tinker, Rad. Effects 74, 1983, p.229.

9. H.T. Sawhill, L.W. Hobbs and M.T. Tinker, Adv. Ceramics 6, 1983, p.128.

10. H.T. Sawhill and L.W. Hobbs, Proc.Int.Congr. on Metallic Corrosion [NRC Canada] 1, 1984, p.21.

11. L.W. Hobbs and S-K. Fan, MRS Symp. Proc. 115, 1988, p.283.

12. S.B. Newcomb, W.M. Stobbs and E. Metcalfe, Phil.Trans.Roy.Soc. (London) A319, 1986, p.191.

13. W.M. Stobbs, S.B. Newcomb and E. Metcalfe, Phil.Trans.Roy.Soc. (London) A319, 1986, p.219.

14. S.B. Newcomb and W.M. Stobbs, Mat. Sci. Techn. 4, 1988, p.384.

15. S.B. Newcomb and L.W. Hobbs, these Proceedings.

16. R.B. Marcus and T.T. Sheng, Transmission Electron Microscopy of Silicon VLSI Circuits and Structures, Wiley, New York, 1983.

17. L.W. Hobbs and H.T. Sawhill, J.Physique 46 Colloque C4, 1985, p.117.

18. A. Atkinson, Harwell Laboratory Report AERE-R12762, 1987.

19. A. Atkinson, F.C.W. Pummery and C. Monty, Transport in Non-Stoichiometric Compounds, Plenum, New York, 1985, p.359.

47 SEQUENTIAL SCANNING ELECTRON MICROSCOPY STUDY OF OXIDE SCALE FAILURE AND REGROWTH

A.T. Tuson and M.J. Bennett

Materials and Manufacturing Technology Division, Building 393, Harwell Laboratory, AEA Technology, Didcot, Oxon OX11 0RA, UK.

ABSTRACT

The oxide scale grown on the 20Cr/25Ni/Nb stainless steel during four oxidation cycles in a carbon dioxide based environment at 875°C, with intermediate furnace cooling to room temperature, has been examined by sequential scanning electron microscopy (SSEM). This enabled investigation of the nature of failure of the initial, uniform, protective chromium rich oxide scale, the re-oxidation of the steel areas exposed involving the formation of iron-rich pitting type scale and also of the spallation of this oxide during subsequent oxidation/cooling cycles.

INTRODUCTION

Due to its technological importance the oxidation behaviour of the 20Cr/25Ni/Nb stabilised stainless steel has been extensively investigated [1-7]. Of particular importance is understanding the growth and failure of the initial, uniform, protective scale comprising an outer Fe, Cr and Mn bearing spinel layer, a middle Cr_2O_3 layer with a thin amorphous SiO_2 layer at the scale/steel interface [2]. When the scale is above a threshold thickness, it can spall on cooling [3] and upon subsequent re-oxidation of the chromium depleted layer a non-protective scale formed, leading to pitting type attack [4,5]. Physical examination of the surface topography at the completion of oxidation by scanning electron microscopy [6] has been quite revealing in furthering understanding of these phenomena. However, because these processes have not been observed directly in-situ there remains some doubt that important mechanistic facets have not yet been exposed fully. The validity of this approach was substantiated by an early hot-stage optical microscopy study [7], although limited severely by the lack of depth of field and spatial resolution. In the absence of a scanning electron microscope fitted with a high temperature environmental cell capable of operation at atmospheric pressure, recourse has been made to an effective alternative technique, namely sequential scanning electron microscopy (SSEM). This is based on following the topographical and compositional changes of the same representative regions of a 20Cr/25Ni/Nb stainless steel surface throughout a defined and relevant oxidation regime.

EXPERIMENTAL

A coupon (10 x 10 x 0.5 mm) of 20Cr/25Ni/Nb stainless steel, with the analysis (w/o) 19.9 Cr, 24.6 Ni, 0.7 Nb, 0.7 Mn, 0.6 Si and 0.04 C, was ground to a 600 grit surface finish, cleaned in Decon 90 solution and, finally, annealed in dry hydrogen for 30 min at 930°C. Two parallel lines were lightly scratched on one side of the coupon in both the X and Y directions to form a grid. The specimen was oxidised at 875°C in flowing gas with the composition CO_2+2% CO+300 vpm CH_4+350 vpm H_2+12 vpm H_2O. After 263h oxidation the specimen was furnace cooled to room temperature, a process taking several hours. This sequence was repeated three further times, with the sample being cooled after cumulative oxidation times of 500, 999 and 1505h. Each period of oxidation followed by cooling to room temperature is described as a thermal cycle, so that the steel was exposed for four such cycles.

The oxidised specimen was examined in an Hitachi S520 Scanning Electron Microscope (SEM) with an Energy Dispersive X-ray Analysis (EDX) facility. After the first thermal cycle a number of areas where the oxide had spalled were identified on the SEM and photographed. EDX analyses were taken using a spot source or a reduced area raster. A semi-quantative value for the composition at these points was obtained after applying a ZAF correction allowing for atomic number, absorption and fluorescence effects. The accelerating voltage of 25 kV meant an analysis depth in the oxide of 2-3 μm. After the second thermal cycle these 'spalled areas' were again located within the SEM using the grid as a guide. They were re-examined at the same magnification and tilt and further EDX analyses were taken. Also, some areas of fresh spallation were studied in a similar way. The process was repeated after the third and the fourth thermal cycles.

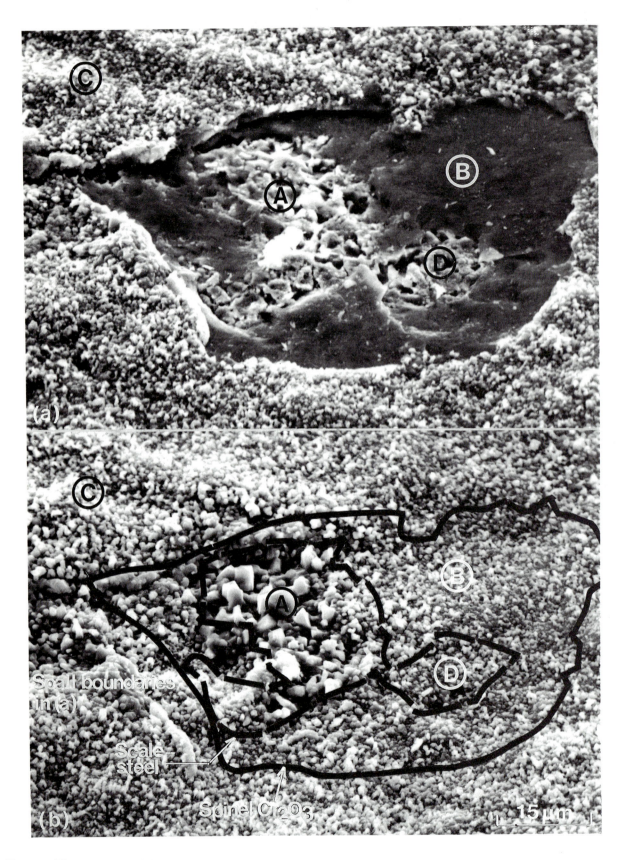

Figure 1: The same area of 20/25Nb steel from which oxide scale had spalled after the second thermal cycle following 500h oxidation (a) and subsequently after the third thermal cycle following 999 h oxidation (b).

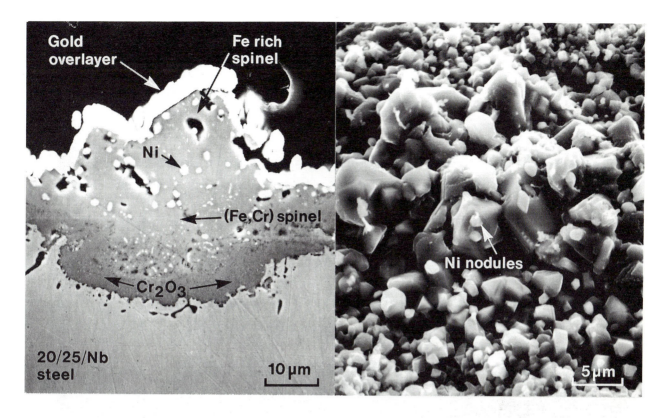

Figure 2: Transverse section through an oxide pit.

Figure 3: External iron rich spinel crystallites with metallic nickel nodules of the pitting attack of 20/25/Nb steel.

RESULTS AND DISCUSSION

The topographical features following loss of the protective uniform scale formed on 20Cr/25Ni/Nb stainless steel after all four thermal cycles were similar. A representative region of the oxidised 20Cr/25Ni/Nb stainless steel surface where a discrete (57 x 40 μm) section of the protective scale had spalled following the second thermal cycle is shown in Figure 1(a). Fracture occurred both through the complete scale section to the underlying steel with decohesion at, or near, the scale-steel interface (position **A**) and also partially through the scale with decohesion at an interoxide interface, probably that between the outer spinel and the inner Cr_2O_3 layer (position **B**), as confirmed by EDX analysis (Table 1). Position C is at the outer surface of intact protective oxide scale. The same region is shown after the third oxidation/cooling cycle in Figure 1(b) where the full black line marks the position of the spinel/chromia interface and the dashed black line the chromia/steel interface in the upper photograph. Analyses again taken at positions **A**, **B** and **C** are given in Table 1. The behaviour of the exposed steel was governed primarily by the chromium depletion profile resulting from the formation of the protective uniform scale and possibly also to a lesser extent by remaining adherent oxide, such as silica and/or chromia. The measured chromium contents of the steel exposed by spallation (about 14 w/o) was below that (~ 16 w/o) needed for the formation of a chromia layer on oxidation, so pitting type attack ensued [4,5]. This is characterised by the initial formation of large Fe_3O_4 crystallites externally (Figure l(b), position **A**) with the development of Fe, Cr bearing spinels beneath, as shown in a transverse section Figure 2. Scale then grows by both outward cation and inward oxidant transport, and as a consequence of the later carbon uptake also occurs into the steel. The spinel at the base of the pit gradually becomes more chromium rich and eventually when the chromium concentration in the steel reaches about 16 w/o a Cr_2O_3 healing layer is formed (Figure 2). Subsequent scale growth is controlled by cation diffusion through this layer. The depth of the pit is determined by the chromium depletion profile and in particular by that of the chromium activity at which the healing layer could form. It is also of further note that the healing layer not only enveloped the pit but also joined up with the Cr_2O_3 layer of the adjoining protective type scale.

Small metallic nickel nodules were apparent on the iron-rich crystal surfaces (Figure 3). These resulted from the reduction of iron-nickel rich spinels developed in the early, rapid stages of pit growth, which were not stable in the CO_2/CO gas mixture. Metallic nickel regions were apparent also within the cross-sectioned pit (Figure 2) where this metal was left unoxidised during pit formation because the oxygen partial pressure was too low.

Figure 4: The same area of 20/25/Nb steel after the third thermal cycle following 999 h oxidation (a) and after the fourth thermal cycle following 1505h oxidation (b).

Table 1. Semi-quantitative analyses (w/o) at the points marked on Fig. 1

Element	Position A		Position B		Position C	
	2*	3*	2*	3*	2*	3*
Si	1.0	0.5	0.6	Nil	Nil	0.1
Nb	2.2	0.9	2.0	0.8	Nil	0.4
Cr	14.2	4.6	75.6	59.6	68.1	58.2
Mn	0.4	15.1	4.2	24.9	20.4	28.0
Fe	58.3	64.2	12.9	13.3	10.0	11.9
Ni	23.9	14.7	4.7	1.4	1.5	1.4

Analysis after thermal cycle number

SSEM and EDX analyses (Table 1) revealed that on re-oxidation the Cr_2O_3 layer at position **B** , Figure 1(b), became covered with Cr, Mn and Fe bearing oxide, similar in composition to the outer spinel of the uniform protective scale (position C). So, essentially having lost only the original spinel layer, this was regenerated by similar processes to those responsible for the continuing growth of the spinel layer of the protective scale and probably involving cation diffusion.

It is interesting to note that regions of this latter type of oxide also grew over part of what was believed to be exposed steel (e.g. at position **D**, Figures 1 (a) and (b)) possibly because a thin residual layer of silica and/or chromia prevented more aggressive attack involving inward oxidant transport. These areas could have been those described as passive in the hot-stage microscopy [7].

Spallation of scale from the 20Cr/25Ni/Nb stabilised stainless steel has always been believed to be attributable solely to loss of uniform protective scale. However, this study showed that pitting type scale can also spall. Figure 4(a) is a region of the oxidised steel surface after the third thermal cycle where there are areas of both pitting and protective-type scale. The former is indicated by large, iron-rich oxide crystallites and the latter by the smaller (Fe, Cr, Mn) spinel grains. Spallation occurred within this region on the next thermal cycle (Figure 4(b)) and the chromia/spinel and steel/scale boundaries delineating fracture are marked in Figure 4(a) again by full and dashed black lines respectively. Some of the iron-rich crystallites apparent in Figure 4(a) were not present in Figure 4(b) indicating that part of the scale formed during pitting attack had indeed spalled. However, spallation was never localised solely to pitting attack but was always part of more extensive spalling taking place in adjacent uniform protective scale. This could indicate that fracture was initiated within the latter and was propogated through the upper pit section. Consistent with this thesis this scale never spalled to the metal but always left an underlying layer of chromium-rich scale. Decohesion of the adjoining areas of protective scale occurred both to the steel-scale and spinel-chromia interfaces, as described previously.

CONCLUSIONS

Sequential scanning electron microscopy proved a rewarding technique for studying oxide scale failure and regrowth on 20Cr/25Ni/Nb stainless steel during an oxidation regime with intermittent furnace cooling. More defined microstructural features were revealed by concentrating on the sequential behaviour of the same surface region than were apparent from the normal SEM procedure of examining random representative areas.

ACKNOWLEDGEMENT

We are most grateful to the Department of Energy and the Electricity Generating Boards in the United Kingdom for funding this study.

REFERENCES

1. M.J. Bennett, 10th International Congress on Metallic Corrosion, Oxford and IBH Publishing Co. PVT Ltd., New Delhi, 4 ,1987, p. 3761.

2. M.J. Bennett, J.A. Desport and P.A. Labun, Proc. Roy. Soc. Lond., A412, 1987, p. 223.

3. H.E. Evans, Mat. Sci. and Tech., 4, 1988, p. 415.

4. M.J. Bennett, M.R. Houlton and R.W.M. Hawes, Corros. Sci., 22, 1982, p. 111.

5. H.E. Evans, D.A. Hilton, R.A. Holm and S.J. Webster, Oxid. Met., 2, 1978, p. 473.

6. R.C. Lobb, CEGB Report TPRD/B/0478/N84 (1984).

7. F.H. Fern and J.E. Antill, Corros. Sci., 10, 1970, p. 649.

48 NUCLEATION AND GROWTH OF SURFACE OXIDES AND SULPHIDES FORMED ON Fe-Cr-Ni ALLOYS DURING EXPOSURE TO H₂-CO-H₂O-H₂S ATMOSPHERES

J. F. Norton, S. Canetoli and K. Schuster

*Commission of the European Communities, Institute of Advanced Materials,
J.R.C. Petten Establishment, 1755 ZG Petten, The Netherlands.*

ABSTRACT

The degradation experienced by metallic components during exposure to process environments of the types encountered in coal conversion plant has prompted considerable research during recent years. These atmospheres generally have low levels of oxygen but high sulphur and carbon activities and therefore sulphidation is one of the primary modes of attack.

In such multi-component environments there is a competition between sulphur, oxygen and carbon to react with the alloy. It is therefore important to establish the factors which govern the formation of a surface oxide and whether this remains intact or, alternatively, becomes disrupted by faster growing sulphides.

The experiments detailed in this paper have been designed to address this problem using a range of H₂-CO-H₂O-H₂S gas mixtures of varying sulphur activity. Several techniques have been used to monitor the nucleation and growth of oxides and sulphides formed on the surfaces of Fe-Cr-Ni alloys during exposures ranging from a few minutes to several hundred hours at 800°C. A sequential SEM technique coupled with cross-sectional metallography has been invaluable in helping establish the corrosion mechanisms involved.

INTRODUCTION

During the conversion of coal into a gaseous form for use in the generation of electricity, corrosion of metallic components such as heat exchangers by the process gas can be significant. In these low oxygen-containing atmospheres, sulphidation can be severe with the extent of attack being dependent upon the metal temperature and the sulphur activity of the environment. There is a competition between the sulphidising, carburising and oxidising species in the gas to react with key elements in the alloy, e.g. Cr, Fe, Ni, Al, Si, etc., and it is therefore important to establish an in-depth understanding of the factors governing such reactions if reliable predictions of materials behaviour are to be made.

A number of investigations have been performed during recent years in an attempt to rationalize the complex reactions occurring in multi-component gaseous environments [1-6]. The studies reported here have utilized H₂-CO-H₂O gas mixtures to which varying amounts of H₂S have been added so that the effect of changing the sulphur activity upon the type and extent of corrosion products formed on the surfaces of selected alloys could be monitored. A sequential SEM technique has been developed whereby identical areas on the specimens can be re-examined with increasing exposure to the corrosive atmosphere thus enabling the nucleation and growth of surface oxides and sulphides to be studied. Exposures have varied from a few minutes to several hundred hours. Cross-sectional metallographic examinations have also been carried out after selected exposures using a vapour deposition preparation method in addition to conventional chemical etching in order to resolve and identify the mixed oxide/sulphide structures.

THERMODYNAMIC ASPECTS

Although thermodynamic considerations can provide a useful insight in indicating which gas/metal reactions *may* occur they cannot be used with certainty to predict what *will* occur. Once corrosion of the alloy has started, local inhomogeneities or changes in alloy composition as well as changes in gas chemistry are possible and thus kinetics play a dominant role. However, such considerations can aid the planning of a co-ordinated series of experiments and can also help in the interpretation of the results subsequently obtained.

The phases which are stable as a result of the reaction of metallic elements with gaseous species can be illustrated in stability diagrams using values derived from thermodynamic calculations. In their simplest form these diagrams depict regions of phase stability for the reaction of an element, e.g. Cr, with two reactants such as O and S. Superimposing diagrams for several elements is more useful and Fig. 1 shows the phase boundaries for the major elements Fe, Cr and Ni and the reactants S and O at the temperature of 800°C selected for these studies. This is a greatly simplified approach. The boundaries shown are for pure metals ($a_m = 1$) whereas metal activities are much less than 1 in alloy systems. The thermodynamic boundaries between elements with $a_m < 1$ and their oxides and sulphides are at higher oxygen and sulphur partial pressures than those shown.

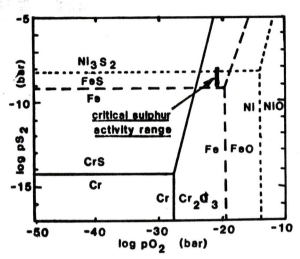

Figure 1: Phase stability diagram for M-S-O at 800°C (M=Cr, Fe, Ni).

EXPERIMENTAL PROCEDURE

Materials and Test Atmospheres

Initially, two commonly used engineering alloys were selected for these studies, i.e. Alloy 800H - an austenitic 20Cr-32Ni-Fe material containing small amounts of Al and Ti which has good high temperature mechanical strength - and a centrifugally cast 25Cr-35Ni-Fe alloy. A laboratory-cast "model" alloy based on this composition but without minor elements was also tested in selected experiments. The chemical analyses of the alloys are given in Table 1.

Alloy	Chemical Composition - wt%						
	Cr	Ni	Fe	C	Mn	Si	Others
800 H	20.1	31.4	Bal.	0.06	0.66	0.37	Al0.28 Ti0.48
HP40Nb	23.7	32.3	Bal.	0.46	1.15	1.29	Nb0.8
Model	25.1	33.7	Bal.	-	-	0.04	-

Table 1. Chemical composition of alloys

The test atmospheres used during these experiments have been designed in terms of the reactant activities, pO_2, pS_2 and a_c using equilibrated gas mixtures. The carbon and oxygen activities have been kept constant whilst the sulphur activity was systematically varied over a critical range which had largely been established during earlier investigations [6]. The compositions and thermodynamic characters of these environments are summarized in Table 2 with their positions being indicated on the phase stability diagram, Fig. 1.

Composition of Mixture				Character (at 800°C)		
H_2	CO	H_2O	H_2S	pO_2(bar)	a_c	pS_2(bar)
Bal.	7%	1.5%	0.06%	10^{-21}	0.3	1×10^{-8}
Bal.	7%	1.5%	0.42%	10^{-21}	0.3	5×10^{-9}
Bal.	7%	1.5%	0.26%	10^{-21}	0.3	2×10^{-9}
Bal.	7%	1.5%	0.20%	10^{-21}	0.3	1×10^{-9}

Table 2. Composition of gas mixtures

Experimental Techniques

Specimens, 10 x 8 x 6 mm, were prepared from each alloy using controlled and reproducible machining and fine grinding techniques and ultrasonically degreased prior to exposure in the autoclave. The correct gas quality was established by evacuating and back-filling the reaction chamber several times with the pre-mixed test gases. The furnace was then heated to 800°C and the gas pressure ($\approx$1.5 bar absolute) and flow ($\approx$15 l/h) controlled before the specimens were rapidly brought to temperature using a plunger device. This device enabled exposures to be varied from a few minutes to several hundred hours and it was thus possible to study the early-stage nucleation and growth processes by intermittent examination and re-examination of the specimen surfaces.

Microstructural Analysis

A simple sequential SEM technique was adopted whereby the positions of corrosion products which had nucleated on the specimen surfaces during the initial exposure to the test gas were accurately indexed and subsequently re-examined after additional exposure periods. This, coupled with EDX analysis, enabled the morphological and compositional changes occurring to be established. At the end of each experiment, cross-sectional examinations were carried out so that the nature and extent of the corrosive attack could be quantified. In this respect, in addition to conventional chemical etching, a ZnSe vapour deposition procedure [7] was adopted as the preparation technique, since this significantly aided phase identification in mixed oxide/sulphide structures.

RESULTS AND DISCUSSION

Previous studies [5,6] indicated that significant changes in the mode and extent of corrosion occurred at 800°C when the sulphur activity was varied from $pS_2=10^{-8}$ bar to 10^{-9} bar, i.e. H_2S levels of 0.6% and 0.2% respectively. Reaction rates were reduced by up to six orders of magnitude. During short-term exposure of the "model" alloy to the 0.6% H_2S containing atmosphere, the surface of the specimen became covered by a thick sulphide scale. After 5 hours the scale consisted of a mixture of $FeCr_2S_4$ and $(Fe_xCr_{1-x})S$ whilst chromium-rich sulphides had precipitated in the alloy substrate. with continuing exposure, appreciable quantities of nickel became incorporated in the outer scale which thickened considerably and the discrete internal sulphide precipitates coalesced to form a continuous inner layer. In contrast, during exposure of this material to the environment containing 0.2% H_2S, specimens became covered by a very thin uniform chromium-rich oxide and the alloy substrate was protected from further attack, even for durations of several thousand hours. There was only a very little evidence of internal sulphide precipitation.

Figure 2 summarises the radical change in corrosion behaviour which accommpanied this comparatively minor reduction in sulphur activity of the environment. In view of these findings, this more detailed study has been carried out using a range of gas mixtures having intermediate H_2S levels with the intention of monitoring both the formation of surface oxides and their breakdown by sulphide nuclei.

Exposure in 0.42% H_2S-containing Gas Mixture (pS_2=5 x 10^{-9} bar)

The nucleation and growth of the surface corrosion products formed on specimens of two commercial alloys, Alloy800H and HP40Nb, exposed for 5, 10, 15 and 25 h to the 0.42% H_2S-containing atmosphere was studied using the sequential SEM technique. An example of the observations made on an Alloy800H specimen is presented in Fig. 3. After 5 h exposure the surface was largely covered by a Cr-rich oxide having a fine, filamentary structure, although small clusters of much coarser sulphide crystallites had already nucleated. Re-examination of the same area after a further 5 h exposure

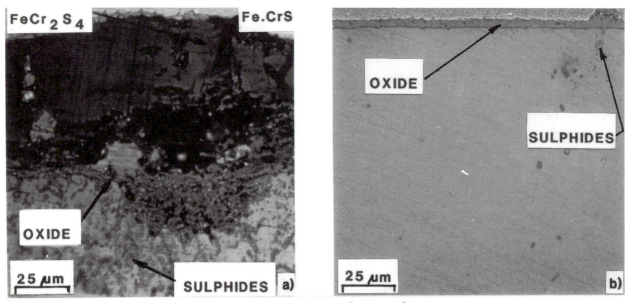

Figure 2: Influence of reducing the sulphur activity from $pS_2 = 10^{-8}$bar to 10^{-9}bar upon corrosion behaviour of "Model" Alloy at 800°C. (a) 5h - 0.6%H_2S, (b) 2000h - 0.2%H_2S.

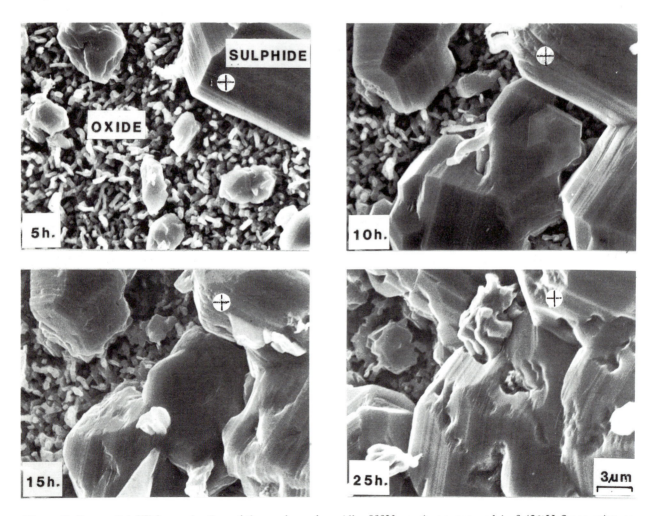

Figure 3: Sequential SEM examination of the surface of an Alloy800H specimen exposed in 0.42%H_2S gas mixture ($pS_2 = 5 \times 10^{-9}$bar) at 800°C.

showed that additional particles had nucleated, and coalescence of the original sulphide particles had also occurred. Elemental EPMA scans for Cr, Fe, Ni and S indicated that the coarser particles were Fe-containing sulphides whilst the background oxide was Cr-rich. With continuing exposure, extensive coalescence and growth of the sulphide nodules occurred and the entire surface became covered by a thick sulphide scale. Cross-sectional microscopy of specimens terminated after short-term exposures has shown that a number of sulphide particles had nucleated on the outer surface of the oxide and that this oxide had been locally penetrated. These sulphide nuclei had a lamellar structure characteristic of mixed Fe-Cr sulphides having varying Fe-Cr ratios. In addition, much smaller sulphides have precipitated in the alloy substrate beneath the oxide. Figures 4a and b illustrate these features for the Alloy800H and HP40Nb.

Exposure in 0.26% H_2S-containing Gas Mixture ($pS_2 = 2 \times 10^{-9}$bar)

Similar studies have been carried out in a 0.26% H_2S-containing gas mixture in which Alloy800H and HP40Nb specimens were exposed for periods of 1, 2, 3, 5, 10, 25, 50 and 100h. A summary of the SSEM examinations of the HP40Nb alloy at the 1, 3, 10 and 100 h. inspections is presented in Fig. 5 along with EDX scans on the features illustrated. The crystalline appearance of the sulphide nodules was in clear contrast to the thinner fine-grained background oxide at the initial SEM examination. This persisted at the 3 h inspection but after 10 h the crystalline morphology had transformed to a more rounded globular structure. After 100 h it was no longer possible to distinguish the sulphide crystals. The filamentary structure depicted in the figure was typical for the area monitored. The semi-quantitative EDX analyses indicated a marked drop in the iron content after 10 h and a significantly reduced S-peak after 100 h exposure. There was also a concomitant increase in chromium level at longer exposures which, when combined with the morphological changes observed, indicate that the sulphide nodules were possibly becoming overgrown by a Cr-rich oxide.

A cross-sectional microstructural examination of the specimen discontinued after 100 h confirmed that overgrowth of the sulphide nuclei by a thin uniform oxide had occurred in a number of areas. Figure 6 illustrates this, for the HP40Nb specimen. Very similar features were observed during the SSEM examination and subsequent cross-sectional study of the Alloy800H specimen. These observations confirm similar results obtained during an earlier investigation [6] when filamentary surface chromium sulphides became completely overgrown by an outer oxide during exposure to a 0.2% H_2S-containing gas mixture. In that study, exposures of up to 500 h were needed before overgrowth by the oxide was complete. In summary, the critical influence of relatively small changes in the sulphur activity of these multi-reactant atmospheres upon corrosion behaviour has been observed to be due to a shift from an oxidation-governed regime to one in which catastrophic sulphidation attack occurs. In both regimes oxide and sulphide particles nucleated but in the lower H_2S-containing atmospheres the Cr-rich sulphides became overgrown by the oxide. Conversely, in the atmospheres with higher H_2S levels the Fe-rich sulphides grew more rapidly resulting in disruption and breakdown of the oxide layer. Figure 7 is a schematic illustration of the principle mechanisms involved.

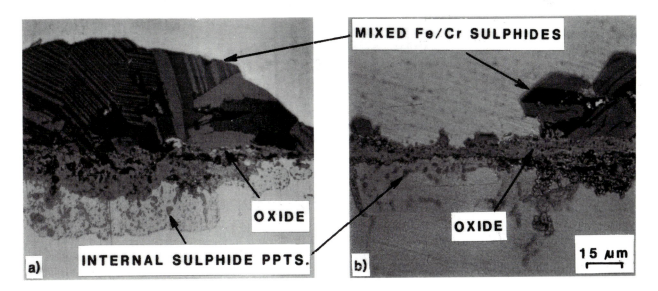

Figure 4: Cross-sectional microstructure of specimens exposed in 0.42% H_2S gas mixture ($pS_2 = 5 \times 10^{-9}$bar) at 800°C. (a) Alloy800H - 10h, (b) HP40Nb - 25h.

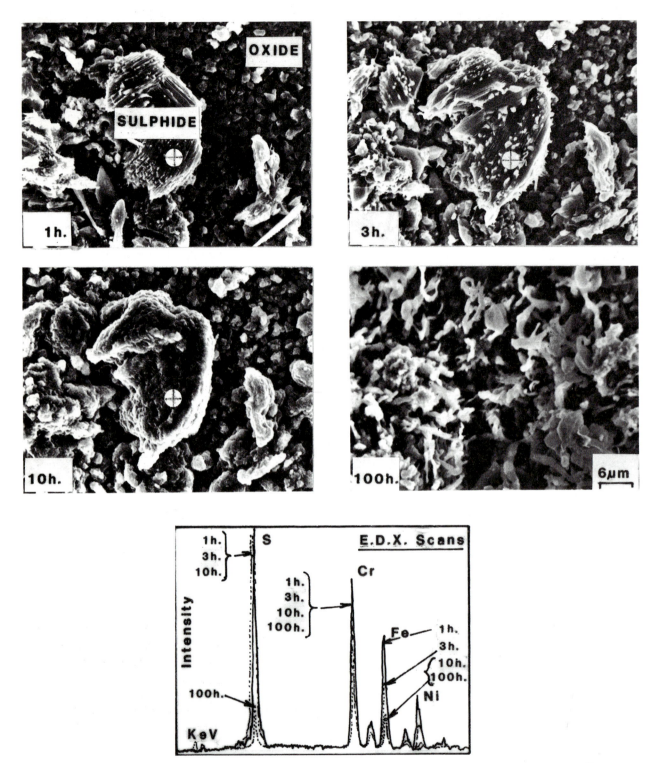

Figure 5: Sequential SEM examination of the surface of a HP40Nb specimen exposed in 0.26%H_2S ($pS_2 = 2 \times 10^{-9}$bar) gas mixture at 800°C.

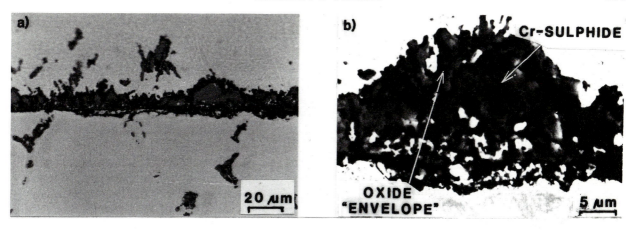

Figure 6: Cross-sectional microstructure of HP40Nb specimen exposed in 0.26%H_2S-containing gas mixture at 800°C. (a) Optical Image, (b) Electron Image.

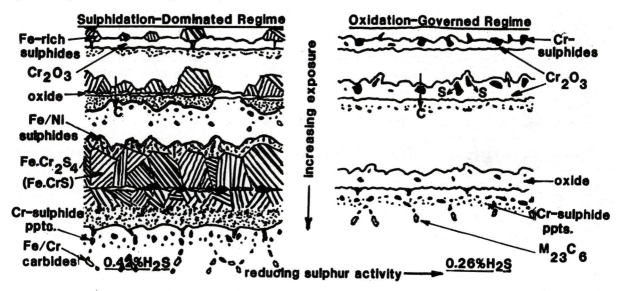

Figure 7: Schematic illustration of factors involved in the simultaneous nucleation and growth of oxides and sulphides on the surface of Fe-Cr-Ni Alloys exposed to H_2-CO-H_2O-H_2S gas mixtures at 800°C.

CONCLUDING REMARKS

The nucleation and subsequent growth of oxides and sulphides on the surfaces of several Fe-Cr-Ni alloys has been monitored using a sequential SEM technique. Representative features have been identified and re-examined intermittently during continuing exposure of the specimens. Cross-sectional microstructural examination of the specimens at the termination of the experiment has confirmed the nature and extent of corrosive attack.

H_2-CO-H_2O-H_2S gas mixtures have been used during these investigations and the effect of varying the sulphur activity over a critical range, i.e. $pS_2 = 10^{-8}$ bar to 10^{-9} bar at 800°C has been studied. Exposures have ranged from a few minutes to several hundred hours.

It has been observed that the sulphides which nucleated on the surface initially were either Cr or Fe-rich, depending on the sulphur activity of the environment. With time, the Cr-rich sulphides became overgrown by a Cr-rich oxide and oxidation-governed kinetics resulted. In the situation where Fe sulphides had formed, these rapidly overgrew the surface oxide eventually leading to catastrophic corrosion due to the incorporation of nickel into the sulphide scale.

For the experimental conditions studied, the critical transition from the oxidation-governed regime to one in which sulphidation attack dominated lies at sulphur activities between 2×10^{-9} bar (0.26% H_2S) and 5×10^{-9} bar (0.42% H_2S). This sulphur activity is lower than that predicted by thermodynamics confirming the presence of the so-called "kinetic" boundary.

ACKNOWLEDGEMENTS

The scientific contribution of J.A. Kneeshaw during his 3-year fellowship at J.R.C.-Petten is gratefully acknowledged. Thanks are also due to M. Van de Voorde (Programme Manager) and E.D. Hondros (Institute Director) for permission to publish this paper.

REFERENCES

1. R.A. Perkins, Alloy800, Eds. Betteridge, Krefeld, Kröckel, Lloyd, Van de Voorde and Vivante, North Holland Publ. Co., 1978, p.213.

2. K. Natesan, Environmental Degradation of High Temperature Materials, The Institution of Metallurgists, Series 3, No. 13, 1980, pp.1/12.

3. J. Stringer and W.T. Bakker, Proc. 8th Int. Cong. on Metallic Corrosion, Dechema, Frankfurt 3, 1981, p.2157.

4. H.J. Grabke, High Temperature Materials Corrosion in Coal Gasification Atmospheres, Ed. J.F. Norton, Elsevier Applied Science Publishers, 1984, p.59.

5. H.J. Grabke, J.F. Norton and F.G. Casteels, in "High Temperature Alloys for Gas Turbines and Other Applications", Eds. Betz, Brunetaud, Coutsouradis, Fischmeister, Gibbons, Kverness, Lindblom, Marriott and Meadowcroft, D. Reidel Publ. Co., 1986, p.245.

6. J.A. Kneeshaw, I.A. Menzies and J.F. Norton, Werkst. u. Korros. 38, 1987, p.473.

7. H.E. Bühler and H.P. Hougardy, Atlas der Interferenzschichten-Metallographie, Deutsche Gesellschaft für Metallkunde, Oberursel, West Germany, 1979.

49 THE USES OF THE FRESNEL METHOD IN THE STUDY OF OXIDATION

S.B. Newcomb, F.M. Ross, D. Ozkaya and W.M. Stobbs

Department of Materials Science and Metallurgy,
Cambridge University, Pembroke Street, Cambridge CB2 3QZ, UK.

ABSTRACT

The Fresnel Method is one of the many recently developed TEM techniques allowing a microscopist to get digital rather than analogue data from an image. The quantitative application of the Method can allow the composition of a flat, edge-on, interface region to be determined to near atomic level spatial accuracy, so it has considerable potential in the study of oxidation reactions. Here we briefly review the principles of the approach and describe a representative range of the qualitative, as well as quantitative, results we are currently obtaining using the Fresnel Method. We then assess some of the critical problems in the study of oxidation which could be practicably tackled in the future.

INTRODUCTION

The conventional approach to TEM remains that the image is used to obtain data on structure and that the analysis of local composition changes requires the use of a probe and either EDX or (P)EELS techniques. However the amplitude and phase changes which occur when an electron is scattered elastically are characteristic of the atomic number. Thus when there is a composition change at an interface viewed in projection, there will be changes in the elastic scattering which are directly related to the form of the projected scattering potential. It is at first sight startling that the digital analysis of a rather low resolution, and coarse, through focal series of images, showing the resultant Fresnel fringe patterns, can yield relatively high resolution data on the form and magnitude of the composional discontinuity which is present. Of course the elements causing the change have to be identified by the (poorer resolution) conventional approach of using a probe and EELS or EDX; in fact the Fresnel Method and probe approaches turn out to be ideally complementary in that, even with parallel recording and a specialised STEM, a probe method becomes somewhat qualitative for the analysis of regions smaller than about 2nm wheras it is beneath this dimension that the Fresnel Method starts to become quantitatively applicable in the characterisation of the profile of an interface.

For those wishing to take up the use of the method we have recently reviewed the basic approach and the methods of quantification [1], but it is probably still helpful to glance at Fig. 1 before proceeding further. Figures 1a and b illustrate schematically the intensity profiles to be expected for a discrete layer of lower scattering potential than the surrounding matrix imaged under- and over-focus respectively, while it is clear from Fig.1c how easy it is in general to determine the thickness of such a layer by the extrapolation of the Fresnel fringe spacings obtained at different defoci to zero defocus. Thickness measurements made in this way represent the first quantitative use of Fresnel contrast data (e.g. [2]), but it is astonishing how much more than simply the thickness of a given layer can be determined from a through-focal image series if the contrast behaviour is more fully characterised, as was pioneered by Ness et al. [3]. The fringe contrast is best defined as in Fig. 1a so as to minimise the sensitivity to uncertainties in the value of the background intensity. In broad terms, the higher the magnitude of the contrast, the bigger the change in the scattering potential (and thus effectively in the atomic number), while the more detailed the Fresnel fringe profile and the more rapidly the contrast develops as the defocus is changed, the sharper the change in scattering potential across the interface.

The sensitivity of the Fresnel Method for the analysis of a thin layer was established in our earliest quantitative work [3], on the characterisation of amorphous grain boundary layers in ceramics, where we suggested that the accuracy of an analysis was limited by the accuracy to which the thickness of the specimen could be measured for thicknesses in the 20 - 40nm range. Weak beam thickness fringe methods allow a thickness to be measured to within 0.5nm, leading to an accuracy of about 5% in the determination of layer composition. However, results obtained since then [4,5] on the variation of fringe contrast with specimen thickness have demonstrated the importance of inelastic scattering processes in altering fringe contrast. From this work it has become clear that there are really two approaches to the measurement of the composition of a thin layer. The first is to extrapolate data at a series of known thicknesses to zero specimen thickness and the second is to include in the analysis a quantitative model of the effects of inelastic scattering. The latter approach is rather difficult [6] while, as is demonstrated in our analysis of the oxides on Si [5], the former method can

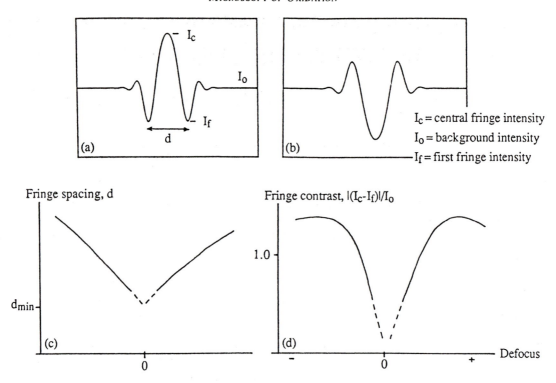

Figure 1: Schematic behaviour of the Fresnel fringes seen in defocused images of a thin layer. The form of the intensity is shown (a) underfocus and (b) overfocus for a region of lower scattering potential than the surrounding matrix. The way the fringe spacing and contrast change with defocus are shown in (c) and (d) respectively (from [1]).

give surprisingly accurate results. While there are thus fundamental, if evadable, difficulties in the quantification of the composition of a thin layer, the evaluation of the sharpness of a composition change at an interface is much less affected by uncertainties in inelastic scattering. It has thus proved susrprisingly easy to characterise the compositional profile of MBE-grown layers of thicknesses from one to a few unit cells of AlAs in GaAs to accuracies approaching the atomic [7]. In this context it is also worth noting that while the quantification of the composition of a layer requires the determination of the thickness of the foil examined in the TEM, this is not the case if it is only the sharpness of the composition change which is of interest. We have thus found it possible to determine the profile of the composition changes in the W(fine polycrystalline)/Si(amorphous) multilayers currently of interest for synchrotron applications without any viable method of measuring the foil thickness [8].

In the following examples of the use of the Fresnel Method in the characterisation of oxidation behaviour, we contrast an example of an area where a fully quantiative approach is required from those in which a qualitative attitude can give useful information.

APPLICATIONS OF THE FRESNEL METHOD IN THE STUDY OF OXIDATION

The area in the study of oxidation in which we have found the application of the Fresnel Method to have been indispensable has been the characterisation of how the manner of oxidation of silicon changes as the scale thickness increases [4]. The critical problem in this area is the determination of both the distance over which the composition of the silica is graded at the Si/SiO_2 interface and the precise position of this SiO_x layer with respect to the underlying Si. Interestingly this required the correlation of high resolution images (which were used to determine the position and structural roughness of the crystalline substrate) with the lower resolution, high defocus, Fresnel contrast images used to characterise the thickness of the layer of intermediate chemistry.

At this point it is perhaps worth explaining why it is that high resolution images are difficult to interpret in terms of the composition near to an interface, whereas a series of lower resolution images can yield high resolution information. Ourmazd et al. [9] have suggested a high resolution recipe for compositional analysis which neglects the effects of the interface in giving Fresnel contrast (as are used directly in the Fresnel Method for interface characterisation). We discuss this in detail elsewhere [10] but here we just note that Fresnel effects recorded at high resolution are sensitive to the precise form of the phase changes caused by the objective lens and these changes themselves are sensitive to a range of parameters which are difficult to quantify uniquely, as has been discussed in relation to such an approach for the characterisation of the rigid body displacement at $\Sigma 3$ grain boundaries [11]. In contrast to these problems in applying a direct high resolution approach, in the Fresnel Method the use of relatively low resolution (circa 1nm) and fairly large defoci makes the individual images insensitive both to the phase contrast associated with contamination and to the details of the high angle transfer behaviour. It is through the relative contrast changes from image to image that a knowledge of the high resolution form of the composition change is regained.

Returning then to the use of the Fresnel Method in the characterisation of the oxidation behaviour of Si, the accuracy of the approach in determing the thickness of the layer of intermediate composition may be assessed from Fig. 2, where experimental values of contrast as a function of defocus are compared with computer simulations for two narrow oxide layers. For both these layers, the thickness of the layer of intermediate composition (which, we note, is seen in projection and is in part due to roughness, as can be characterised from the high resolution image) can be determined with atomic level sensitivity. The summarised data we obtained for a variety of oxide thicknesses, prepared both by "wet" and dry oxidation, are shown in Fig. 3, and the significant result is that we can see that as oxidation proceeds the layer of

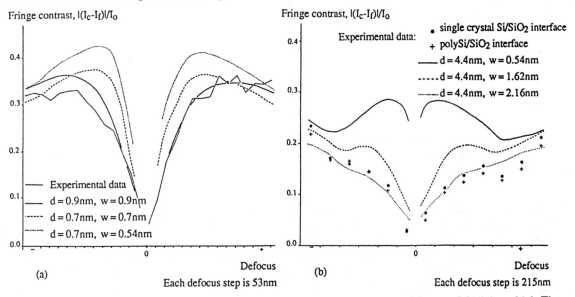

Figure 2: Measured and simulated values of fringe contrast for two silica layers: (a) 0.9nm and (b) 4.4nm thick. These oxides were formed by chemical oxidation and after growth were capped with polysilicon. The simulations assumed an oxide layer width and spread in composition (i.e. interface diffuseness) indicated by d and w respectively. The best fitting simulations had interface diffusenesses of 0.9 and 1.2nm respectively. For details see [4].

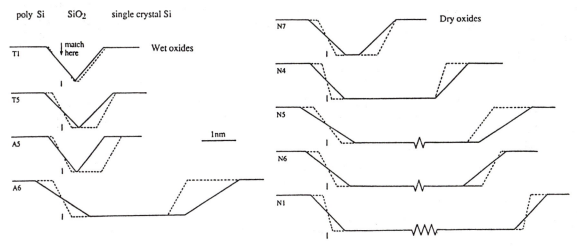

Figure 3: A comparison, for two sets of oxides, of structural measurements of oxide width and interface diffuseness (dotted lines) obtained from high resolution images and compositional measurements (solid lines) obtained from the best fitting Fresnel simulations. In comparing the profiles, the polySi/SiO₂ interfaces have been matched. For details see [4].

intermediate composition moves from the amorphous side of the interface into the crystalline region. The characterisation of this effect has allowed us to reconsider the wealth of models that have been proposed for Si oxidation and thence to formulate an essentially new model based on a rate-controlled mechanism change when the oxide is between about 2 and 3nm in thickness [4].

An evaluation of the type described above, while being rewarding, is time-consuming, and we should emphasise how effective qualitative approaches can be. In Fig. 4 we show a more technological application in the same field in which it is probable that the thickness of an oxide layer between the Si substrate and a superposed polysilicon layer above a critical lower limit was responsible for the poor performance of devices incorporating the overlayer as a contact. Here the presence of the amorphous oxide layer about 1.5nm thick is best demonstrated by taking a dark field image (such as that in Fig. 4a) using an aperture on the silica halo in a direction perpendicular to the interface normal. The thickness of the layer is then however best evaluated by the simple expedient of extrapolating the Fresnel fringe profile peaks (in images such as those shown in Fig. 4b and c) to their positions at zero defocus while remembering that there can be small systematic errors in this procedure [4]. We include this example here partially because it is important to emphasise that even perfectly clean interface would exhibit rather strong Fresnel contrast if it were also flat because of the rigid body displacement which causes a local discontinuity in the scattering potential.

A contrasting qualitative application is to be found in work we describe elsewhere in these proceedings on the oxidation behaviour of Al alloy based metal matrix composites containing SiC particles [12]. Since in general these carbides develop only rather thin scales of silica, the problem was often to differentiate a layer of oxide upon them from the overlayers of araldite used in the preparation of the edge-on specimens. The difference in scattering potential for silica and silicon carbide, when compared with that to be expected between silica and the polymeric overlayer, lent a characteristic appearance to the Fresnel fringe image series of oxide coated particles. This may be seen in Fig. 5, which shows such images for the outer surface of an oxidised SiC particle which is shown under differing image conditions in [12].

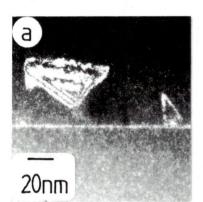 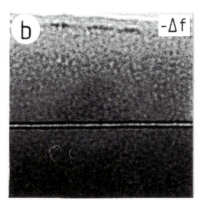 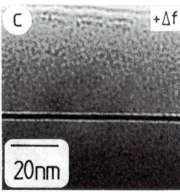

Figure 4: A polysilicon layer superposed on a Si substrate. The presence of the oxide is demonstrated by the d.f. silica halo image in (a) while its thickness is easily evaluated by the extrapolation of fringe peak positions in images such as those shown in (b) and (c) to zero defocus.

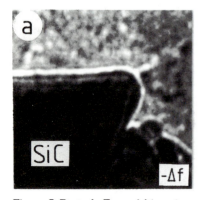 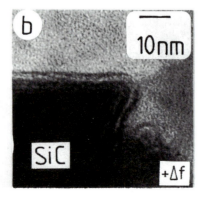 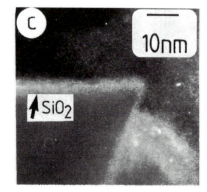

Figure 5: Part of a Fresnel fringe image series used in the characterisation of the superposed layer seen on the SiC particle. See [12] for a further description of the oxidation behaviour.

THE FUTURE POTENTIAL OF THE FRESNEL METHOD IN OXIDATION

The most exciting result we have recently obtained with respect to the future use of the Fresnel Method is that it appears to be possible to evaluate both the rigid body displacement and the compositional profile of a species segregating to a grain boundary, provided (as is normally the case for a segregating element) that its atomic number is substantially different from that of the matrix. Sets of Fresnel fringe profiles are shown in Fig. 6 for a "clean" grain boundary in Al and for quenched and annealed Al-Pb and Al-Sn alloys. The details of the interpretation of these images are described elsewhere [13], but the important point to note is that we have two unknowns (the rigid body displacement and the segregational profile) and two features of the Fresnel image series (the contrast changes and the absorption) to use in their evaluation. Even a qualitative evaluation of the profiles shown in Fig. 6 demonstrates that, whereas Sn segregates in a non-equilibrium fashion, Pb behaves in an equilibrium manner. Given, then, the ability to quantify this type of internal boundary, we can start to consider a whole range of useful applications of the Fresnel Method. We should, for example, be able to evaluate the effects of corrosion on a segregating species at a grain boundary. In this context we were struck by the way the grain boundaries on the Al-Sn sample described above showed high Fresnel contrast when the surface of the sample was

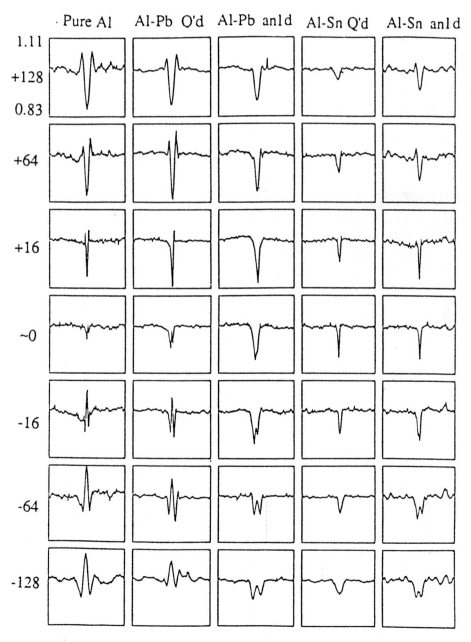

Figure 6: Fresnel fringe profiles for grain boundaries in (a) pure Al, (b) quenched and (c) annealed Al-Pb, and (d) quenched and (e) annealed Al-Sn alloy. The defocus step between images is 65nm.

anodised (Fig. 7a), and accordingly we examined the intersection of a grain boundary in this material in the quenched state (with nonequilibrium Sn segregation) with the anodised surface in an edge-on sample (Fig. 7b). While the Fresnel contrast from the boundary near to, as well as far from (circa 7μm), the anodised surface was now more indicative of the segregation being nearer to equilibrium (Figs.7c, d) than was observed some six months earlier for the same specimen (Fig. 6), this is unfortunately probably explained by room temperature ageing (as is evidenced by the grain boundary precipitation of Sn particles (Fig. 7b) which was not seen in the previous work). Furthermore the Fresnel contrast of the boundaries was qualitatively the same both near to and far from the surface, so it is unlikely that the effects on the segregation we are seeing are due to the anodisation.

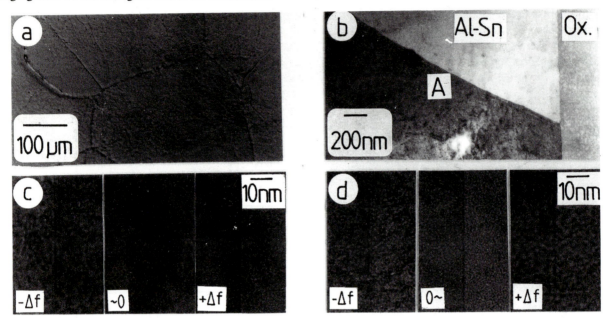

Figure 7: Grain boundary near to the anodised Al-Sn surface: (a) optical of surface; (b) edge-on image of g.b. intersecting surface; (c) and (d) Fresnel pairs of g.b. from A and from 7μm below this.

There are of course many possible applications of classical importance in the modelling of oxidation phenomena. It will, for example, be a challenge to see whether or not there are differences in grain boundary structure and composition in the Fe-Cr spinels in the composition range over which their diffusivity changes.

ACKNOWLEDGEMENTS

We are grateful to Prof. D. Hull for the provision of laboratory facilities and to the SERC for support (S.B.N.) as well as the Turkish Government (D.O.).

REFERENCES

1. W.M. Stobbs and F.M. Ross, in NATO ASI Series B (Physics), 203, Ed. D. Cherns, Plenum, London, 1989, p183.

2. D.R. Clarke, Ultramicrosc. 4, 1979, p.33.

3. J.N. Ness, W.M. Stobbs and T.F. Page, Phil. Mag. A54, 1986, p.679.

4. F.M. Ross and W.M. Stobbs, A study of the initial stages of oxidation of Si using the Fresnel Method, Phil. Mag., in press.

5. F.M. Ross and W.M. Stobbs, Computer Modelling for Fresnel Contrast Analysis, Phil. Mag., in press.

6. W.M. Stobbs and W.O. Saxton, J.Microsc. 151, 1988, p.171.

7. F.M. Ross, E.G. Britton and W.M. Stobbs, Proc. AEM, Manchester, 1987 (Ed. G. Lorimer), IOM Publications, 1988, p.205.

8. W-C. Shih and W.M. Stobbs, The use of the Fresnel Method for the profiling of W/Si multilayers, Ultramicrosc., in press.

9. A. Ourmazd, D.W. Taylor, J. Cunningham and C.W. Tu, Phys. Rev. Lett., 62, 1989, p.933.

10. M.J. Hytch, C.B. Boothroyd, F.M. Ross, E.G. Bithell and W.M. Stobbs, Proc. EMAG 89, Inst. Phys. Conf. Ser. Ed. P.J.Goodhew, 98, 1990, p.345.

11. G.J. Wood, W.M. Stobbs and D.J. Smith, Phil.Mag. A50, 1984, p.375.

12. P.B. Prangnell, S.B. Newcomb and W.M. Stobbs, these proceedings, p.245.

13. D. Ozkaya, W-C. Shih, S.B. Newcomb and W.M. Stobbs, Proc. EMAG 89, Inst. Phys. Conf. Ser. Ed.P.J.Goodhew, 98, 1990, p.199.

50 OXIDATION OF A FERRITIC STAINLESS STEEL CONTAINING 17% Cr STUDIED BY RAMAN LASER SPECTROSCOPY AND NUCLEAR REACTION ANALYSIS

H. Chaudanson, F. Armanet, G. Beranger and J. Godlewski

Division Matériaux - UA 849 CNRS- Université de Technologie de Compiègne, BP 649, F-60206 Compiègne CEDEX, France.

and

A. Hugot-Legoff, S. Joiret and M. Lambertin

Laboratoire de Physique des Liquides et Electrochimie - LP 15 CNRS-Université P et M Curie Tour 22, 4 place Jussieu 75252 Paris CEDEX 05, France.

ABSTRACT

The oxidation behaviour of a ferritic stainless steel containing 17%Cr in dry or wet air was examined during exposures of less than one minute at 1200°C. The extent of attack was assessed by direct weight measurement complemented by nuclear reaction analysis of the oxygen uptake. The phases produced were identified by Laser Raman and Glow Discharge spectroscopy. The oxidation mechanism appeared to be a competition between chromia formation, its evaporation as CrO_3 and spinel oxide formation.

INTRODUCTION

Knowledge of the oxides formed on the surface of stainless steels during hot forming are of great importance, as is the understanding of their growth. The oxides studied were produced in different conditions, (particularly the humidity of the atmosphere), and during short exposures. The present work deals with the nature and the growth processes of such oxides.

EXPERIMENTAL

The oxidation runs to simulate the growth of these oxides were short. Consequently the oxidation kinetics were difficult to record. Two different techniques were used to assess the extent of attack: measurement of the weight change and determination of the oxygen taken up by the sample by using Nuclear Reaction Analysis (NRA). The oxide scales formed were analyzed by X-ray diffraction (XRD), EDX analysis, Glow Discharge Optical Spectroscopy (GDOS) and Laser Raman Spectroscopy (LRS).

Weight Gain Measurement

The apparatus used to oxidize the samples is shown on Fig. 1. The specimen was placed in the silica tube which was evacuated (10^{-10} Pa = 10^{-5} atm). Then the furnace, which was previously heated, was moved around the sample. When the temperature was stabilized in the tube, the oxidizing atmosphere (dry or wet air) was introduced. At the completion of oxidation, the tube was evacuated and the furnace was moved from the sample. The gain in the specimen mass during oxidation was measured, the mass change was also determined during the heating up period. Several experiments were carried out without introducing the oxidizing gas (oxidation duration t = 0) and these results proved to be very important.

Nuclear reaction analysis allows the amount of various elements (O, C, N, P, S, ...) in a different matrix to be determined with high accuracy [1]. For this purpose, charged particles (deuterons) were accelerated (900 KeV) in a Van De Graaff accelerator to induce the nuclear reaction: $^{16}O (d,p)^{17}O*$. The emitted charged particles (protons) were detected by mean of semiconductor detectors and their energy spectrum was recorded. The interpretation of these spectra allows identification of the nuclei in the target. A standard reference target permitted the amount of oxygen taken up by the sample to be calculated. It has to be noted that the thickness analysed depends on the matrix. In the present work it can be reasonably presumed that the entire oxide scale was examined.

Glow Discharge Optical Spectroscopy

This method is well-known but quantitative analyses are difficult to perform, particularly when the specimen is non-homogeneous. However, it is possible to obtain an approximate image of the element distribution even in thin scales and the substrate.

Laser Raman Spectroscopy

Laser Raman Spectroscopy (LRS) is recognized as a powerful technique for the analysis of reaction product films on metal surfaces [2-7]. Interest in the technique is based on the physical nature of the Raman scattering process [7]. Light of energy $h\nu_0$ is scattered from a sample at energy $h(\nu_0 - \nu_1)$ (Stokes shifted) and $h(\nu_0 + \nu_1)$ (anti-Stokes shifted), where $h\nu_1$ is a Raman

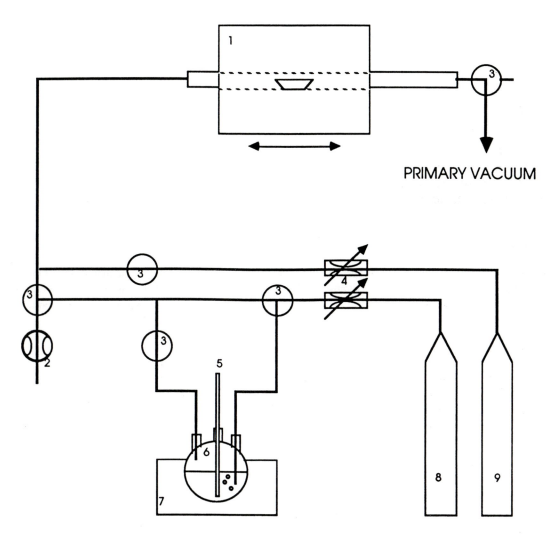

PRIMARY VACUUM

1: FURNACE
2: FLOW -METER
3: SLUICE
4: MANOMETER
5: CONTACT THERMOMETER
6: WATER VAPOUR
7: BALLON-HEATER
8: AIR
9: NITROGEN

Figure 1: Experimental apparatus for oxidation testing.

allowed transition energy corresponding to rotational, vibrational or electronic molecular energy difference. The energy loss from the incident photons are chemical compound specific and correspond to wavelength shifts in the scattered light. Detailed analysis of band positions and widths allows information of more physical nature to be obtained, such as stoichiometry, stress intensity and degree of crystallinity [8].

The oxidized specimen was positioned under an optical microscope. An argon laser beam (514,53 nm, 200 mW) was focused onto the specimen. The scattered light was directed into a monochromator. Comparison of the observed wave number shift with standards enabled the actual phases to be identified [8-9]. LRS readily permitted the micro compositional examination of most scales formed during short exposures. Its important advantage is that it is a non-destructive technique.

Samples and Oxidation Conditions

The alloy studied in this work was a ferritic stainless steel containing 17% chromium with the composition given in Table 1.

Table 1. Alloy composition (wt%)

	Cr	Ni	Si	Mn	C	Fe
Fe-17%Cr	16.50	0.38	0.22	0.45	0.04	Bal

Oxidation runs were performed at 1200°C during which the reaction durations were less than 1 minute. The oxidising atmosphere was dry or wet air. The water partial pressure varied between 0 and 31150 Pa (highest dew point: 70°C). Kinetic oxidation tests were performed only in dry air and in wet air with 2330 Pa and 7370 Pa of water vapour (dew points 20°C and 40°C).

RESULTS

Oxidation Kinetics

The oxidation curves obtained using both experimental techniques, shown in Fig. 2, indicate that whatever measurement method is used, the reaction kinetics did not fit any simple law. The irregularities in these curves became more significant with increasing humidity levels. Comparison of both series of measurements show that they are in good accordance: minima and maxima of the irregularities (waves) were obtained for the same times (Fig.3).

Some details are important. Firstly, the specimen weight had a "negative" value at the beginning of the run (t = 0) which means that a mass loss occurred during heating under vacuum. Secondly, the mass variations which are directly measured and those which are calculated using NRA are different by an order of magnitude. Thirdly, the oxygen uptake at the beginning of the run (t = 0) is not zero. It corresponds to that of the natural room temperature formed oxide film or more probably due to a film formed during the heating under vacuum.

Oxide Scale Analysis

Glow discharge optical spectroscopy and laser Raman spectroscopy were mainly used. X-ray diffraction and EDX analysis were more difficult to employ due to the thinness of the scales grown. Analysis of a taper cross-section was attempted but the scales were brittle and poorly adherent. It is important to note that the analyses were performed after short oxidation exposures (15 s) when mass charge curves were continuous so that the scales were coherent.

The results (Fig. 4) are as follows:

(a)   In dry air, the Raman bands corresponded to Cr_2O_3 and to a spinel ($Fe_xCr_{3-x}O_4$) with the main corrosion product being chromia.

(b)   When water vapour was introduced into the atmosphere (dew point 20°C or 40°C), the Raman bands corresponding to a spinel oxide (690 cm^{-1}) increased, while the Raman band corresponding to Cr_2O_3 (555 cm^{-1}) remained unchanged.

(c)   When increasing humidity (dew points 50 and 60°C), the spinel oxide bands become even more intense, whilst the chromia band decreased.

(d)   Finally, if the humidity content of the atmosphere was increased again (dew point 70°C), only iron oxides were formed on the alloy.

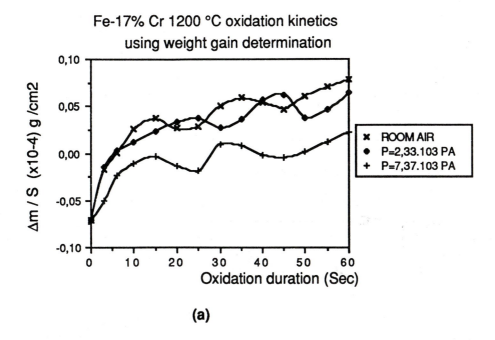

(a)

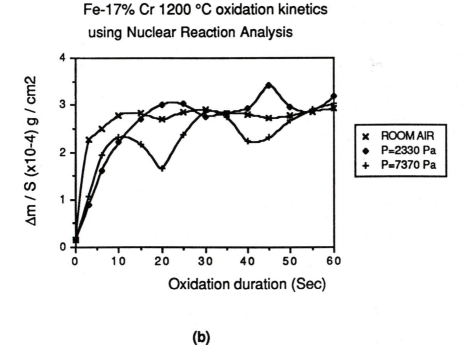

(b)

Figure 2: Fe-17% Cr: oxidation kinetics (a) weight gain determination, (b) using Nuclear Reaction Analysis.

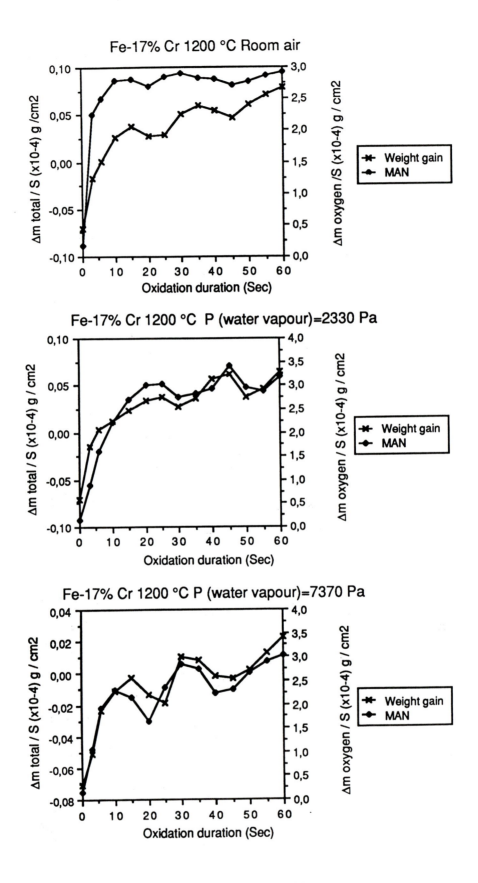

Figure 3: Fe-17% Cr oxidation kinetics determined by weight gain and Nuclear Reaction Analysis for different water vapour pressures.

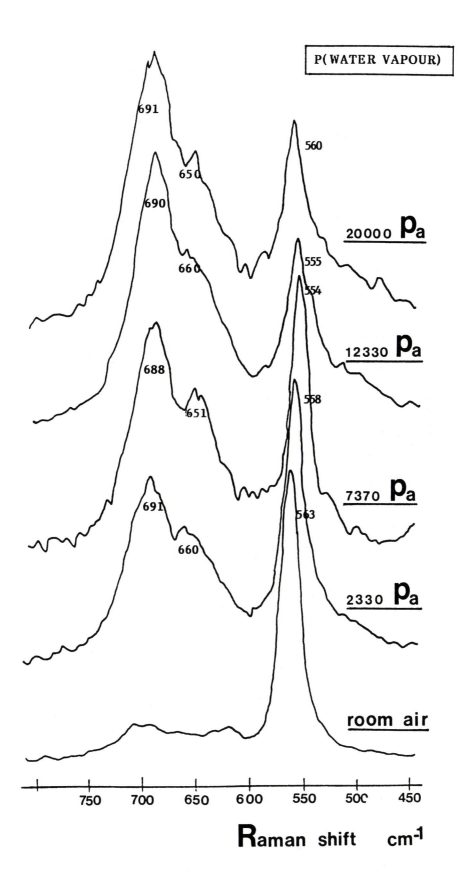

Figure 4: Fe-17% Cr oxidized during 15 sec at 1200°C. Raman spectras evolution versus water vapour pressure.

GDOS analyses (Fig.5) seem to confirm these results. These suggest that only Cr_2O_3 was present with dry air; chromia and spinel formed in air with a low dew point, and only spinel in air with the highest humidity. It is difficult to quantify the LRS results but an approximation of the scale composition can be obtained from a ratio of the intensities of the different Raman bands. For example, the chromia content can be calculated using the expression:

$$\%Cr_2O_3 = 100\ I(Cr_2O_3)\ /\ I(spinel) + I(Cr_2O_3)$$

Results obtained are reported in Table 2. These calculations confirm that chromia disappears when the humidity of the environment increases, whilst spinel oxide becomes increasingly important.

Table 2: Estimated composition of oxide scales formed after 15 s

P H$_2$O (Pa)	Cr$_2$O$_3$ %	Spinel %
Room air	82	18
2330	47	53
7370	41,5	58,5
12330	29	71
20000	32	68

DISCUSSION

The first point of interest is the oxidation kinetics. It was mentioned that a mass loss occurred during heating, which means that one of the alloying elements was being evaporated. The steel oxidised contains chromium (17%). Even if vacuum was established in the apparatus, sufficient oxygen remained to oxidize the specimen and preferentially form chromium oxide (Cr_2O_3). However, it is well-known that volatile oxides (like CrO_3) may form at very high temperatures [10-12]. For that reason it is probable that the weight loss during the heating was due to the chromium evaporation as this volatile oxide. A proof of this pre-oxidation is given by the amount of oxygen (determined by NRA) at the beginning of the exposure which is not negligible.

Chromium evaporation continued after the introduction of the oxidizing atmosphere. When the curves obtained by both experimental techniques are compared, it can be seen that the weight gain measured by weighing the sample is 15 to 20 times lower than the mass gain calculated by NRA. A simple calculation established that a large part of the scale formed disappeared. This result is to be expected as chromium oxide evaporation is favoured thermodynamically by an oxygen pressure above $2 . 10^{+4}$ Pa = 0.2 atm.

However it is more difficult to understand the observation of the "waves" in the curves. It can only be concluded that abnormal kinetic processes occur. Some authors have observed similar results and explained these by local thermal fluctuations during the scale growth, favouring alternatively two competitive processes. In this case one of them could be the Cr_2O_3 formation and the other CrO_3 evaporation.

It may be noted that an increase of the humidity in the environment was accompanied by a simultaneous increase in the amplitude of the waves. This is in accord with results reported by different authors where the introduction of water vapour in the atmosphere favours both the Cr_2O_3 formation and the volatile CrO_3 formation.

Turning to the results of the different analyses performed on the scales, increasing the dew point in the atmosphere resulted in the preferential growth of the spinel oxide, whilst Cr_2O_3 disappeared. This can be explained by two factors. The first, as previously explained, was that chromium oxide evaporation was favoured by the presence of humidity in the atmosphere. At the same time, the plasticity of Cr_2O_3 increases: so it may be supposed that more defects are produced. Generally the spinel oxides are not formed directly on the metallic substrate but grow by cationic diffusion and solid-solid reaction. In this case, growth could be favored by an increase of point defects in the remaining Cr_2O_3. Finally, at the highest dew point, chromium oxide evaporation becomes so fast that a large Cr depletion is obtained and the formation of iron oxide occurs.

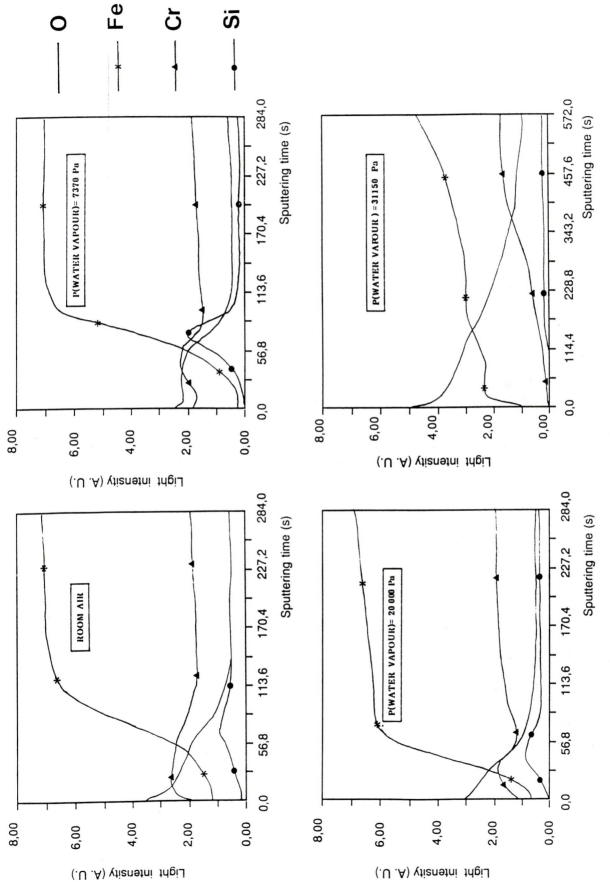

Figure 4: Fe-17% Cr oxidized during 15 s at 1200°C. Raman spectras evolution versus water vapour pressure.

CONCLUSION

This work has shown that various oxides are formed by high temperature oxidation of a ferritic stainless steel during short times, their composition being essentially related to the humidity in the atmosphere.

Difficulties were experienced due to the short exposure times in the oxidizing atmosphere, so the scales formed were thin, brittle and poorly adherent to the substrate. Moreover, oxide evaporation occurred during these exposures. It was necessary therefore to employ several different techniques to get information. The extent of attack was assessed by direct weight measurements complemented by nuclear reaction analysis of oxygen; while Laser Raman Spectroscopy allowed identification of oxide phases formed. All the results permitted the postulation of an oxidation mechanism, based on a competition between chromia formation, its evaporation (as CrO_3) and spinel oxide formation.

ACKNOWLEDGEMENTS

The authors thank Ugine-Savoie-Society for financial support and are grateful to J.P. Moreau and J.P. Cazet of IRSID St Germain en Laye who performed Glow Discharge Optical Spectroscopy analysis. Nuclear Reaction Analysis was supported by GRD 86 CNRS.

REFERENCES

1. G. Amsel, J.P. Nadai, E. D'Artemare, D. David, E. Girard and J. Moulin, Nuclear instruments and methods, 92, 1971, p.481.

2. D.J. Gardiner, C.J. Littleton, K.M. Thomas and K.N. Strafford, Oxid. Met., 27, 1987, p.57.

3. R.L. Farrow, P.L. Mattern and A.S. Nagelberg, Appl.Phys.Lett., 36, 1980, p.212.

4. R.L. Farrow and A.S. Nagelberg, Appl.Phys.Lett. 36, 1980, p.945.

5. M.J. Bennett, P.R. Graves and R.W.M. Hawes. High Temperature Alloys. Their exploitable potential. Eds. J.B. Marriott, M. Merz, J. Nihoul and J. Ward. (1985) Elsevier Applied Science, p.205.

6. A. Hugot-Le Goff. Méthodes usuelles de caractérisation des surfaces, Eds. D.David and R.Caplain, Eyrolles, Paris, 1988, p.102.

7. W.A. England, M.J. Bennett, D.A. Greenhalgh, S.N. Jenny and C.F. Knights, Corros. Sci., 26, 1986, p.537.

8. K.F. McCarty and D.R. Boehme, J. Solid State Chem., 79, 1989, p.19.

9. N. Boucherit, A. Hugot-Le Goff, S. Joiret, G. Beranger and H. Chaudanson, Thin Solid Films, 174, 1989, p.111.

10. J.A. Menzies and D. Mortimer, Corros. Sci., 6, 1966, p.517.

11. F. Armanet, "Influence comparée de la plasticité des couches d'oxydes sur la résistance à l'oxydation à haute température du nickel, du chrome et de certains de leurs alliages: rôle de la vapeur d'eau". Thesis, University of Compiegne, 1984.

12. L. Uller, "Oxydation des aciers inoxydables dans les mélanges vapeur d'eau oxygène à hautes températures". Thesis, University of Pierre et Marie Curie, Paris VI, 1980.

51 REAL TIME STUDIES OF URANIUM DIOXIDE OXIDATION

J.B. Price, M. J. Bennett and F. L. Cullen

Materials and Manufacturing Technology Division, AEA Industrial Technology, Building 393, Harwell Laboratory, Didcot, Oxon OX11 0RA, UK.

and J. F. Norton and S. R. Canetoli

Joint Research Centre, Petten Establishment, 1755ZG Petten, The Netherlands.

ABSTRACT

Three series of real time experiments have been undertaken concerned with the disintegration of UO_2 during oxidation in air at temperatures in the range 225 to 550°C. Gravimetric measurements, using a controlled atmosphere microbalance, have established the time at which powdering first occurred and the rate of subsequent continuing spallation. In-situ observations have been made also using environmental cells with hot stages fitted respectively to an X-ray diffractometer and an optical microscope. The former have examined the chemical and the latter the topographical changes during UO_2 oxidation. The results have both confirmed and extended current understanding of the mechanisms of the chemical and physical processes involved.

INTRODUCTION

In the majority of nuclear power plants the fuel is solid uranium dioxide (UO_2) pellets contained in a protective metal cladding, e.g. of Zircaloy for Light Water Reactors and of 20Cr/25Ni/Nb stainless steel for the Advanced Gas-cooled Reactors (AGR). Possible fault scenarios, which can be envisaged in the respective fuel cycles during both reactor operation and post reactor discharge, transport and storage, could involve exposure of defective fuel elements to air. Therefore, the oxidation behaviour of UO_2 in this environment has been the subject of continuing studies [1-4]. In a low temperature regime up to ~ 550°C, the reaction sequence initially involved the formation of U_3O_7, while the final reaction product was the higher oxide U_3O_8. These transformations involved substantial volume changes. As a consequence oxidation caused disintegration of the UO_2 pellet with concurrent spallation of oxide particulates. In the context of the behaviour of defective reactor fuel elements, oxidation of fuel could result in the splitting open of the protective metal cladding and the release of radioactive dust, as well as of gaseous fission products.

At 200-550°C the kinetics of oxidation comprised of an induction period with a linear growth rate followed by an increased constant rate, until after ~80% oxidation the rate steadily decreased as UO_2 was consumed [1, 2]. As might be expected, increasing temperature was marked by a decreasing induction period and an increasing constant oxidation rate. For these measurements the UO_2 specimens were contained in crucibles, which consequently retained spalled oxide particulates. The end of the induction period has been assumed to coincide with the onset of spallation and values derived from the kinetic curves [2] agreed well with the limited direct measurements available for the time to powdering [4]. In the current study the first series of real time experiments, therefore, using freely suspended specimens aimed to extend these data by thermogravimetric measurement of the time for onset of spallation during air oxidation to cover the complete temperature range of interest (225-550°C). Additionally for the first time the rates of continuing spallation were determined.

The chemical changes during the low temperature oxidation of UO_2 have been studied by X-ray photoelectron spectroscopy and X-ray diffraction [3], while the corresponding physical changes have been followed by optical and scanning electron microscopy [1, 3]. These have led to the postulation of a possible mechanism for oxidation and spallation [3]. All these observations, however, were made on specimens cooled to room temperature, so that there always remains questions whether these exactly reproduced the situation at temperature or if some modification, particularly of a physical nature, e.g. cracking, occurred on cooling.

To consolidate current understanding of the mechanism of oxidation of UO_2 in the low temperature regime two further series of real time observations have been made using environmental cells with hot stages fitted to an X-ray diffractometer and to an optical microscope. The former examined the compositional changes and the latter the physical processes which occurred during UO_2 oxidation in air at 250-550°C. These results, together with the gravimetric measurements of the time for onset of spallation and of the continuing spallation rate, form the basis of this paper.

EXPERIMENTAL

Slices (approximately 4 mm x 5 mm x 7.5 mm) of UO_2 were cut using a diamond saw from pellets of current AGR fuel (2.21% enriched). The pellets had a nominal density of 10.81 g cm^{-2} (99% theoretical density) and a grain size of 14-16 μm (mean linear intercept). For hot stage microscopy one side of the UO_2 slice was mechanically polished down to 3 μm diamond to produce a mirror finish. Oxidation was carried out using a CI Electronics controlled atmosphere microbalance. The specimen was supported within a simple gold wire cradle, which did not impede the ejection of fine particles. The oxidising gas, laboratory air, at 1 atm pressure flowed over the specimen at 400 cc min^{-1}.

X-ray diffraction studies on the UO_2 pellet were performed using an APEX goniometer incorporating a GTP high temperature environmental chamber. The radiation employed was Cu Kα and detection of the diffracted X-rays was by an MBRAUN position sensitive detector with a beta filter. The detector was centred at 30° 2~ giving an effective measuring range between 25° and 35° 2u. The sample was mounted in a stainless steel holder using aluminium foil to wedge it in place. Heating of the sample was effected in static air by a tantalum element and the furnace temperature was measured and controlled by a Ni-Cr/Ni-Al thermocouple situated above the sample. The furnace was heated to 250°C and the diffraction data were collected after periods of 5 minutes and 55 minutes, using a count time of 5 minutes. After 1 h at 250°C the temperature was raised to 300°C and during a similar exposure duration diffraction data were obtained after the same times (5 and 55 mins). The procedure was then repeated successively at 325 and 350°C with the total oxidation times being 85 and 55 mins respectively.

Topographical changes on the polished UO_2 surface during oxidation in 1 atm air at 450, 500 and 550°C were studied using a conventional optical microscope attached to a small environmental chamber containing an alumina hot stage heated by Pt-20% Rh windings. The furnace temperature was measured and controlled by a Pt/Pt-Rh thermocouple attached to the stage. The environmental chamber was water-cooled and had side ports for gas inlet and outlet, electrical supply and the control thermocouple. The chamber was sealed with a transparent quartz top plate, which acted as a viewing window. The microscope was fitted with special lenses to enable changes in the UO_2 surface to be followed during oxidation, normally at x320 magnification. The image was enhanced using a video camera coupled to a recording/display system, which enabled magnification to x1300, whilst retaining excellent resolution. Oxidation was observed by time lapse recording 10 s of each 60 s exposure to the oxidant, thereby allowing rapid topographical changes to be followed.

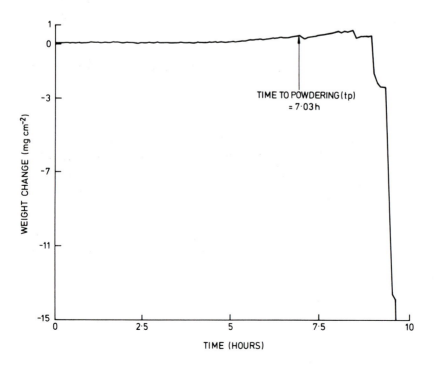

Figure 1: Variation with time of the weight of UO_2 during oxidation in air at 300°C.

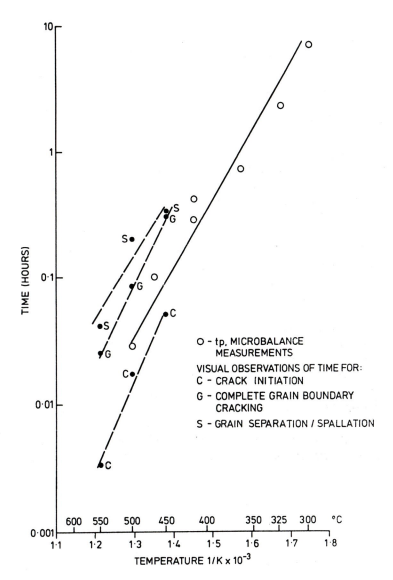

Figure 2: The effect of temperature upon visual observations of cracking and spallation and of gravimetric measurements of the time (tp) to powdering.

RESULTS AND DISCUSSION

During air oxidation at all temperatures the UO_2 specimen weight varied with time, as shown in Fig. 1 for oxidation at 300°C. Initially the specimen weight increased until at a time designated as tp , the time to powdering, a small drop of weight occurred due to the spallation of oxide particulate. For a short while afterwards the specimen weight continued to increase overall, although with occasional small losses, until further oxidation resulted in a continuing loss at a steady rate, which was measured as that of spallation (R_s). The weight gains at the onset of spallation ranged between 0.48 and 1.65 mg/cm² and appeared to be temperature independent. By contrast wlth increasing temperature tp decreased and R_s increased, as indicated by the respective Arrenhius plots (Figs. 2 and 3). Values of tp at 225 and 250°C were 697 and 161 h and are not shown in Fig. 2. Linear regression analysis would suggest that the best estimates of these parameters between 225 and 550°C may be represented by the following expressions,

$$\log tp = -9.028 + \frac{5804}{T}$$

$$\log R_s = 9.246 - \frac{4561}{T}$$

where tp = hours, T = degrees Kelvin and R_s = mg/cm².

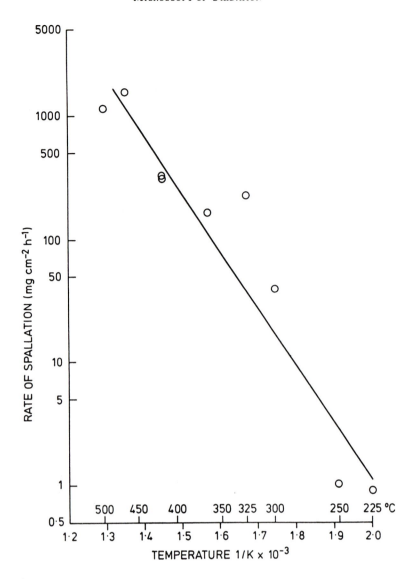

Figure 3: Arrhenius plot of the rate of spallation during UO₂ oxidation in air at 225 to 500°C.

The temperature of the UO₂ sample in the X-ray diffractometer environmental cell was increased in a step wise mode. Successive exposures were 1 h at 250°C, 1 h at 300°C, 85 min at 325°C and finally 55 min at 350°C. Therefore the cumulative exposure, equivalent to 2.84 h at 350°C just exceeded the anticipated tp (2.0 h) for isothermal exposure at this temperature. X-ray patterns were taken at regular intervals. As shown in Fig. 4 increasing exposure time and temperature was marked by a gradual reduction in the intensity of all UO₂ peaks and a corresponding increase in the peaks from a new tetragonal phase, U_3O_7. During oxidation at the highest temperature (350°C), concurrent with the expected tp, a faint U_3O_8 line also became apparent. Direct XRD observations at this stage were hampered by the upward movement of the UO₂ surface as a result of oxidation. Detailed XRD on cooling the specimen to room temperature confirmed the presence of U_3O_7 and U_3O_8 as reaction products.

Optical hot stage microscopy revealed that the same sequence of topographical changes occurred during the oxidation of UO₂ in air at all three temperatures (450, 500 and 550°C). The UO₂ surface was not changed visually by any slight oxidation occurring during heating up to temperature (450°C) (Fig. 5(a)) and was characterised by small depressions which could be attributed primarily to material removal on polishing, together with some remnant inherent porosity. No grain structure was apparent. The visual observations identified three sequential stages during oxidation. The first was initiation of cracking within the UO₂ grain boundaries and also within the grains, with some associated minor distortion of the surface, as is shown in Fig. 5(b), which is the same area as in Fig. 5(a). During the second stage (Fig. 5(c), another area of the surface) cracking and surface distortion became more pronounced with reaction product, probably U_3O_8, formation within the cracks. Finally spallation of complete grains or grain sections occurred (compare Fig. 5(d) with same

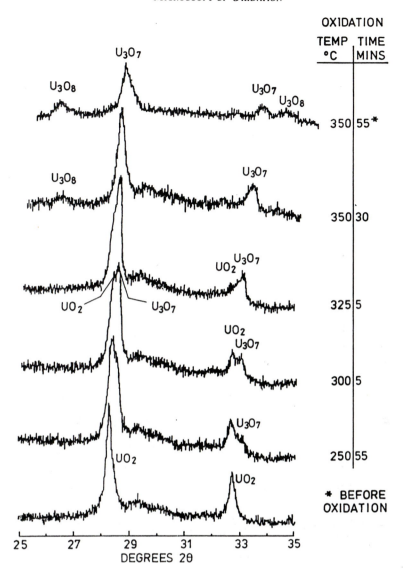

Figure 4: Change in X-ray diffraction peaks between 25° and 35° 2θ during the oxidation of UO₂ at successively increasing temperatures between 250° and 350°C. Traces marked with an asterisk were measured at room temperature.

area Fig. 5(c)). The times for each successive stage at the three temperatures were obtained from the video recording and decreased with increasing temperature. These values are plotted on Fig. 6 and appear to be entirely consistent with the gravimetric tp measurements; each obeying a similar Arrhenius type relationship. The times to crack initiation were shorter than tp, while those for complete grain boundary cracking and spallation were in reasonable agreement with tp.

The present results were entirely consistent with the postulated understanding of the low temperature oxidation behaviour of UO₂ in air [1-3]. Because these were real time studies, they provided substantial validation of the proposed mechanism by the elimination of possible uncertainties associated with the discontinuous observations on which this mechanism was originally based. The oxidation of UO₂ within this temperature regime is shown schematically in Fig. 6. Initially during the incubation period, oxidation of UO₂, via inward oxidant transport, resulted in the formation of a sub-stoichiometric U_3O_7 layer, several microns thick. The rate determining step is believed to be reaction at the U_3O_7-UO_2 interface, involving the incorporation of oxygen ions into stable interstitial clusters in the underlying UO_2 matrix. The overall contraction of the UO_2 fluorite lattice associated with the incorporation of extra oxygen atoms resulted in intergranular and transgranular cracking within the surface oxide layer. Optical microscopy showed this also involved fracture within grains beneath the surface [1], which would not have been apparent looking down on the surface. At this stage of oxidation disruption of the surface was only minor and there was no spallation. A critical local U_3O_7 thickness might have been needed before the growth stress exceeded the fracture strength of a grain boundary or surface. U_3O_8 nucleated at high surface energy sites generated by the intergranular and transgranular microcracks. The 30% increase in volume associated with the phase

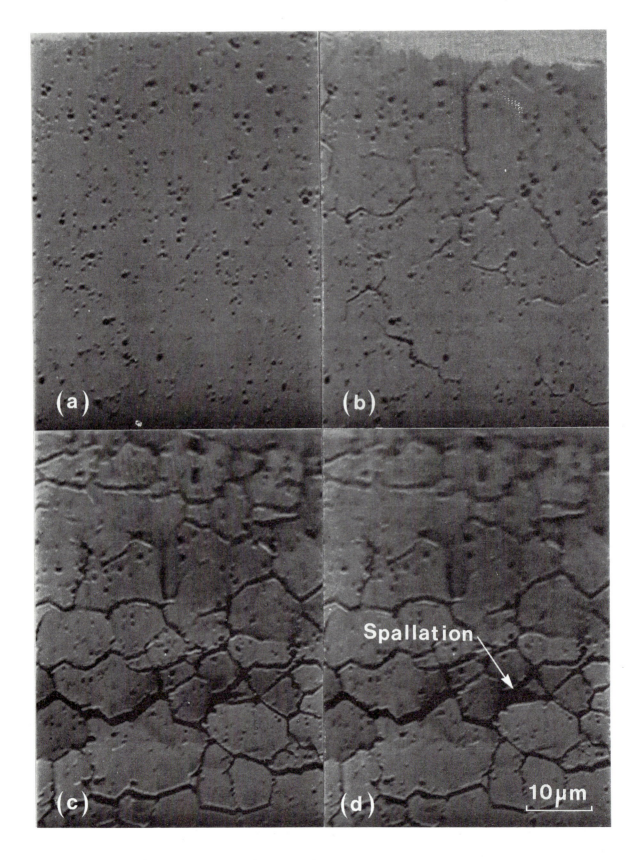

Figure 5: Optical micrographs of the UO$_2$ surface on reaching the oxidation temperature (450°C) (a) and following subsequent 10 min (b), 20 min 30 s (c) and 20 min 31 s (d) oxidation in air at this temperature.

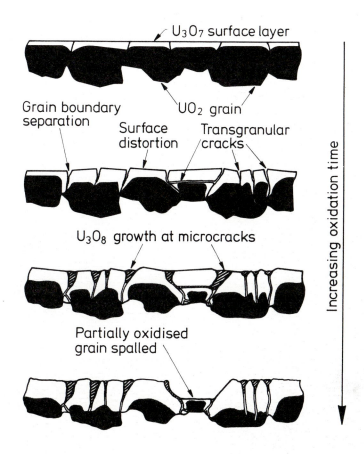

Figure 6: Schematic representation of the oxidation of UO_2 in air at temperatures between 225 and 550°C.

change from U_3O_7 to U_3O_8 and resultant compressive stresses caused more pronounced surface distortion, propagation of microcracks, an enhanced rate of oxidation and eventually spallation of oxide particulates. In the main these would probably have been smaller than the original UO_2 grains and would not necessarily have been completely oxidised to U_3O_8. Consequently, the particulates would have comprised of U_3O_8, U_3O_7 and possibly in some cases also of unoxidised UO_2. The acceleration of uranium dioxide oxidation to the higher constant rate reflected the increased effective reaction surface area. It also involved the continuous formation and breakdown of a thin barrier U_3O_7 layer, together with the associated concurrent spallation of oxide particulates.

CONCLUSIONS

1. This study has confirmed the value of direct *in situ* observation of both chemical and physical changes during oxidation as they happened. It has also established that relevant observations can be obtained by the attachment of an environmental cell fitted with a hot stage to an X-ray diffractometer and to an optical microscope.

2. Results obtained by these techniques and thermogravimetry have both confirmed and extended current mechanistic understanding of the chemical and physical processes involved in the distintegration of UO_2 during oxidation in air at temperatures in the range 225 to 550°C.

ACKNOWLEDGEMENT

The study was funded by the United Kingdom Department of Energy, which is gratefully acknowledged.

REFERENCES

1. Chemical Reactivity of Oxide Fuel and Fission Product Release, Proceedings of a Workshop held at the Berkeley Nuclear Laboratories, April 7-9, 1987, Eds. K.A. Simpson and P. Wood, CEGB, London.

2. M.J. Bennett, J.B. Price and P. Wood, Nuclear Energy, 27, 1988, p. 49.

3. P.A. Tempest, P.M. Tucker and J.W. Tyler, J. Nucl. Matls., 151, 1988, p. 251.

4. G. Skyme, P. Wood and G.A. Brown, Proc. of the Third Int. Spent Fuel Storage Sympsoium, Seattle, 1986, CONF-860417.

SECTION EIGHT
The Future

52 THE FUTURE OF THE MICROSCOPY OF OXIDATION

J.B.Castle

The Department of Materials Science and Engineering,
University of Surrey,
Guildford GU2 5XH, U.K.

ABSTRACT

In this final lecture of the programme I shall endeavour, as I was asked, to look forward to the imaging techniques which might become coupled with microscopy in the future. A review of the titles preceding this presentation might indicate that all has already been said. Nevertheless, starting from the viewpoint that this years spectroscopy is next years microscopy, there are points to be made.

Experimental techniques can be grouped, conveniently, in terms of their resolution in each of four modes: depth, chemical state, microstructural, and position; the latter being of most importance for techniques having any claim to be the basis of a microscopy. Metal oxidation, if a significant commercial problem, presents a thick film for analysis and it could be felt that depth resolution is of lesser interest. However, as examples in the literature have already shown, depth resolution is important in the analysis of spalling interfaces, of thin barrier layers, or of grain boundaries revealed in fracture cross-sections. The established techniques of AES (SAM), SIMS and EPMA between them supply all that is needed for the identification of elements present in the above situations. They serve chemical state determination and structure less well however.

Of the techniques available for chemical state resolution, photoelectron spectroscopy may be most appropriate. It is, for example, able to reveal the difference between mixed oxides, e.g., spinel, and single oxides; between valence states; and between cations associated with different anions (chlorides and oxides). Developments in imaging XPS will be reviewed and future possibilities outlined. In addition the basis of chemical state recognition by means of Auger spectroscopy will be outlined and an indication given of the situations in which high (energy) resolution would give worthwhile information in Auger images.

Whilst great advances in the preparation of cross-sections for TEM have been made it still remains that structural information associated with the SEM would be a great advantage. Techniques capable of giving structural information in the reflected mode are not common and it may be that the direct use of the STM on cross-sections will make this need redundant. However, the observation of EXAFS-like structure in the Auger spectrum is exciting and could assist in the study of grain boundary structure.

INTRODUCTION

The exciting feature of the oxidation of metals is that the evidence, of reaction, remains for study in its entirety in the form of the oxide layer. As we have seen there is no shortage of ideas for the means of studying this forensic pile. Thus, in considering the future, two key questions emerge: firstly, what features of oxidation are left outstanding from the present methods of oxidation; and, secondly, what are the signals that such features send, which might enable their detection and study?

MICROSCOPY OF OXIDATION: THE NEEDS

Consideration of the first question posed in the introduction shows that it centres on the need to deduce, from the oxide remains, something of the rate determining mechanism of the reaction. Corrosion studies after all are undertaken in order to either predict the rate of attack for the purposes of the design engineer, or in order that an unsatisfactory rate of attack should be suppressed. It is necessary, also, to overcome the disadvantage of the solid oxide which is that, in contrast to aqueous corrosion for example, neither the reaction interface nor the transport path can be studied 'in vitro'. This is why marker experiments are so important and thus in particular why imaging forms of SIMS [1], in which ^{18}O is used as a marker, have such a role to play. It often happens that this year's spectroscopy is next year's microscopy and the development of imaging SIMS is a nice illustration of the point. There have been some excellent contributions on this topic at this conference.

What then remains to study? Topography, including the presence of porosity, is well catered for by the SEM; elemental concentrations are revealed by the EPMA; the chemistry of exposed interfaces by XPS and local pitting or grain boundary attack by the scanning Auger microscope; and, as we have seen, detailed, localised, marker studies are possible by SIMS. Diffraction studies of oxide, either detached or in situ in the TEM, have enabled important structures to be obtained. specific examples of the used of each of these techniques in the context of corrosion may be found in an earlier article by the author [2] whilst details of the techniques themselves are found in a summary published in the E-Mag series [3]. Figure 1 provides a succinct view of many of the current techniques and of the particular focus which they have; i.e., analysis as a function of position, or depth, or in terms of chemical state or structure. Clearly there are overlaps and those which concern this conference are the overlaps which include positional register since this is the essential attribute of microscopy.

In answer to the question posed in the previous paragraph, we may note that it is unusual for any of the techniques to reveal the transported species or to provide more than circumstantial evidence of the path that they take. Oxidation theory requires that they be known for a complete description of the process. The participation of a vacancy or an interstitial ion in the mass action balance at an interface determines the rate of transport under that gradient. Yet could it be that microscopy, and its success, has turned our thoughts from this central question? Carl Wagner [4] had demonstrated the presence of copper vacancies and the concomitant positive hole by direct titration of dissolved oxide as early as 1938 and showed too how their concentration varied with temperature. Which technique might permit an oxide to be examined for this property by physical means? In the next section some of the emerging microscopies are reviewed in this light.

FUTURE MICROSCOPIES

The SIMS experiment is the outcome of some 30 years of work on the determination of transport numbers and diffusion coefficients by isotopic tracers and might logically be extended now to in situ studies of the decay of surface activity. Measurement of the changes in concentration of a tracer ion with respect to time at oxidation temperature and position would provide direct evidence for variation in self diffusion coefficients with respect to the surface phase. The destructive nature of the SIMS technique might be cited as a drawback for such work. It would require static SIMS but in its time-of-flight mode there is every hope that this will be combined with imaging. Briggs [5], for example, has shown images of polymer types under static conditions (using ion doses of the order 10^{-11}).

ATOM PROBE MICROSCOPIES

Pickering [6] has shown that ion probe microscopy can be used for elegant experiments in which the low concentration of nickel ions in copper oxide formed on a copper-nickel alloy can be measured and concentration gradients at dopant levels traced across interfaces. This is perhaps the ultimate in both spacial and concentration resolution, but the work establishes that concentrations, which are significant in terms of the expected numbers of point defects involved in oxidation, can be detected.

IMAGING ELECTRON SPECTROSCOPIES

Whilst the mass spectroscopies are increasingly able to give accurate stoichiometries by direct counting, the electron spectroscopies are able to show the influence of changes in electron density of states. This is less direct, for mass transport considerations, than the determination of stoichiometries. Nevertheless considerations of electroneutrality often require that one is accompanied by the other. The preeminent technique for speciation is XPS - X-ray photoelectron spectroscopy, sometimes known as ESCA. The title ESCA stands for electron spectroscopy for *chemical* analysis and was coined by Siegbahn in his early papers as a result of his excitement in showing that copper(l) oxide could be distinguished from copper(ll) oxide by this then new (1958) method. The socalled, chemical shift giving rise to the possibility of distinguishing chemical state is still the best known feature of XPS and is much used in oxidation studies. Further effects are also helpful however: the shake-up satellite on the divalent states of copper is a very effective indicator for Cu(ll) as was pointed out in 1971[7], in the context of the oxidation of copper-nickel alloys; whilst the chemical shift on the photoexcited Auger line enables copper(l) to be identified unequivocally in the presence of copper metal [3] a fact that was used in discussing the oxidation of copper nickel alloys in [9]. Since the early work it has become possible to separate the majority of oxidation states by means of one of the above spectroscopic techniques. It has also become possible to

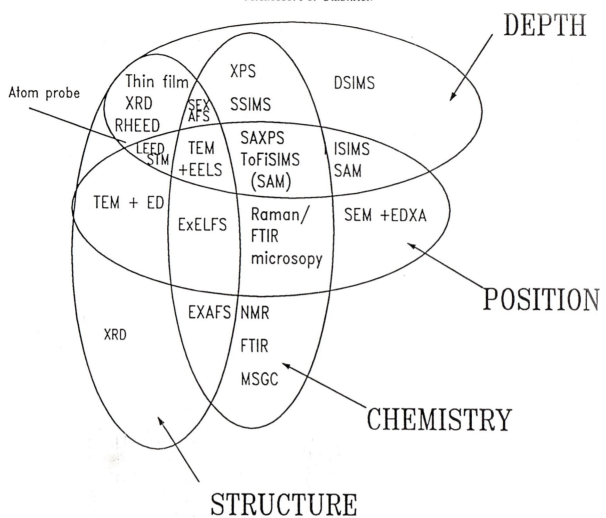

Figure 1: Venn diagram showing techniques according to resolution in depth, chemical state, structure and position. This review concerns those which have positional-chemical state sensitivity.

distinguish, within the formal oxidation state, between ions in differing local environments [10] and as a function of the strain within the oxide [11]. Each of these effects stems from the relaxation of electrons associated with the photo-ionised ion. The relaxing electrons maybe outer orbitals of the ion itself, intra-atomic relaxation, or a displacement of the charge associated with the ligands, extra-atomic relaxation. In its extreme case charge maybe completely transferred into unfilled levels created by the core ionisation and it is this which gives rise to the satellites of the divalent transition metals, such as the copper(ll) satellite mentioned above. Since its inception as a commercially available technique in 1970 XPS has been widely used for the study of oxidised surfaces and for the study of interfaces exposed by a variety of fracture methods [13]. The spectroscopic methods summarised above have each been applied where appropriate. The problem has been that XPS is not a microscopic technique. However a number of efforts to give it a spatially resolved dimension now appear to be bearing fruit and a brief account of these may be useful.

Conventional XPS utilises a rather simple form of X-ray source to illuminate an area of several square millimetres of the sample. Until the early 1980s there was no possibility of distinguishing detail within this excitation footprint. This conference does not need reminding of the topographic and morphological complexity of the typical oxidised surface and information from the whole of this region is averaged out in the XP spectrometer. A number of developments appeared in the latter part of the 1980s which have begun to change this. Firstly X-ray sources became more sophisticated, enabling point illumination with a spot of 150 μm², and secondly, lens and analyser combinations were developed which enabled the examination of a selected area within the illuminated zone. Lens design did not permit a spatial resolution of better than 150 μm² with this technique either. However the simple addition of scan plates did enable the production of images by the latter method [13]. Although hardly a microscopy this technique did give a resolution which is better than the first Auger scans published in the late 1970's.

The first imaging XPS instrument did not employ scanning but produced a real image by a projection method. The instrument known as the spectromicroscope was developed from a concept discovered by Turner at Oxford. Photoelectrons produced in a strong magnetic field become trapped in the field lines, spiralling with a helical radius which depends on the field strength and the kinetic energy of the electron. Turner and his coworkers showed that the field line trapped the full hemisphere of emission and that this enhanced the signal strength over techniques which collected only a given solid angle, determined by the size of the detector. With a relatively simple analyser images with a resolution of 10 μm have been produced. The disadvantage of the spectromicroscope is the need to locate the sample at the centre of a superconducting solenoid in order to provide the high field necessary. Furthermore, reasonable magnification requires that the field lines be intercepted some way from the point of emission, a flight path of about 1 metre was used in the original instrument. The method of imaging does however make this form of microscope very suitable for use with high brightness beams from a synchrotron or other source and it is understood that an instrument will be commissioned on the high brightness source at the free electron laser project at Stanford. There it is anticipated that 1 μm will be achieved and the project goal is 0.1 μm. The possible production of synchrotrons as stand-alone sources for X-ray lithography may mean that this type of technique will have a more broadly-based future than it appears at the moment. Note however that the spacial resolution, even with exotic sources, is still determined by the magnetic field. It might therefore be necessary to tune the photon energy so that each photoelectron is produced at an optimum low energy in order to achieve the best resolution in an image. Alternative a method will have to be found to scan the beam as a fine probe.

More recently, two further types of imaging spectrometers have been produced and these are now commercially available. The instrument produced by Scienta [14] uses the dispersion plane of the hemispherical analyser to give a one dimensional image of an illuminated line on the surface of the sample. The other dimension being kept for the orthogonal, energy dispersive, plane of the analyser which thus repeats the line scan for a number of differing energies. The instrument operates with an advanced X-ray source, giving high brightnesses and the spatial resolution has been recorded at 23 μm and the energy resolution at 0.27 eV. This form of display is ideal for the study of interfaces, revealing any change in oxidation state at the interface in a very direct way. In forming a two-dimensional image the energy dispersion has to be sacrificed and this is what is done in the second instrument, the ESCASCOPE produced by VG Scientific [15]. This uses a lens to construct a true image of the photoelectron emission, incorporating the hemispherical filter as a 1:1 component in the lens train. Measurements suggest that the present lens permits a resolution of 10 μm in the image. This instrument is perhaps worth a much brighter X-ray source than the conventional source usually called for. Nevertheless it is the first to produce worthwhile images at anything like microscopic levels of spatial resolution.

The images available, in company literature [16], are not of topics in oxidation but they do show the extent to which information might be obtained by the method.

SCANNING AUGER MICROSCOPY

Scanning Auger microscopy, like SIMS, is now part of the standard repertoire of the microscopist. Here we shall consider the extent to which chemical state, as distinct from elements, can be imaged. When an Auger spectrum is excited by a scanning electron beam, the secondary electron corresponding to the creation of the core hole does not have a characteristic energy.

However, the electron resulting from the de-excitation of the hole does have a characteristic value, sufficient to identify the element in all cases but not necessarily sufficient to identify the chemical state. This depends on the width of the line in relation to its chemical shift from the relaxation effects described above. For the steel forming transition elements this relationship is completely wrong; the line is too broad, because of the involvement of the 3d levels in the Auger transition for the chemical state to be resolved. For the post transition elements however the lines become very narrow and the chemical state is easily revealed. Thus useful work can be done with copper and zinc and with the semiconductor elements, Ga,Ge and As. For similar reasons the chemical shift in Mg,Al and Si compounds is good, as it is on Cd through to Te. Not many of these elements are of interest to oxidation scientists however. The situation is not likely to be improved because of the fundamental nature of the limitation.

One technique which is already improving the use of scanning Auger microscopy for the study of complex surfaces is that of data analysis by use of scatter diagrams. This method has been pioneered by Prutton at York and has recently been implemented on our own Link AN 10000 system at Surrey. In constructing a scatter diagram the intensities of two given elements are plotted against each other for each pixel of the map. Particular phases show up then as clusters on the diagram and the pixels represented by these intensities can then be highlighted in a map. The mean ratio of the intensities constituting this cluster then can be interpreted as a phase. The method is sensitive and makes use of the available data in a statistically significant way. Concentration gradients can be revealed in the scatter diagram and this could be a very useful method for the examination of cross-sections, possibly for the determination of gradients in stoichiometry, although again oxidation crosssections have yet to be examined in this way.

Recently EXAFS-like structure has been shown to exist in the Auger spectrum [17]. The acronym EXFAS has been adopted (Extended Fine Auger Structure) and it is of interest because of its easy access, its association with a technique of high spacial resolution, and its usefulness on solid surfaces in the reflection mode. The modulation of the emitted signal stems from the variation in energy of the Auger electron by filling of the core hole with secondary electrons in the continuum provided by the primary beam. The signal is weak, but just as the plasmon has proved a useful signal with which to gauge thickness, so an EXFAS wiggle may one day be used to gauge local structure and order.

THE SCANNING TUNNELLING MICROSCOPE

This is one technique which did not start as a spectroscopy. Much has been written in the past year or two and it is not appropriate for me to add to this. However, two exciting possibilities emerge which are relevant to this conference. The first is that direct observation of vacancies at surfaces has been demonstrated, albeit on the surface of a catalyst. These give rise to the possibility that the mass action equilibrium at the surface could be directly observed by counting the vacancies is very real. Secondly it has been shown that both Auger and X-ray analysis can be performed using the STM tip when it is retracted from the surface and used then as an electron emitter at a potential of several thousand volts [18]. Thus the structure and chemical information associated with EXFAS could ultimately be available at STM resolution. This would certainly enable the goal of identifying those layers or interfaces acting as resistances to oxidation to be achieved.

SUMMARY

Several suggestions for new directions in the microscopy of oxidation have emerged from this review of the new microscopies. The aim has been to find the technique capable of determining local non-stoichiometry, local concentration gradients, diffusion coefficients and the presence of vacancies and or positive holes. the following conclusions have been reached:

1. SIMS might be extended from its present use in conducting marker experiments to a use in the determination of diffusion coefficients by decease in surface concentration of initial deposits or by depth profiling.

2. The atom probe is already demonstrating that the mass spectroscopies have sufficient sensitivity to enable stoichiometric gradients to be observed.

3. The imaging forms of XPS may permit the direct observation of chemically significant electronic structures, but their resolution is likely to be confined to around 10 μm except at national centres of exotic sources.

4. Data analysis by methods such as scatter diagrams might permit chemical phases or concentration gradients to be observed in the Auger microscope.

5. The scanning tunnelling microscope offers the best opportunity for the direct measurement of defect concentration at the oxide/gas interface.

REFERENCES

1. A. Brown and J. C. Vickerman, Surf.Interface Anal., 6, 1984, p. 1.

2. J. E. Castle, Br. Corros. J.,22, 1987, p. 77.

3. P. J. Goodhew and J. E. Castle, Proc. EMAG '83, Inst. Phys. Conf. Series No 68, 1984, p. 515.

4. C. Wagner and K. Grunewald, Z. Phys. Chem., 40B , 1938, p. 455.

5. D. Briggs, Int. Conf. Interfaces in Composite Materials, Ed. F. R. Jones, Pergamon (1989).

6. H. W. Pickering, Private communication

7. J. E. Castle, Nature Physical Science, 234, 1971, p. 93.

8. J. E. Castle and D. C. Epler, Proc. Roy. Soc., A339, 1974, p. 49.

9. J. E. Castle and M. Nasserian-Riabi, Corros. Sci., 15, 1975, p. 537.

10. R. H. West and J. E. Castle, Surf. Interface Anal., 4, 1982, p. 68.

11. M. J. Edgell, S. C. Mugford, J. E. Castle and N. A. Pirie, J. Electrochem. Soc., 137 No 1, 1990, p. 201.

12. A. D. Brooker and J. E. Castle, Surf. Interface Anal., 8, 1986, p. 13.

13. AXIS - see commercial literature published by KRATOS Ltd, Barton Dock, Manchester.

14. U. Geilius et al., Proc. Int. Conf. Elec. Spectros. 4, Eds. C. R. Brundle, G. E. McGuire & J. J. Pireaux, Elsevier, 1990, p. 747.

15. S. M. Hues and R. J. Colton, Surf. Interface. Anal., 14 No 3, 1989, p. 101.

16. See commercial literature published by VG Scientific.

17. B. P. Hollebone et al., Proc. Int. Conf. Elec. Spectros. 4, Eds. C. R. Brundle, G. E. McGuire & J. J. Pireaux, Elsevier, 1990, p. 661.

18. J. Gimzewski, Physics World (UK), 2, No 8, 1989, p. 25.

INDEX